METHODS IN MOLECULAR BIOLOGY™

Series Editor
John M. Walker
School of Life Sciences
University of Hertfordshire
Hatfield, Hertfordshire, AL10 9AB, UK

For further volumes:
http://www.springer.com/series/7651

Pluripotent Stem Cells

Methods and Protocols

Edited by

Uma Lakshmipathy

Primary and Stem Cell Systems, Life Technologies, Carlsbad, CA, USA

Mohan C. Vemuri

Primary and Stem Cell Systems, Life Technologies, Frederick, MD, USA

Editors
Uma Lakshmipathy
Primary and Stem Cell Systems
Life Technologies
Carlsbad, CA, USA

Mohan C. Vemuri
Primary and Stem Cell Systems
Life Technologies
Frederick, MD, USA

ISSN 1064-3745 ISSN 1940-6029 (electronic)
ISBN 978-1-62703-347-3 ISBN 978-1-62703-348-0 (eBook)
DOI 10.1007/978-1-62703-348-0
Springer New York Heidelberg Dordrecht London

Library of Congress Control Number: 2013934004

Printed on acid-free paper

Humana Press is a brand of Springer
Springer is part of Springer Science+Business Media (www.springer.com)

Preface

Human pluripotent stem cells such as human embryonic stem cells (hESC) and induced pluripotent stem cells (iPSC) with their unique developmental plasticity hold immense potential as cellular models for drug discovery and in regenerative medicine as source for cell replacement. While hESC are derived from a developing embryo, iPSC are generated with forced expression of key transcription factors to convert adult somatic cells to ESC-like cells, a process termed reprogramming. This is a rather appealing choice for generating pluripotent stem cells since it overcomes ethical issues concerning the use of developing embryos and iPSC can be generated from patient-specific or disease-specific cells for downstream applications. The purpose of the book "Pluripotent stem cells: Methods and Protocols" is to highlight the best methods and systems for the entire work flow.

A key concern with iPSC is that randomly integrated transcription factors used during reprogramming can lead to genomic instability and altered epigenome. In order to generate safer iPSC, several advances have been made in the development of footprint-free systems for the delivery of transcription factors. However, some of the currently available footprint-free reprogramming methods are more efficient or consistent than others in producing iPSC from diverse somatic sources. The advantage in having the capability to generate iPSC from diverse somatic cells is that it offers a choice to utilize convenient sources of patient material such as peripheral blood or altered somatic cell sources in case of disease patients where peripheral blood may not be an ideal choice. Part I focuses on this aspect with reviews and articles on ESC derivation and iPSC generation.

An additional key consideration for the culture of safer, therapy-compliant cells is the choice of media and matrix used for the generation and expansion of cells. Traditionally, pluripotent stem cells are cultured on murine embryonic fibroblasts in media systems that contain serum or serum replacement. The biggest challenge in routine and large-scale culturing of these cells is the effort and technical expertise required to maintain the pluripotent stem cells in their undifferentiated and karyotypically normal state. In order to distill the art of culturing these cells to a predictable science, several media systems that are serum free, feeder free, and chemically defined have been developed and successfully used. This, in combination with matrices that eliminate the use of feeder cells, paves the path towards more defined and regulated culture systems. Further, use of different reprogramming methods from diverse somatic cell types cultured under varying conditions can often lead to iPSC clones that differ in their differentiation potentials. Identification of methods and streamlined protocols for directed differentiation into highly specialized cells is critical for efficient conversion of the pluripotent stem cells into the desired functional cell type. Part II summarizes media systems for expanding ESC and iPSC, with detailed protocols for directed differentiation into specific lineages.

A key unanswered question that is currently the focus of several studies is whether iPSC are truly pluripotent and equivalent to ESC. Traditional characterization have relied on cellular methods of pluripotent stem cell characterization such as observing morphology, differential dye staining or specific antibody labeling. Furthermore, the teratoma assay is currently considered as the in vivo gold standard for pluripotency confirmation. However, the combination of transcriptome and epigenome analysis using new technology platforms is rapidly evolving and will enable high-throughput characterization and standardization of pluripotent stem cells. Some of the commonly used cellular and molecular characterization methods are compiled under Part III.

Finally, pluripotent stem cells are amenable to genetic labeling and can thus be used for dissecting basic biology, and for creating disease models and screening systems. Successful gene delivery, however, depends on easy cloning systems to rapidly generate vectors, gene delivery methods to deliver these vectors into pluripotent stem cells, and engineering platforms that meet the need for sustained context-specific gene expression. Part IV reviews the potential application of labeled stem cells with specific methods for cloning, gene delivery, and cell engineering.

Integrated work flow, familiarization to available methods and detailed protocols are critical for simplifying the complex technical challenges associated with pluripotent stem cell. Our intended goal and the objective of this book will be met when both novice and expert users can use this book as a comprehensive guide for the successful generation, culture, characterization, and differentiation of pluripotent stem cells.

Carlsbad, CA, USA *Uma Lakshmipathy*
Frederick, MD, USA *Mohan C. Vemuri*

Contents

Contributors

Ronald V. Abruzzese • *Molecular And Cell biology, Life Technologies, Austin, TX, USA*
Michal Amit • *Department of Obstetrics and Gynecology, Technion Israel Institute of Technology, Rambam Medical Center, Haifa, Israel*
Janice Au-Young • *Genetic Analysis, Life Technologies, Forter City, CA, USA*
Hossein Baharvand • *Department of Stem Cells and Developmental Biology, Cell Science Research Center, Royan Institute for Stem Cell Biology and Technology, ACECR, Tehran, Iran; Department of Developmental Biology, University of Science and Culture, ACECR, Tehran, Iran*
Yun Bao • *Genetic Analysis, Life Technologies, Forter City, CA, USA*
Shayne Boucher • *Primary and Stem Cell Systems, Life Technologies, Frederick, MD, USA*
Paul W. Burridge • *Institute for Cell Engineering, Johns Hopkins University School of Medicine, Baltimore, MD, USA*
Richard Eckert • *Department of Biochemistry and Molecular Biology, University of Maryland, Baltimore, MD, USA*
Catharina Ellerström • *Cellectis Stem Cells, Cellartis AB, Göteborg, Sweden*
Richard A. Fekete • *Molecular and Cell Biology, Life Technologies Austin, TX, USA*
Andrew Fontes • *Primary and Stem Cell Systems, Life Technologies, Carlsbad, CA, USA*
Elena Grigorenko • *Genetic Analysis, Life Technologies, Forter City, CA, USA*
Ronald P. Hart • *Rutgers Stem Cell Research Center, The W.M. Keck Center for Collaborative Neuroscience, Rutgers University, Piscataway, NJ, USA; Department of Cell Biology and Neuroscience, Rutgers University, Piscataway, NJ, USA*
Seyedeh-Nafiseh Hassani • *Department of Stem Cells and Developmental Biology, Cell Science Research Center, Royan Institute for Stem Cell Biology and Technology, ACECR, Tehran, Iran; Department of Developmental Biology, University of Science and Culture, ACECR, Tehran, Iran*
Kejin Hu • *Wisconsin National Primate Research Center, University of Wisconsin-Madison, Madison, WI, USA*
Johan Hyllner • *Cellectis Stem Cells, Cellartis AB, Göteborg, Sweden*
Jasmeet Kaur • *Primary and Stem Cell Systems, Life Technologies, Carlsbad, CA, USA*
Uma Lakshmipathy • *Primary and Stem Cell Systems, Life Technologies, Carlsbad, CA, USA*
Pauline T. Lieu • *Primary and Stem Cell Systems, Life Technologies, Carlsbad, CA, USA*
Muluye E. Liku • *Diabetes Center, University of California San Francisco, San Francisco, CA, USA*
Ying Liu • *Department of Neurosurgery, UT Health, Houston, TX, USA*
Chad C. MacArthur • *Primary and Stem Cell Systems, Life Technologies, Carlsbad, CA, USA*
Nasir Malik • *NIH CRM, National Institute of Arthritis and Musculoskeletal and Skin Disorders (NIAMS), National Institutes of Health (NIH), Bethesda, MD, USA*
Ricardo Mancebo • *Genetic Analysis, Life Technologies, Forter City, CA, USA*
Jennifer C. Moore • *Rutgers NIMH Stem Cell Center and Department of Genetics, Rutgers, The State University of New Jersey, Nelson Biological Laboratory, Piscataway, NJ, USA*

SUNALI N. PATEL • *Genetic Analysis, Life Technologies, Forter City, CA, USA*
MARK J. POWERS • *Primary and Stem Cell Systems, Life Technologies, Frederick, MD, USA*
RENE H. QUINTANILLA JR. • *Primary and Stem Cell Systems, Life Technologies, Carlsbad, CA, USA*
MAHENDRA S. RAO • *NIH CRM, National Institute of Arthritis and Musculoskeletal and Skin Disorders (NIAMS), National Institutes of Health (NIH), Bethesda, MD, USA*
CHRISTOPHER L. RICUPERO • *Rutgers Stem Cell Research Center, The W.M. Keck Center for Collaborative Neuroscience, Rutgers University, Piscataway, NJ, USA; Department of Cell Biology and Neuroscience, Rutgers University, Piscataway, NJ, USA*
PHILIP ROELANDT • *Stem Cell Institute, Katholieke Universiteit Leuven, Leuven, Belgium*
DAVID RUFF • *Life Technologies, South San Francisco, CA, USA*
ALEXANDRIA SAMS • *Primary and Stem Cell Systems, Life Technologies, Frederick, MD, USA*
IGOR SLUKVIN • *Wisconsin National Primate Research Center, University of Wisconsin-Madison, Madison, WI, USA*
RAIMUND STREHL • *Cellectis Stem Cells, Cellartis AB, Göteborg, Sweden*
MAVIS R. SWERDEL • *Rutgers Stem Cell Research Center, The W.M. Keck Center for Collaborative Neuroscience, Rutgers University, Piscataway, NJ, USA; Department of Cell Biology and Neuroscience, Rutgers University, Piscataway, NJ, USA*
MARY LYNN TILKINS • *Primary and Stem Cell Systems, Life Technologies, Frederick, MD, USA*
DENNIS VAN HOOF • *Diabetes Center, University of California San Francisco, San Francisco, CA, USA*
JOLIEN VANHOVE • *Stem Cell Institute, Katholieke Universiteit Leuven, Leuven, Belgium*
MOHAN C. VEMURI • *Primary and Stem Cell Systems, Life Technologies, Frederick, MD, USA*
CATHERINE VERFAILLIE • *Stem Cell Institute, Katholieke Universiteit Leuven, Leuven, Belgium*
KATE JUDD • *Primary and Stem Cell Systems, Life Technologies, Grand Island, NY, USA*
YALEI WU • *Genetic Analysis, Life Technologies, Forter City, CA, USA*
ELIAS T. ZAMBIDIS • *Division of Pediatric Oncology, Institute for Cell Engineering, Johns Hopkins University School of Medicine, Sidney Kimmel Comprehensive Cancer Center at Johns Hopkins, Baltimore, MD, USA*

Part I

Generation of Pluripotent Stem Cells

Chapter 1

Sources and Derivation of Human Embryonic Stem Cells

Michal Amit

Abstract

Human embryonic stem cells (hESCs) are pluripotent cells derived from the inner cell mass (ICM) of the developing embryo. hESCs culture as cell lines in vitro and possess great potential in such research fields as developmental biology and cell-based therapy, as well as such industrial purposes as drug screening and toxicology. When ESCs were first derived by Thomson and colleagues, traditional methods of immunostaining and culturing, using primary mouse embryonic fibroblasts and medium supplemented by serum were used. Considerable efforts have since led to improved methods for isolating new lines in defined and reproducible conditions. This chapter discusses sources for embryos for ESC isolation, commonly used methods for deriving hESC lines, and a number of possible culture systems.

Key words Human, Pluripotent stem cells, Derivation, Sources

1 Introduction

The history of culturing pluripotent cell lines dates back to Stevens and Little's description of testicular teratomas formed by mouse strain "pluripotential embryonic cells," which were shown to divide asymmetrically and give rise to both differentiating and undifferentiating cells (1). This lay the basis for the embryonic stem cell definition. A few years later, similar cells were isolated and cultured, and shown to differentiate in vitro (2, 3). These cells, named embryonal carcinoma (EC), were demonstrated to form teratocarcinoma in vivo, after being isolated and cultured in vitro from a single cell (4). This paved the way for the exploration of pluripotent stem cells in vitro. The field was further advanced by the creation of chimeras, first by injecting cells isolated from inner cell mass (ICM), and later by injecting EC cells into blastocytes, and demonstrating integration of injected cells into the three embryonic germ layers of the developed mouse embryos (5, 6), and even creating normal chimeric mice (7). During the following decades, many studies disrobed the capabilities of EC cells to differentiate

Uma Lakshmipathy and Mohan C. Vemuri (eds.), *Pluripotent Stem Cells: Methods and Protocols*, Methods in Molecular Biology, vol. 997, DOI 10.1007/978-1-62703-348-0_1, © Springer Science+Business Media New York 2013

in vitro (8) [reviewed by Andrews, 2002], including via embryoid bodies (EBs) (9). It has become clear that processes of EC cell differentiation in vitro, within EBs, mimic early embryonic events.

In 1981, an additional type of pluripotent cells was derived from the ICM of mouse blastocyst, and grown as a cell line in culture—embryonic stem cells (ESCs) (10, 11). ESCs demonstrate similar features to EC cells; they form EBs in vitro and teratoma (not teratocarcinoma like EC cells) in vivo, create normal chimeric mice and proliferate indefinitely as undifferentiated if cultured under appropriate culture conditions. Unlike EC cells, ESCs exhibit normal and stable karyotype.

The capacity to isolate and culture pluripotent cells necessitated the development of suitable culture techniques. The traditional method, based on a supportive layer of fibroblasts, was adapted for EC cells, enabling the culture of clonal lines as undifferentiated cells (12). In that pioneering work chick fibroblasts were used as feeder layers at the initial passages. Later studies used mouse embryonic fibroblasts (MEFs), primary or from a cell line, to support the culture of undifferentiated pluripotent cells (12). The same culture methods were used for isolating ESCs (from mouse, primate, and human sources), and later for the first isolation of induced pluripotent stem cell (iPSC) lines (13).

2 Overview

2.1 Sources

By means of similar culture and isolation methods as those used for the derivation of mouse ESCs, primate ESCs were derived 14 years later (14, 15), and human ESC lines only 17 years later, in 1998 (16). The 17-year gap between the isolation of mouse ESC and hESC lines is partly explained by the scarceness of high quality blastocysts for research purposes. The availability of high quality blastocysts for research was dependent on the following developments: (1) in vitro fertilization (IVF) methods, (2) controlled ovarian stimulation protocols, and (3) methods for culturing human embryos to the blastocyst stage, and for freezing human embryos. Most human embryos available for derivation of ESCs or for other research purposes are surplus embryos from IVF programs. The first successful human IVF was achieved by Steptoe and Edwards, who, in 1978, retrieved one oocyte following a spontaneous cycle, and transferred one resultant embryo, leading to the birth of baby girl Louise Brawn (17). Since the first pioneering work of the Nobel Prize winner Edwards, millions of people have benefitted from technologies for assisted reproduction. Protocols for IVF have improved dramatically since the first IVF cycle, particularly from the development of controlled ovarian stimulation, which is now routinely performed, and which results in the collection of an average of ten oocytes per cycle.

Due to 60% fertilization rates in vitro, and the usual retrieval of only two embryos per cycle (To avoid the risks of multiple pregnancies), many cycles produce surplus embryos, which are routinely frozen. Since many couples do not use all their embryos for reproductive purposes, thousands of surplus embryos are estimated to be frozen in liquid nitrogen worldwide, most of which may never be claimed by their parents.

Culturing techniques for human embryos, including culturing to the blastocyst stage, have developed alongside advances in assisted reproductive technologies. Embryo culturing is not trivial since, in vivo, environmental changes during early embryo development, are followed by physiological changes (18). To adapt culture conditions to the needs of developing embryos, a stepwise culture media was developed, with different formulations for different stages of development, and suitability to clinical applications (defined serum-free medium) (19). The development of methods for culturing human embryos from fertilization to the blastocyst stage has made the derivation of hESC lines possible, since the use of early stage embryos for isolation is problematic.

Most of the existing hESC lines were derived from surplus embryos from IVF programs (16, 20). However, other options are available, such as embryos of low quality that were excluded from clinical uses (21) or genetically abnormal embryos that were discarded after pre-implantation genetic diagnosis (PGD) (22, 23) [and Amit and Itskovitz-Eldor, unpublished data]. An additional source is embryos produced for research purposes from donated gametes, though in some countries the production of embryos for nonreproductive purposes is not allowed (24). Embryo production can also be achieved by oocyte activation and by nuclear transfer. In the derivation of ESCs from nonhuman primate activated oocytes, as demonstrated by Cibelli and colleagues, most ESC features were exhibited (25). Later, ESC lines were isolated from embryos that were developed from activated human oocytes (26, 27). One research team derived hESC lines from blastocytes that resulted from nuclear transfer; however, the validity of that study has been questioned (28).

A different approach to deriving hESC lines uses a single blastomere rather than an intaier blastocyte. Mouse ESC lines have been derived by mixing a single blastomere with an already-established cell line; after propagation the newly derived line is isolated by means of a selective tag (29, 30). If adapted to hESC line isolation, one or two blastomeres can be removed from a 3-day-old human embryo for ESC line isolation, similar to the technique used for PGD, and the embryo can be used for reproductive purposes. This was demonstrated in a recent study, in which three hESC lines were derived from a blastomere biopsy, using foreskin fibroblast as feeders (30).

Using embryos from different sources, and using different derivation methods, at least 600 human ESC lines have been derived and characterized worldwide (the human European Embryonic Stem Cell Registry, http://www.hescreg.eu/), and are now available for research purposes. However, the use of embryos for research, industrial and clinical applications has raised ethical questions. Countries have addressed such issues by publicizing specific guidelines for the use of embryos for hESC studies. The guidelines published and used by the National Institute of Health in the United States are widely used as a reference. The availability of iPSC lines can serve to bypass the derivation of pluripotent stem cells, while avoiding the need to discard a human embryo. However, some questions were recently raised regarding the epigenetic memory of these lines, which may limit their differentiation potential (31), their karyotype stability (32), and tumor formation (33).

2.2 Derivation Methods

As mentioned earlier in this chapter, hESC lines were derived using the culture methods that were developed earlier for EC cell lines and for mouse ESC line derivation. To derive ESC lines, the ICM should be selectively removed and propagated in culture. ICM can be isolated by means of an immunosurgical or mechanical method. Alternatively, ICM can be isolated during the second passage, after culturing intact embryo (following removal of the zona pellucid, ZP), thus enabling ICM cell outgrowth.

Immunosurgery, the most commonly used method for ESC line derivation, was developed in the 1970s by Solter and Knowles for studying early embryonic development (34). The procedure aims to selectively isolate the ICM from the blastocyst. To this end, the blastocyst is exposed to anti-human whole serum antibodies, which attach to the trophoectoderm. The ICM cells remain intact due to prevention of antibody penetration into the blastocyst by the outer layer of the trophoectoderm (connected by tight junctions). Next, using the capability of the immune system to lyse cells marked by an antibody, the embryo is incubated with guinea pig complement-containing medium, which lyses the trophoectoderm antibody-marked cells. The intact ICM is exposed and cultured with a suitable supportive layer, like mitotically inactivated mouse primary embryonic fibroblasts (MEFs) (16). Pictures of embryos at different stages of derivation are summarized in Fig. 1.

The trophoectoderm layer can be selectively removed by an alternative method, such as mechanically, by means of fine needles under stereoscope, or by laser (35, 36). In both methods the isolated ICM is further cultured and expanded on the supportive layer.

To selectively isolate the ICM cells during the second passage, the ZP is removed and the whole embryo is plated in culture. 24–48 h after plating, the exposed embryo attaches to the feeder layer and usually flattens. The ICM continues to grow with surrounding trophoblasts as a monolayer. An example of plated

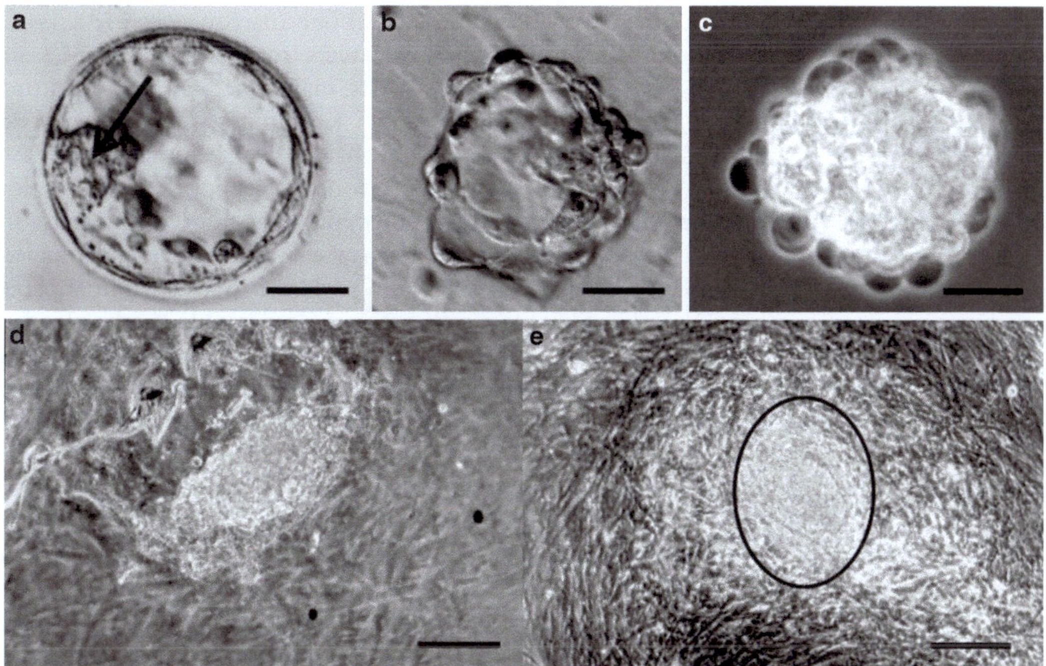

Fig. 1 Derivation of hESCs. ICM of the blastocyst is selectively removed and further propagated on MEF either by means of immunosurgery or by using the whole embryo approach. (**a**) A 6-day-old human blastocyst. The ICM is marked with an *arrow*. Bar 40 μM. (**b**) An exposed embryo incubated in anti-human whole serum antibodies. Bar 50 μM. (**c**) ICM after incubation with guinea pig complement-containing medium, which lyses the trophoectoderm antibody-marked cells. Bar 30 μM. (**d**) ICM outgrowth after plating it with MEFs. Bar 60 μM. (**e**) ICM (marked by a *circle*) outgrowth after plating a whole embryo on MEFs. Bar 90 μM. Picture (**e**) was taken with permission from (59)

embryo is illustrated in Fig. 1. When the ICM reaches a sufficient size, it is mechanically removed under a stereoscope and transferred to a fresh plate. This method is regarded as less efficient than immunosurgery, due to the risk of ICM differentiation.

The number of available hESC lines is testament that methods of derivation are a reproducible procedure. Success rates vary from less than 10% (24) to over 40%, depending on the quality of the embryos and the technique involved (16, 20, 35).

2.3 Derivation of hESCs Using Different Culture Conditions

Traditional methods for isolating and culturing ESCs include MEFs as feeder layer and medium supplemented with fetal bovine or calf serum. Possible future uses of hESCs for clinical and industrial purposes will require a culture system that will be free from animal products, defined, and most importantly reproducible. The first step toward this aim is replacing the MEFs with a human supportive layer, preferably while using a cell line. Several alternatives have been suggested as human feeders, including human embryo-derived fibroblasts or fallopian tube epithelial cells (37), foreskin fibroblasts (38, 39), human placenta fibroblasts (40, 41), adult human fibroblasts (42), and adult bone marrow cells (43). While using these cells as supportive layers for culturing hESCs for prolonged culture, the stem cells maintained all ESC features including pluripotency and stable karyotypes. Human embryo-derived fibroblasts, placenta fibroblasts, adult human fibroblasts and foreskin fibroblasts were also found to support the derivation of new hESC lines under animal product-free conditions (37, 39–42, 44). In most studies, human serum was used as a replacement for FBS or knockout serum replacer, to avoid animal products. Foreskin fibroblast cells are most commonly used as feeders for ESC line derivation. From the lines derived in the cited studies, about 90% were derived using foreskin fibroblast (37, 40–42, 45–50). For clinical or industrial applications, some of these cell lines were derived while using clinical-grade materials in a clean room (under good manufacturing practice, GMP). The reported clinical-grade cell lines were all isolated and cultured using foreskin fibroblasts as feeders (49, 51). We were able to derive GMP-grade cell lines while using NutriStem™ as culture medium (serum-free, animal-free commercially available defined medium), which demonstrated all ESC features and required sterility tests. One cell line of foreskin fibroblast, which was derived and cultured in a clean-room using animal-free medium was used as a supportive layer for culturing and banking this ESC line (Amit et al., unpublished data). Thus, the use of a culture system based on human cell lines as feeders, and animal-free culture medium, proved to be a reliable system for derivation, culture and banking of hESCs for research, clinical and industrial uses.

To establish a defined environment for culturing ESCs, removal of the supportive layer from the culture and the use of a defined animal- and serum-free medium and matrix are recommended. The first study reported the derivation of new hESC lines without use of a supportive layer, while using a culture system consisting of a MEF-produced matrix and a medium supplemented with bFGF (16 ng/ml), LIF, serum replacement, and plasmanate (52). The six new hESC lines that resulted demonstrated ESC features. Despite the use of animal products and non-defined materials, this pioneering study demonstrated the feasibility of a feeder-layer free derivation of hESC lines. Later, hESC lines were derived under defined conditions. In the study of Ludwig and colleagues a defined serum- and animal-free medium and a matrix containing a combination of recombinant human proteins were used to derive 2 hESC lines (53). The newly derived cell lines were cultured for prolonged culture using these defined conditions, while maintaining most hESC features. Thus, for the first time, defined, animal-, serum- and feeder-free culture conditions were established for hESCs. However, the cell lines did not exhibit karyotype stability. One of the isolated hESC lines was reported to harbor 47, XXY and the other one exhibited chromosome 12 trisomy. Since the karyotype analysis was conducted several months post derivation, it is unclear whether the embryos were originally defected or whether the abnormality was acquired later during the culture (53).

In recent years, a new concept for a feeder-layer free culture of hESCs was reported, whereby cells are cultured in suspension (54–58). This new culture system is not only defined and reproducible, but also a system for mass production of ESCs (55). In one study, new cell lines were derived directly in suspension using animal-free medium (58). Three new hESC lines were derived, exhibiting normal hESC features (58). This study demonstrated that supportive layers or direct contact to extracellular matrices are not necessarily required for hESC culture and isolation, provided that suitable conditions are employed (58).

Thus, ESC lines can be isolated and cultured in a system that is animal and serum free, reproducible, and scalable.

Acknowledgments

The author thanks Mrs. Hadas O'Neill for editing the manuscript. The research conducted was partly supported by Technion Research and Development Foundation (TRDF).

References

1. Stevens LC, Little CC (1954) Spontaneous testicular teratomas in an inbred strain of mice. Proc Natl Acad Sci USA 40:1080–1087
2. Pierce GB, Dixon FJ (1959) Testicular teratomas. II. Teratocarcinoma as an ascetic tumor. Cancer 12:584–589
3. Pierce GB, Verney EL (1961) An in vitro and in vivo study of differentiation in teratocarcinomas. Cancer 14:1017–1029
4. Kleinsmith LJ, Pierce GB (1964) Multipotentiality of single embryonal carcinoma cells. Cancer Res 24:1544–1551
5. Gardner RL (1998) Contribution of blastocyst micromanipulation to the study of mammalian development. Bioessays 20:168–180
6. Brinster RL (1974) The effect of transferred into to the mouse blastocyst on subsequent development. J Exp Med 140:1049–1056
7. Mintz B, Illmensee K (1975) Normal genetically mosaic mice produced from malignant teratocarcinoma cells. Proc Natl Acad Sci USA 72:3585–3589
8. Andrews P (2002) From teratocarcinomas to embryonic stem cells. Philos Trans R Soc Lond B Biol Sci 357:405–417
9. Martin GR, Evans MJ (1975) Multiple differentiation of clonal teratocarcinoam stem cells following embryoid body formation in vitro. Cell 6:467–474
10. Evans MJ, Kaufman MH (1981) Establishment in culture of pluripotential cells from mouse embryos. Nature 292:154–156
11. Martin GR (1981) Isolation of a pluripotent cell line from early mouse embryos cultured in medium conditioned by teratocarcinoma stem cells. Proc Natl Acad Sci USA 78:7634–7638
12. Evans MJ (1972) The isolation and properties of a clonal tissue culture strain of pluripotent mouse teratocarcinoma cells. J Embryol Exp Morphol 28:163–196
13. Takahashi K, Yamanaka S (2006) Induction of pluripotent stem cells from mouse embryonic and adult fibroblast cultures by defined factors. Cell 126:663–676
14. Thomson JA et al (1995) Isolation of a primate embryonic stem cell line. Proc Natl Acad Sci USA 92:7844–7848
15. Thomson JA et al (1996) Pluripotent cell lines derived from common marmoset (Callithrix jacchus) blastocysts. Biol Reprod 55:254–259
16. Thomson JA, Itskovitz Eldor J, Shapiro SS, Waknitz MA, Swiergiel JJ, Marshall VS, Jones JM (1998) Embryonic stem cell lines derived from human blastocysts. Science 282:1145–1147

17. Steptoe PC, Edwards RG (1978) Birth after the reimplantation of a human embryo. Lancet 2:366
18. Gardner DK, Lane M, Calderon I, Leeton J (1996) Environment of the preimplantation human embryo in vivo: metabolite analysis of oviduct and uterine fluids and metabolism of cumulus cells. Fertil Steril 65:349–353
19. Gardner DK (1994) Mammalian embryo culture in the absence of serum or somatic cell support. Cell Biol Int 18:1163–1179
20. Reubinoff BE, Pera MF, Fong C, Trounson A, Bongso A (2000) Embryonic stem cell lines from human blastocysts: somatic differentiation in vitro. Nat Biotechnol 18:399–404
21. Mitalipova M, Calhoun J, Shin S, Wininger D, Schulz T, Noggle S, Venable A, Lyons I, Robins A, Stice S (2003) Human embryonic stem cell lines derived from discarded embryos. Stem Cells 21:521–526
22. Verlinsky Y, Strelchenko N, Kukharenko V, Rechitsky S, Verlinsky O, Galat V, Kuliev A (2005) Human embryonic stem cell lines with genetic disorders. Reprod Biomed Online 10:105–110
23. Mateizel I, De Temmerman N, Ullmann U, Cauffman G, Sermon K, Van de Velde H, De Rycke M, Degreef E, Devroey P, Liebaers I, Van Steirteghem A (2006) Derivation of human embryonic stem cell lines from embryos obtained after IVF and after PGD for monogenic disorders. Hum Reprod 21:503–511
24. Lanzendorf SE, Boyd CA, Wright DL, Muasher S, Oehninger S, Hodgen GD (2001) Use of human gametes obtained from anonymous donors for the production of human embryonic stem cell lines. Fertil Steril 76:132–137
25. Vrana KE, Hipp JD, Goss AM, McCool BA, Riddle DR, Walker SJ, Wettstein PJ, Studer LP, Tabar V, Cunniff K, Chapman K, Vilner L, West MD, Grant KA, Cibelli JB (2003) Nonhuman primate parthenogenetic stem cells. Proc Natl Acad Sci USA 100:11911–11916
26. Revazova ES, Turovets NA et al (2007) Patient-specific stem cell lines derived from human parthenogenetic blastocysts. Cloning Stem Cells 9:432–449
27. Revazova ES et al (2008) HLA homozygous stem cell lines derived from human parthenogenetic blastocysts. Cloning Stem Cells 10:11–24
28. Hwang WS et al (2005) Patient-specific embryonic stem cells derived from human SCNT blastocysts. Science 308:1777–1783
29. Chung Y, Klimanskaya I, Becker S, Marh J, Lu SJ, Johnson J, Meisner L, Lanza R (2006) Embryonic and extraembryonic stem cell lines derived from single mouse blastomeres. Nature 439:216–219
30. Giritharan G, Ilic D, Gormley M, Krtolica A (2011) Human embryonic stem cells derived from embryos at different stages of development share similar transcription profiles. PLoS One 6:e26570
31. Kim K, Zhao R, Doi A, Ng K, Unternaehrer J, Cahan P, Huo H, Loh YH, Aryee MJ, Lensch MW, Li H, Collins JJ, Feinberg AP, Daley GQ (2011) Donor cell type can influence the epigenome and differentiation potential of human induced pluripotent stem cells. Nat Biotechnol 29:1117–1119
32. Mayshar Y, Ben-David U, Lavon N, Biancotti JC, Yakir B, Clark AT, Plath K, Lowry WE, Benvenisty N (2010) Identification and classification of chromosomal aberrations in human induced pluripotent stem cells. Cell Stem Cell 7:521–531
33. Okita K, Ichisaka T, Yamanaka S (2007) Generation of germline-competent induced pluripotent stem cells. Nature 448(7151): 313–317
34. Solter D, Knowles BB (1975) Immunosurgery of mouse blastocyst. Proc Natl Acad Sci USA 72:5099–5102
35. Amit M, Itskovitz-Eldor J (2002) Derivation and spontaneous differentiation of human embryonic stem cells. J Anat 200:225–232
36. Turetsky T, Aizenman E, Gil Y, Weinberg N, Shufaro Y, Revel A, Laufer N, Simon A, Abeliovich D, Reubinoff BE (2007) Laser-assisted derivation of human embryonic stem cell lines from IVF embryos after preimplantation genetic diagnosis. Hum Reprod 23:46–53
37. Richards M, Fong CY, Chan WK, Wong PC, Bongso A (2002) Human feeders support prolonged undifferentiated growth of human inner cell masses and embryonic stem cells. Nat Biotechnol 20:933–936
38. Amit M, Margulets V, Segev H, Shariki C, Laevsky I, Coleman R, Itskovitz-Eldor J (2003) Human feeder layers for human embryonic stem cells. Biol Reprod 68:2150–2156
39. Hovatta O, Mikkola M, Gertow K, Stromberg AM, Inzunza J, Hreinsson J, Rozell B, Blennow E, Andang M, Ahrlund-Richter L (2003) A culture system using human foreskin fibroblasts as feeder cells allows production of human embryonic stem cells. Hum Reprod 18:1404–1409
40. Simón C, Escobedo C, Valbuena D, Genbacev O, Galan A, Krtolica A, Asensi A, Sánchez E, Esplugues J, Fisher S, Pellicer A (2005) First derivation in Spain of human embryonic stem cell lines: use of long-term cryopreserved embryos and animal-free conditions. Fertil Steril 83:246–249

41. Genbacev O, Krtolica A, Zdravkovic T, Brunette E, Powell S, Nath A, Caceres E, McMaster M, McDonagh S, Li Y, Mandalam R, Lebkowski J, Fisher SJ (2005) Serum-free derivation of human embryonic stem cell lines on human placental fibroblast feeders. Fertil Steril 83:1517–1529
42. Tecirlioglu RT, Nguyen L, Koh K, Trounson AO, Michalska AE (2010) Derivation and maintenance of human embryonic stem cell line on human adult skin fibroblast feeder cells in serum replacement medium. In Vitro Cell Dev Biol Anim 46:231–235
43. Cheng L, Hammond H, Ye Z, Zhan X, Dravid G (2003) Human adult marrow cells support prolonged expansion of human embryonic stem cells in culture. Stem Cells 21:131–142
44. Inzunza J, Gertow K, Stromberg MA, Matilainen E, Blennow E, Skottman H, Wolbank S, Ahrlund-Richter L, Hovatta O (2005) Derivation of human embryonic stem cell lines in serum replacement medium using postnatal human fibroblasts as feeder cells. Stem Cells 23:544–549
45. Ström S, Holm F, Bergström R, Strömberg AM, Hovatta O (2010) Derivation of 30 human embryonic stem cell lines. In Vitro Cell Dev Biol Anim 46:337–344
46. Aguilar-Gallardo C, Poo M, Gomez E, Galan A, Sanchez E, Marques-Mari A, Ruiz V, Medrano J, Riboldi M, Valbuena D, Simon C (2010) Derivation, characterization, differentiation, and registration of seven human embryonic stem cell lines (VAL-3, -4, -5, -6 M, -7, -8, and -9) on human feeder. In Vitro Cell Dev Biol Anim 46:317–326
47. Ilic D, Giritharan G, Zdravkovic T, Caceres E, Genbacev O, Fisher SJ, Krtolica A (2009) Derivation of human embryonic stem cell lines from biopsied blastomeres on human feeders with minimal exposure to xenomaterials. Stem Cells Dev 18:1343–1350
48. Valbuena D, Galánm A, Sánchez E, Poo ME, Gómez E, Sánchez-Luengo S, Melguizo D, García A, Ruiz V, Moreno R, Pellicer A, Simón C (2006) Derivation and characterization of three new Spanish human embryonic stem cell lines (VAL 3 -4 -5) on human feeder and in serum-free conditions. Reprod Biomed Online 13:875–886
49. Ellerström C, Strehl R, Moya K, Andersson K, Bergh C, Lundin K, Hyllner J, Semb H (2006) Derivation of a xeno-free human embryonic stem cell line. Stem Cells 24:2170–2176
50. Crook JM, Peura TT, Kravets L, Bosman AG, Buzzard JJ, Horne R, Hentze H, Dunn NR, Zweigerdt R, Chua F, Upshall A, Colman A (2007) The generation of six clinical-grade human embryonic stem cell lines. Cell Stem Cell 1:490–494
51. Prathalingam N, Ferguson L, Young L, Lietz G, Oldershaw R, Healy L, Craig A, Lister H, Binaykia R, Sheth R, Murdoch A, Herbert M (2012) Production and validation of a good manufacturing practice grade human fibroblast line for supporting human embryonic stem cell derivation and culture. Stem Cell Res Ther 3:12
52. Klimanskaya I, Chung Y, Meisner L, Johnson J, West MD, Lanza R (2005) Human embryonic stem cells derived without feeder cells. Lancet 365:1636–1641
53. Ludwig TE, Levenstein ME, Jones JM, Berggren WT, Mitchen ER, Frane JL, Crandall LJ, Daigh CA, Conard KR, Piekarczyk MS, Llanas RA, Thomson JA (2006) Derivation of human embryonic stem cells in defined conditions. Nat Biotechnol 24:185–187
54. Amit M, Chebath J, Marguletz V, Laevsky I, Miropolsky Y, Shariki K, Peri M, Revel M, Itskovitz-Eldor J (2010) Suspension culture of undifferentiated human embryonic and induced pluripotent stem cells. Stem Cell Rev 6:248–259
55. Amit M, Laevskym I, Miropolsky Y, Shariki K, Peri M, Itskovitz-Eldor J (2011) Dynamic suspension culture for scalable expansion of undifferentiated human pluripotent stem cells. Nat Protoc 6:572–579
56. Olmer R et al (2010) Long term expansion of undifferentiated human iPS and ES cells in suspension culture using a defined medium. Stem Cell Res 5:51–64
57. Singh H, Mok P, Balakrishnan T, Rahmat SN, Zweigerdt R (2010) Up-scaling single cell-inoculated suspension culture of human embryonic stem cells. Stem Cell Res 4:165–179
58. Steiner D et al (2010) Derivation, propagation and controlled differentiation of human embryonic stem cells in suspension. Nat Biotechnol 28:361–364
59. Amit M, Itskovitz-Eldor J (2012) Atlas on human pluripotent stem cells—derivation and culturing. Series: stem cell biology and regenerative medicine. In: Turksen (ed), Chapter 1, Fig 1.11 a, Humana Press

Chapter 2

A New Chemical Approach to the Efficient Generation of Mouse Embryonic Stem Cells

Hossein Baharvand and Seyedeh-Nafiseh Hassani

Abstract

Here, we present a highly efficient and reproducible method for the establishment of mouse embryonic stem cells (mESCs) from embryonic day 3.5 (E3.5) whole blastocysts. This protocol involves the use of small molecules SB431542 and PD0325901, which inhibit transforming growth factor-β (TGF-β) and extracellular signal-regulated kinases (ERK1/2), respectively. This protocol is universal in the derivation of mESC lines from NMRI, C57BL/6, BALB/c, DBA/2, and FVB/N strains, which have previously been considered refractory or non-permissive for ESC establishment. The efficiency of mESC lines generation is 100%, regardless of genetic background.

Key words Mouse, Embryonic stem cells derivation, Small molecule, ERK, TGF-β

1 Introduction

Mouse embryonic stem cells (mESCs) are derived from the inner cell mass (ICM) of murine embryos and have the long term capability to self-renew in culture conditions as well as differentiate into derivatives of three embryonic germ layers by chimera or teratoma formation (1, 2). These potential capabilities make mESCs the best model for studying early mammalian developmental events, particularly by knocking out various genes and making germ-line chimeras (3). Furthermore, in recent years, extensive efforts have been undertaken with mESCs to develop efficient differentiation protocols for the derivatives of the three embryonic germ layers to be potentially used in regenerative medicine.

Another area of research interest in the field of mESCs involves understanding the molecular basis of pluripotency, particularly because of the considerable properties these cells have as naïve pluripotent stem cells (4). Naïve mESCs by having a highly efficient clonogenicity from single cells, lower rate of doubling time, efficient contribution to chimeric embryos, and the absence of differentiation

Uma Lakshmipathy and Mohan C. Vemuri (eds.), *Pluripotent Stem Cells: Methods and Protocols*, Methods in Molecular Biology, vol. 997, DOI 10.1007/978-1-62703-348-0_2,

bias have demonstrated the ground state of pluripotency (4, 5). Some reports have been published that converted human ESCs (hESCs) as primed stem cells into their naïve state in order to resolve their sensitivity to single cell dissociation or to prevail over their developmental propensity among different lines. These initial attempts, however, have been unable to produce long-term maintenance of naïve hESCs without the expression of ectopic transgenes (6). Therefore, more research is necessary to recognize the mechanisms of naïve pluripotency.

We have recently reported a new, highly efficient route for the generation of mESCs from different mouse strains such as NMRI, BALB/c, C57BL/6, DBA/2, and FVB/N which have previously been considered as refractory or non-permissive (7). We have shown that the dual inhibition of ERK and TGF-β signaling pathways by using a combination of PD0325901 (PD03) and SB431542 (SB43) small molecules, respectively, during the line derivation procedure could overcome observed differences between various strains. Recent publications show that the use of PD032 with GSK3 inhibitor CHIR99021 (CHIR) could support the derivation and maintenance of the pluripotency ground state in different strains (8). However, CHIR or other GSK3 inhibitors induce chromosomal instability (9, 10). It seems that inhibition of TGF-β which induce differentiation in mESCs by an autocrine process has preference over the inhibition of the more complicated GSK3. When compared to the other available protocols for the derivation of mESCs, our protocol differs as follows: a new signaling inhibition by small molecules, technical simplicity, high efficiency and reproducibility, and universality for various genetic backgrounds.

2 Materials

2.1 Chemicals

All reagents and materials should be sterile.

1. EGTA.
2. L-glutamine, 100× (i.e., 200 mM), Store aliquots at −20°C. Use 1:100.
3. Mineral oil.
4. Ethanol 70% (v/v).
5. Mouse fetus, embryonic day 12.5 (E12.5) or E13.5 mouse embryos from NMRI mouse strain.
6. Sodium bicarbonate.
7. Ultrapure deionized water.

2.2 Disposables

1. 15 and 50 ml conical tubes.
2. Plastic disposable pipettes (5, 10, 25 ml).

3. 0.22-μm vacuum filtration (500 ml).
4. 0.22-μm pore size filter.
5. Glass Pasteur pipettes, 9 in. Sterilize by autoclave.
6. Center-well plate.
7. 24-well plates.
8. 12-well plates.
9. 6-well plates.
10. Flask T-25.
11. Flask T-75.
12. 60 mm non-adhesive bacterial plate.
13. Syringes (1, 5, 20, 50 ml).
14. Filter pipette tips (0.5–10, 5–100, 50–1,000 μl).
15. Cryovials.

2.3 Equipment

1. Stereomicroscope.
2. Inverted phase contrast microscope (4, 10, 20, and 40× objectives).
3. Micropipette (1–10, 10–100, 100–1,000 μl)
4. Pipettor.
5. Laminar flow hood (Class I & II).
6. Hemocytometer.
7. Tissue culture incubator, with humidity and gas control to maintain 37°C and 95% humidity in an atmosphere of 5% CO2 in air.
8. Water bath.
9. Liquid nitrogen tank.
10. Foot-operated Venturi pump.
11. Cryovial storage rack.
12. Freezers −20°C and −70°C.

2.4 Reagent Setup

1. *LIF solution*: Dilute 10^7°U/ml Leukemia inhibitory factor (LIF) ten times Phosphate-buffered saline (PBS) without Ca^{2+}/Mg^{2+} (PBS). Store in 1 ml aliquots at −20°C (even though Chemicon recommends storage at 4°C).
2. *β-Mercaptoethanol*: Add 70 μl β-mercaptoethanol in 10 ml PBS– to 100 mM (1,000×). Sterilize through a 0.22-mm filter. Solution should be maintained in a tube with a dark cover. b-Mercaptoethanol is toxic; avoid inhalation, ingestion, or contact with eyes, skin, or mucous membranes.
3. Mouse *ES cell culture medium*: To one, 500 ml bottle of Knockout DMEM, add 75 ml fetal bovine serum ES-qualified

(ES-FBS), 5 ml nonessential amino acid solution (NEAAs), 5 ml penicillin/streptomycin, and 50 μl β-mercaptoethanol. Store in 50 ml aliquots at 4°C. Prior to use, add 500 μl L-glu and 500 μl LIF to give a final concentration of 1,000 U/ml LIF.

4. *MEF* culture medium: Dissolve one pack of DMEM in 800 ml deionized water and add 10 ml Pen/Strep, 7 μl b-mercapto-ethanol (original stock, undiluted), and 3.4 g sodium bicarbonate. Add 2.5–3 ml HCl (1 N) and then water to reach 850 ml. Sterilize through a 0.22-mm vacuum filter. Add 150 ml fetal bovine serum to the medium.
5. Mitomycin C: Prepare a 2 mg/ml stock solution by dissolving 2 mg of Mitomycin C in 1 ml of deionized water and store at 4°C. Solution should be maintained in a tube with a dark cover. Mitomycin C is very toxic if inhaled or swallowed, and poses a danger of cumulative effects.
6. Freezing medium: Mix 10% dimethyl sulfoxide, 50% ES-FBS, and 40% mESC medium. Always prepare fresh on ice. Keep DMSO away from sources of ignition and electrostatic charge.
7. EGTA: To make 100× (50 mM) EGTA solution, dissolve 951 mg EGTA in 49.4 ml deionized water. Add 600 μl of 10 M NaOH (pH = 8) to solute. Sterilize through a 0.22-μm filter. Store at room temperature. For a working solution, add 500 μl of 100× stock in 49.5 ml PBS^-.
8. Acid Tyrode's solution: Mix 8 μg/ml NaCl, 0.2 μg/ml KCl, 0.24 μg/ml $CaCl_2 \cdot 2H_2O$, 0.1 μg/ml $MgCl_2 \cdot 6H_2O$, 1 μg/ml glucose, 4 μg/ml polyvinylpyrrolidone, pH 2.5. Alternatively, the solution can be purchased as acidic Tyrode's solution.
9. Trypsin: 10× trypsin (2.5%) containing 2.5 g/l of trypsin (1:250), 8.5 g/l NaCl, and 9 μl/ml 1 M NaOH in Dulbecco's PBS/$MgSO_4$.
10. Trypsin/EDTA: (0.05%/0.53 mM). Store 10 ml aliquots at -20°C and, after thawing at 4°C, use within 2 weeks. Do not leave at room temperature.
11. PD0325901 solution: For a 10 mM concentrated stock solution, reconstitute 1 mg of PD0325901 by adding 207.4 μl of DMSO. Store aliquots at -20°C. For making media, add 1 μl of stock solution to 10 ml of prewarmed media (1 μM final concentration).
12. SB431542 solution: For a 10 mM concentrated stock solution, reconstitute 1 mg of SB431542 by adding 1.3 ml of DMSO. Store aliquots at -20°C. For making media, add 1 μl of stock solution to 1 ml of prewarmed media (10 μM final concentration).

13. *0.1%* gelatin: Dissolve 1 g of gelatin in deionized water and sterilize by autoclaving 30 min at 120°C. Store at room temperature.
14. Gelatin-coated plates: Cover the bottoms of the dishes with a 0.1% gelatin solution and incubate for at least 5 min. Remove excess gelatin solution. Coated plates can be stored at 4°C for up to 6 months.

3 Methods

All tissue culture reagents (medium, trypsin, etc.) should always be at least hand warmed and, ideally, prewarmed to 37°C.

3.1 Preparation of Mouse Embryonic Fibroblasts

1. Prepare mouse embryonic fibroblasts (MEFs), and generate and inactivate feeder cells following standard procedures. In general, we prepare MEFs as previously described (11–13) from a 12.5–13.5 day post-coitum outbred NMRI mouse strain in accordance with relevant guidelines and regulations (see Note 1). All media are prewarmed at 37°C. We inactivate confluent MEFs with mitomycin C for 1.5–3 h (10 mg/ml).
2. Prepare gelatin-coated center-well plates.
3. After mitomycin C inactivation, wash MEFs twice with PBS, trypsinize, and replate onto gelatinized center-well plates. We split the cells based on the area of the original and desired dishes. For example, mitotically inactivated MEFs in a small T flask which has a 25 cm^2 area is split into one 12-well plate (see Note 2). Alternatively, the MEFs are counted and replated 30,000–35,000 cells/cm^2 (60,000–70,000 cells/center-well).

3.2 Blastocyst Recovery and ICM Expansion

1. Replace MEF medium with mESC medium, including 1 μM PD03/10 μM SB43, 1–3 h before blastocyst recovery.
2. Prepare mESC medium drops under mineral oil in a 60-mm plate and place at 37°C, 5% CO_2.
3. Kill time-mated females at E3.5 and immediately dissect their uteri into 5 ml of mESC in 60-mm plate preheated to 37°C in the incubator (see Note 3).
4. Transfer the uteri into 2 ml of fresh preheated mESC in a 60-mm plate.
5. Flush the blastocysts out of the uterine horn under a stereomicroscope using a 1-ml syringe with a 26-G needle.
6. Collect blastocysts under a stereomicroscope using mouth-controlled pipette, transfer to a new culture dish containing microdrops of pre-equilibrated mESC medium and wash.

7. Remove zona pellucida from the blastocysts by subjecting them to pre-incubated acid Tyrode's solution as described before (14) (Fig. 1). Monitor until the zona pellucida disappear; this will take only seconds to a few minutes, depending on how much medium is carried over (see Note 4).
8. Under a dissecting microscope (ideally within a tissue culture hood) transfer the zona free-blastocysts onto mitotically arrested MEFs in the center-well plates (maximum ten blastocysts per plate) which have been prepared one day earlier and contain 700 μl mESC medium that consists of PD03/SB43 pre-equilibrated at 37°C and 5% CO_2.
9. Incubate at 37°C and 5% CO_2. Do not disturb for ~48 h to allow the blastocysts to partially attach to the culture dish. mESC medium can be then renewed every other day while the ICM expands.
10. Prepare and label appropriate numbers of 24-well plates with confluent layers of mitotically inactivated MEFs 1 day in advance.
11. After 7–9 days, under a stereomicroscope, make cuts around the outgrowth by a thin Pasteur pipette (mechanical passage, passage 0) to remove the surrounding trophoectoderm and pick up the individual full-grown ICMs (see Note 5) (Fig. 1). If the outgrowth is large (more than 200 mm in diameter) cut into pieces before pick up.
12. Culture the isolated outgrowths individually on a fresh 24-well dish covered with a mitotically arrested MEF. Make sure that the blastocyst has been successfully transferred by monitoring for its presence under a stereomicroscope.

3.3 First Trypsinization and mESC Proliferation

1. After 2 days, aspirate the mESC medium including SMs from the wells containing expanded secondary outgrowths on MEFs, wash with PBS^- and add 100 μl of EGTA to the wells for 3 min (see Note 6).
2. Remove EGTA and add 100 μl 2.5× trypsin (0.625% w/v) solution to the well and incubate for 3–5 min in an incubator (see Note 7).
3. Remove the trypsin with a pipette and add 500 μl of mESC medium without SMs to submerge the outgrowth.
4. Dissociate the outgrowths into individual cells and small cell clumps, by pipetting up and down 10–20 times with a 1,000 μl pipette tip.
5. Transfer all the cell suspensions into a well of a 24-well plate that contains fresh MEF feeder and mESC medium. After this step, the SM is removed.
6. Renew the medium every day.

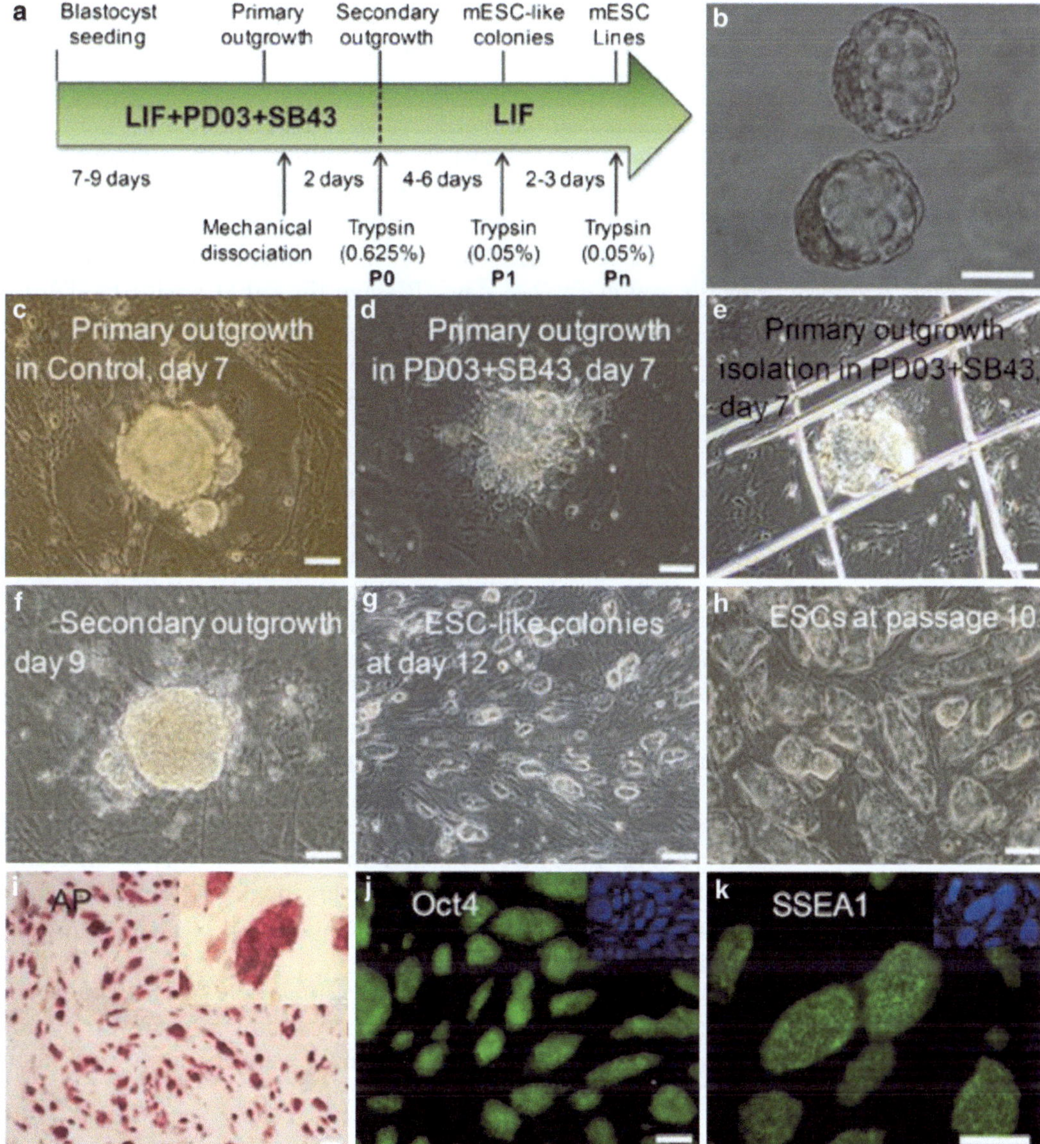

Fig. 1 The procedure of mESC line generation in the presence of PD03 + SB43 and the characterization of a derived line. (**a**) Schematic of the mESC line derivation procedure. (**b**) Zona-free blastocysts from DBA/2 mouse strain. (**c**) A 7-day outgrowth of a whole blastocyst without small molecules (control). In this condition, the outgrowths were large with numerous surrounded trophectodermal and differentiated cells. (**d**) A 7-day outgrowth of a whole blastocyst in presence of PD03 + SB43. The ICM outgrowths surrounded by a few trophectodermal cells. (**e**) Mechanical cutting of primary outgrowth by a thin Pasteur pipette (Passage 0), followed by replating on fresh MEF to make secondary outgrowth (**f**) after 48 h. (G) The secondary outgrowths were treated with 1× EGTA, and subsequently trypsinized (0.625% w/v) (Passage 1). (**h**) A representative picture of passage 10. The expression of (**i**) Alkaline phosphatase, (**j**) Oct4, and (**k**) SSEA1 in generated mESC line. Scale bar: 100 μm

7. 4–6 days after trypsinization, typically packed dome mESC colonies will be detected in all wells (passage 1) (see Notes 8 and 9) (Fig. 1).
8. Dissociate colonies with trypsin/EDTA (0.05% w/v) and passage them into 12-well dishes, followed by 6-well and then a small T flask, respectively (see Note 10).

3.4 Expansion and Freezing/Thawing of mESCs

Passage mESCs when they reach confluency. Usually, they are subcultured every 2–3 days. Passage the mESCs with EGTA and then trypsin/EDTA (0.05% w/v). Mix the resulting single-cell suspension with mESC medium and split the cells 1:3 to 1:6. The mESC medium should be changed daily. mESCs are frozen/thawed as previously described (12, 13).

3.5 Anticipated Results

The above described protocol was applied to different mouse strains of various genetic backgrounds, which were previously considered refractory or non-permissive for ESC establishment. We observed that dual inhibition of ERK1/2 and TGF-β hijacked the strain type in mice during mESC line derivation. The established mESC lines showed dome-shaped colony morphologies, high nuclear-cytoplasmic ratios, the ability to propagate following trypsin digestion and clonal growth from single cells. They expanded in mESC medium over multiple passages and maintained their pluripotency over long-term culture as determined by the expression of alkaline phosphatase (AP) activity and SSEA-l, as well as Oct4 (Fig. 1). The established lines had the ability to differentiate into lineages of all three germ layers in vitro by embryoid body formation. The evaluated lines exhibited high rates of chimerism when injected into blastocysts.

We believe that the protocol presented here offers a simple and efficient derivation of mESCs from strains of various genetic backgrounds which may be valuable for the generation of ESCs in various mammalian species.

4 Notes

1. NMRI-derived MEFs support mouse ESCs better than Balb/c-derived MEFs.
2. We usually use MEF passages 2–4, because the later-passage MEFs do not support ESC derivation to the same extent.
3. Blastocysts from NMRI strain were recovered by superovulation. For the other strains (C57BL/6, BALB/c, DBA/2, and FVB/N), embryos were obtained by natural mating.
4. The culturing of zona-free blastocysts increases their attachment to MEF. Zona-free blastocysts are very sticky. It is helpful

to pipette with an up and down motion and to always have some medium in the pipette before collecting the zona-free embryos.

5. In the presence of PD03/SB43 the occurrence of trophoblast-derived cells is significantly reduced and the initial outgrowth typically looks more like a mESC colony.
6. We found this procedure increased the efficiency of mESC derivation compared to trypsinization of whole ICM and trophectoderm outgrowths as reported by Meissner and colleagues (12).
7. We used different concentrations of trypsin and observed that 2.5× trypsin (0.625% w/v), has the most effective single cell dissociation of outgrowths with the lowest death rates.
8. In the presence of PD03/SB43, the mESC derivation efficiency decreased when 4.5 dpc blastocysts (based on Batlle-Morera et al. (15)) were used instead of 3.5 dpc blastocysts.
9. FBS functions as a trypsin inhibitor. One milliliter of mESC medium can inactivate a maximum of 20 µl of 2.5% trypsin. Make sure that the ratio of 2.5% trypsin and mESC medium is 1:50 or lower (16).
10. The dual inhibition of Erk1/2 and TGF-β by PD03/SB43 promotes the derivation of mESC lines from NMRI, C57BL/6, BALB/c, DBA/2, and FVB/N strains, which previously have been considered refractory or non-permissive for ESC establishment (17–20).

Acknowledgments

This study was funded by grants provided from Royan Institute, the Iranian Council of Stem Cell Technology, and the Iran National Science Foundation (INSF).

References

1. Martin GR (1981) Isolation of a pluripotent cell line from early mouse embryos cultured in medium conditioned by teratocarcinoma stem cells. Proc Natl Acad Sci USA 78:7634–7638
2. Evans MJ, Kaufman MH (1981) Establishment in culture of pluripotential cells from mouse embryos. Nature 292:154–156
3. Yu J, Thomson JA (2008) Pluripotent stem cell lines. Genes Dev 22:1987–1997
4. Nichols J, Smith A (2009) Naive and primed pluripotent states. Cell Stem Cell 4:487–492
5. Hanna J, Markoulaki S, Mitalipova M, Cheng AW, Cassady JP, Staerk J, Carey BW, Lengner CJ, Foreman R, Love J, Gao Q, Kim J, Jaenisch R (2009) Metastable pluripotent states in NOD-mouse-derived ESCs. Cell Stem Cell 4: 513–524
6. Buecker C, Geijsen N (2010) Different flavors of pluripotency, molecular mechanisms, and practical implications. Cell Stem Cell 7:559–564
7. Hassani SN, Totonchi M, Farrokhi A, Taei A, Larijani MR, Gourabi H, Baharvand H (2012) Simultaneous suppression of TGF-beta and ERK signaling contributes to the highly efficient and reproducible generation of mouse embryonic stem cells from previously considered refractory and non-permissive strains. Stem Cell Rev 8(2):472–481

8. Ying QL, Wray J, Nichols J, Batlle-Morera L, Doble B, Woodgett J, Cohen P, Smith A (2008) The ground state of embryonic stem cell self-renewal. Nature 453:519–523
9. Acevedo N, Wang X, Dunn RL, Smith GD (2007) Glycogen synthase kinase-3 regulation of chromatin segregation and cytokinesis in mouse preimplantation embryos. Mol Reprod Dev 74:178–188
10. Tighe A, Ray-Sinha A, Staples OD, Taylor SS (2007) GSK-3 inhibitors induce chromosome instability. BMC Cell Biol 8:34
11. Michalska E (2007) Isolation and propagation of mouse embryonic fibroblasts and preparation of mouse embryonic feeder layer cells. Curr Protoc Stem Cell Biol Chapter 1: Unit1C 3
12. Meissner A, Eminli S, Jaenisch R (2009) Derivation and manipulation of murine embryonic stem cells. Methods Mol Biol 482:3–19
13. Wong ES, Ban KH, Mutalif R, Jenkins NA, Copeland NG, Stewart CL (2010) A simple procedure for the efficient derivation of mouse ES cells. Methods Enzymol 476:265–283
14. Hogan B (1994) Manipulating the mouse embryo: a laboratory manual. Cold Spring Harbor Laboratory Press, Plainview, NY, USA
15. Batlle-Morera L, Smith A, Nichols J (2008) Parameters influencing derivation of embryonic stem cells from murine embryos. Genesis 46(12):758–767
16. Bryja V, Bonilla S, Arenas E (2006) Derivation of mouse embryonic stem cells. Nat Protoc 1:2082–2087
17. Baharvand H, Matthaei KI (2004) Culture condition difference for establishment of new embryonic stem cell lines from the C57BL/6 and BALB/c mouse strains. In Vitro Cell Dev Biol Anim 40:76–81
18. Umehara H, Kimura T, Ohtsuka S, Nakamura T, Kitajima K, Ikawa M, Okabe M, Niwa H, Nakano T (2007) Efficient derivation of embryonic stem cells by inhibition of glycogen synthase kinase-3. Stem Cells 25:2705–2711
19. Schoonjans L, Kreemers V, Danloy S, Moreadith RW, Laroche Y, Collen D (2003) Improved generation of germline-competent embryonic stem cell lines from inbred mouse strains. Stem Cells 21:90–97
20. Cinelli P, Casanova EA, Uhlig S, Lochmatter P, Matsuda T, Yokota T, Rulicke T, Ledermann B, Burki K (2008) Expression profiling in transgenic FVB/N embryonic stem cells over expressing STAT3. BMC Dev Biol 8:57

Chapter 3

A Review of the Methods for Human iPSC Derivation

Nasir Malik and Mahendra S. Rao

Abstract

The ability to reprogram somatic cells to induced pluripotent stem cells (iPSCs) offers an opportunity to generate pluripotent patient-specific cell lines that can help model human diseases. These iPSC lines could also be powerful tools for drug discovery and the development of cellular transplantation therapies. Many methods exist for generating iPSC lines but those best suited for use in studying human diseases and developing therapies must be of adequate efficiency to produce iPSCs from samples that may be of limited abundance, capable of reprogramming cells from both skin fibroblasts and blood, and footprint-free. Several reprogramming techniques meet these criteria and can be utilized to derive iPSCs in projects with both basic scientific and therapeutic goals. Combining these reprogramming methods with small molecule modulators of signaling pathways can lead to successful generation of iPSCs from even the most recalcitrant patient-derived somatic cells.

Key words Induced pluripotent stem cells, Reprogramming human somatic cells, Footprint-free iPSCs, Reprogramming with small molecules

1 Introduction

The discovery that somatic cells could be reprogrammed to a pluripotent state has profoundly altered the landscape in which stem cell research is conducted. Since the original demonstrations that mouse (1) and human (2–4) fibroblasts could be reprogrammed to become induced pluripotent stem cells (iPSCs) by viral overexpression of specific transcription factors, numerous methods have been developed to generate iPSCs. These methods have yielded increases in efficiency of reprogrammed cells and/or the generation of foot print-free iPSC lines that lack integration of any viral vector sequences into their genomes. This chapter will review the current state of iPSC reprogramming methods focusing on the reprogramming of human cells with occasional references to mouse cell reprogramming.

The original method of reprogramming murine fibroblasts by Yamanaka (1) utilized retroviral transduction of Oct4, Sox2,

Uma Lakshmipathy and Mohan C. Vemuri (eds.), *Pluripotent Stem Cells: Methods and Protocols*, Methods in Molecular Biology, vol. 997, DOI 10.1007/978-1-62703-348-0_3, © Springer Science+Business Media New York 2013

Klf4, and c-myc into mouse embryonic fibroblasts (MEFs) or tail-tip fibroblasts (TTF) derived from mice expressing β-galactosidase-neomycin fusion protein at the Fbx15 locus, which is specifically expressed in pluripotent stem cells and serves as an excellent marker for pluripotency. Drug selection with G418 after transduction of the four factors resulted in reprogramming of 0.02% of the MEFs or TTFs 14–21 days post-transduction. Wernig et al. (5) also succeeded in reprogramming MEFs by utilizing the same strategy with a drug selection marker knocked-in to the Nanog or Oct4 locus. This resulted in reprogramming in 0.05–0.08% of MEFs 20 days after transduction of the four Yamanaka reprogramming factors. Reprogramming of adult human dermal fibroblasts (HDFs) was first reported by the Yamanaka group to occur at an efficiency of ~0.02% at ~30 days after transducing the four reprogramming factors (2). A second group achieved reprogramming of ~0.1% fetal fibroblasts 14 days after infection of the four factors but was unable to reprogram adult fibroblasts with this reprogramming cocktail (4). Upon adding hTERT and SV 40 large T antigen to the reprogramming cocktail, ~0.25% of adult fibroblasts were reprogrammed. A lentiviral expression system was employed to deliver Oct4, Sox2, Nanog, and Lin28 to fetal MRC5 lung fibroblasts and newborn BJ-1 foreskin fibroblasts (3). iPSC colonies appeared 20 days post-transduction at an efficiency of 0.03–0.05% for fetal fibroblasts and 0.01% for newborn fibroblasts. After this initial discovery phase in the reprogramming field, modifications were made to redesign the reprogramming factor expression vectors and new modes of delivery were attempted to increase efficiency and minimize or remove vector sequences that were integrated into the reprogrammed iPSC genome.

2 Reprogramming Methods

2.1 Single Cassette Reprogramming Vectors with Cre-Lox Mediated Transgene Excision

As lentivirus, unlike retrovirus, can infect nondividing and proliferating cells it became the preferred delivery vehicle for expressing reprogramming factors in somatic cells. One of the concerns about reprogramming with four or more individual vectors was the suboptimal stoichiometry achieved using such a method (6). There were also worries about incorporation of the lentiviral vector sequences into the iPSC genome. The first concern was addressed with the development of a single cassette reprogramming vector in which each of the reprogramming factors was separated by a self-cleaving peptide signal (7–9). Some of these vectors were also engineered with loxP sites so that any integrated sequence could be excised by the overexpression of Cre-recombinase (8, 9). The first publication documenting Cre-mediated excision of integrated sequences from iPSCs was by episomal expression of a Cre-puromycin plasmid in iPSCs generated with four loxP containing lentiviral reprogramming vectors from patients with Parkinson's disease (10). After recombination and removal of transgene sequences the patients' iPSCs retained pluripotency and could be differentiated to dopaminergic neurons. It was later shown that adenoviral Cre was able to mediate excision of integrated transgene sequences in single cassette vectors (8, 9). A humanized version of one of these vectors with high reprogramming efficiency in mouse cells (0.5%) was constructed. The vector termed STEMCCA is now in wide use with reported reprogramming efficiencies of 0.1–1.5% (11).

2.2 Reprogramming by Nonintegrating Viruses

2.2.1 Adenovirus

Because adenovirus is a non-integrating virus it appears to be an excellent expression vehicle for generating iPSCs. However, the reprogramming efficiency of this method is only 0.001–0.0001% in mouse (12) and 0.0002% in human cells (13). For adenovirus to have any useful applications in reprogramming, considerable work has to be performed to optimize expression and increase reprogramming efficiencies.

2.2.2 Sendai Virus

Sendai virus has the advantageous property that it is an RNA virus that does not enter the nucleus and is therefore diluted out of cells ~10 passages after infection. A second desirable characteristic of Sendai virus is that it can produce large amounts of protein. When Sendai-based reprogramming vectors were made and used to generate iPSCs it was found that Sendai can reprogram neonatal and adult fibroblasts as well as blood cells (14–16). Cells are reprogrammed in ~25 days at an efficiency of 0.1% for blood cells and 1% for fibroblasts. Sendai is more difficult to work with than lentiviruses, but there are commercially available viral extracts for the Yamanaka factors that are ready to use. One disadvantage of Sendai-based reprogramming is that it takes ~10 passages for the virus to be completely lost from recently reprogrammed iPSCs and cells may need to be cultured at a higher temperature (39°C) to fully remove virus.

2.2.3 Protein

Expression of reprogramming factors as proteins would be an ideal method to generate footprint-free iPSCs. Unfortunately, it has been technically challenging to synthesize large amounts of bioactive proteins that can cross the plasma membrane. Two groups were able to make enough bioactive proteins in an *E. coli* expression system to reprogram 0.006% of mouse fibroblasts (17) and 0.001% of human fibroblasts (18). The low efficiency, technical challenges, and lack of published studies in non-fibroblast cell types suggest that much work needs to be done before protein reprogramming is a viable method.

2.3 Nonviral Reprogramming Methods

2.3.1 mRNA Transfection

The ability to express reprogramming factors as mRNA offers another method to make footprint-free iPSCs. Warren et al. were able to overcome several hurdles to transcribe mRNAs to efficiently express reprogramming factors (19). They were able to reprogram human fibroblasts at an efficiency of 1.4% within 20 days. By adding Lin28 to the Yamanaka reprogramming factor protocol, culturing at 5% O_2, and including valproic acid in the cell culture medium, the efficiency could be increased to 4.4%. Although reprogramming factor mRNAs are commercially available, this method suffers from the fact that it is labor intensive, requires addition of mRNA daily for 7 days, and has not been validated in cells other than fibroblasts.

2.3.2 miRNA Infection/ Transfection

Several miRNA clusters are strongly expressed in ESCs. When synthetic mimics of the mature miR-302b and/or miR-372 plus the four lentiviral Yamanaka factors were added to MRC5 and BJ-1 fibroblasts there was a 10- to 15-fold increase in reprogramming efficiency in comparison with the four lentiviral factors alone (20). It was found that certain miRNAs could reprogram cells at high efficiency without the presence of the Yamanaka factors. Expression of the seed sequences for the miR302/367 sequence as lentivirus particles generated iPSCs in ~10% of BJ-1 fibroblasts 12–14 days after infection (21). Another miRNA reprogramming paper found that transfection of certain miRNAs could reprogram human cells. The mir-200c, mir-302s, and mir-369s were transfected into HDFs and adipose stromal cells four times over a 6-day period and reprogrammed 0.002% of the cells 20 days after the first transfection (22). There have been no published reports replicating results with any of the three variations of miRNA reprogramming so it is difficult to determine if this is a robust reprogramming method. If the efficiency of transfection of miRNAs could be improved and a canonical set of reprogramming miRNAs were identified, this could be a promising tool for reprogramming iPSCs.

2.3.3 PiggyBac

PiggyBac is a mobile genetic element (transposon) that in the presence of a transposase can be integrated into chromosomal TTAA sites and subsequently excised from the genome footprint-free upon re-expression of the transposase. When cloned into a piggyBac vector and co-transfected into MEFs the Yamanaka factors can reprogram 0.02–0.05% of cells 14–25 days post-transfection (23, 24). The piggyBac vector could be cleanly excised from the iPSCs upon re-expression of the transposase. Human mesenchymal stem cells (MSCs) were reprogrammed at an efficiency of 0.02% using piggyBac with the addition of sodium butyrate; however, this was 50-fold less efficient than retroviral-mediated reprogramming of MSCs (25). PiggyBac appears to be a promising method for reprogramming mouse iPSCs but there is no published study showing that the vector can be excised from human iPSCs or that cell types other than MSCs can be adequately reprogrammed.

2.3.4 Minicircle Vectors

Minicircle vectors are minimal vectors containing only the eukaryotic promoter and cDNA(s) that will be expressed. A Lin28, GFP, Nanog, Sox2, and Oct4 minicircle vector expressed in human adipose stromal cells was able to reprogram 0.005% of the cells in ~28 days (26). The method was even less efficient at reprogramming neonatal fibroblasts and there are no published reports of successful reprogramming of other somatic cells. Therefore, more validation is necessary before this method can be widely used.

2.3.5 Episomal Plasmids

Transient expression of reprogramming factors as episomal plasmids would allow for the generation of footprint-free iPSCs. However, transient transfection with a standard plasmid vector does not result in expression for a long enough period of time to reprogram cells unless transfections are repeated daily and even then reprogramming efficiency is unacceptably low (27). OriP/EBNA-based plasmids are stably expressed for a longer period of time than standard plasmids, and therefore have been used to express reprogramming factors for the generation of iPSCs. A single transfection of three oriP/EBNA plasmids containing Oct4, Sox2, Nanog, Klf4; Oct4, Sox2, SV40 large T antigen; and c-myc and Lin28 into human foreskin fibroblasts resulted in 0.0003–0.0006% of cells being reprogrammed ~20 days post-transfection (28). Only one-third of the subclones from two of the original iPSC lines lost the episomal plasmid. Another study found similarly low levels of reprogramming in cord blood cells, although addition of thiazovivin enhanced the process tenfold. However, 0.035% of cord blood mononuclear cells were reprogrammed by day 12 (29). Unlike the study of Yu et al. (28), this study found that all iPSC lines lost the plasmid by passage 15. Co-expression of EBNA mRNA with the reprogramming vectors and reprogramming in defined E8 media under hypoxic conditions significantly improved reprogramming efficiency in fibroblasts to 0.006–0.1% with the variation attributable to the characteristics in the parental fibroblast lines that were being reprogrammed (30).

New oriP/EBNA vectors were constructed with the Yamanaka factors plus Lin28 in one cassette and another oriP/EBNA vector containing SV40 large T antigen (31). These vectors were expressed in CD34+ cord blood, peripheral blood, and bone mononuclear cells in media supplemented with sodium butyrate, resulting in iPSC colonies in 14 days in 0.02, 0.009, and 0.005% of the cells, respectively. The transfected plasmids were lost by passage 12 with subsequent whole genome sequencing confirming loss of the plasmids in the bone marrow-derived iPSCs (32). Overall episomal reprogramming appears to be an effective strategy to generate footprint-free iPSCs with the only negative being an inability to reprogram fibroblasts at an acceptable efficiency without modifications to the way cells are cultured.

3 Oocyte Reprogramming as an Alternative Technology

The reprogramming of a somatic cell genome by transfer to enucleated oocytes has been shown to occur quickly and with nearly 100% efficiency in several species including mouse (33). However, ethical considerations and technical challenges have slowed progress of somatic cell nuclear transfer in humans. A recent report achieved positive results with reprogramming patient fibroblasts by transferring the fibroblast nucleus into an oocyte in which the nucleus was not removed (34). The resulting reprogrammed cells had expression profiles and epigenetic signatures that were very similar to pluripotent stem cell lines. Significant improvements still need to be made to technique as the reprogrammed cell types are not competent for therapeutic use because they are triploid. Future advances in this technology may further broaden the methods available to routinely reprogram somatic cells.

4 Enhancing Reprogramming Efficiency for Recalcitrant Cells

Even when using the same method there can be great variability in iPSC efficiency—especially in patient-specific disease lines. This variability is most likely attributable to the parental line that is to be reprogrammed, and could be due to disease-specific mutation(s) or an issue related to how the tissue source was collected, expanded, and stored long term. Various small molecules have been shown to enhance reprogramming efficiency (Table 1). Several known mechanisms enable these molecules to facilitate reprogramming including inhibition of histone deacetylation (25, 35), blockade of the TGFβ and MEK signaling pathways (36, 37), enhancement of function of epigenetic modifiers (38), inhibition of the ROCK pathway (34), and induction of glycolysis (39). Amongst these small molecules the histone deacetylatase inhibitors valproic acid and sodium butyrate are the most commonly used in reprogramming protocols. It should also be noted that culture of cells in 5% oxygen during the reprogramming process can also increase efficiency of iPSC derivation ~5-fold in mice and threefold in human cells (40). For samples that are particularly difficult to reprogram, the addition of a small molecule and culture in hypoxic conditions may yield enough improvements to generate iPSC clones (Table 2). The immense datasets from whole transcriptome studies of pluripotency networks are also likely to yield new transcription factors that can be added to the combinations that are currently in use or be used in new combinations to improve reprogramming efficiency (41–43).

One intriguing alternative to reprogramming difficult somatic cell lines may be to use embryonic stem cell-conditioned medium (ESCM) to induce expression of endogenous reprogramming factors.

Table 1
Compounds increasing iPSC reprogramming efficiency

Treatment	Process affected
Valproic acid	Histone deacetylase inhibition
Sodium butyrate	Histone deacetylase inhibition
PD0325901	MEK inhibition
A-83-01	TGFβ-inhibition
SB43152	TGFβ-inhibition
Vitamin C	Enhances epigenetic modifiers, promotes survival by antioxidant effects
Thiazovivin	ROCK inhibitor, promotes cell survival
PS48	PI3K/Akt activation, promotes glycolysis
5% Oxygen	Promotes glycolysis

Table 2
Increased efficiency of human iPSC generation with various treatment regimens

Treatment	Reprogramming method	Fold enhancement
Valproic acid	3-factor retroviral	>100 (33)
Sodium butyrate (NaB)	Episomal, piggyBac	15–50 (25)
SB431542 + PD0325901	4-factor retroviral	100 (34)
SB + PD + thiazovivin	4-factor retroviral	>200 (34)
A-83-01 + PD	4-factor retroviral	~90 (37)
PS48	4-factor retroviral	15 (37)
NaB + PS48	4-factor retroviral	~25 (37)
5% oxygen	4-factor retroviral	3 (38)

This strategy was used to reprogram rat limbal progenitors to iPSCs at an efficiency of 0.002% without the exogenous expression of any reprogramming factors (44). The efficiency improved to 0.008% with the addition of valproic acid. The limbal progenitors have endogenous expression of Klf4, Sox2, and c-myc which was further upregulated by culture in ESCM with Oct4 also being induced after 10 days of ESCM culture. Such a strategy may not be effective in more differentiated cell types but the use of conditioned media along with small molecules may enhance the ability of exogenously expressed reprogramming factors to increase reprogramming efficiency, particularly in cells that are otherwise difficult to reprogram.

5 Choosing a Reprogramming Method

The primary factors to consider when deciding on a reprogramming method to produce iPSCs are what cell is being reprogrammed, the capacity of the reprogramming method to adequately reprogram this cell type, and whether the presence of integrated sequences in the iPSCs will hinder downstream application. How these factors are weighted will be dependent on the goals of the project. The reprogramming methods can be divided into six groups based upon efficiency, footprint, and number of different somatic cell types known to be reprogrammed by the method (Fig. 1). If there are no long-term translational goals for the iPSCs, a viral infection with the STEMCCA will suffice as this method works for many cell types and offers the option of excising the integrated sequences with Cre-recombinase at a later time point. Projects with translational aspirations should utilize a completely footprint-free method with consideration of whether fibroblasts or blood cells will be reprogrammed. Sendai virus works well with all cell types but requires ~10 passages for the generated iPSCs to be footprint-free. Reprogramming using the episomal method is excellent for blood cells but requires modification of standard culture conditions for fibroblasts. There are also differences in the amount of time it takes to lose the footprint between episomal- and Sendai-based methods. There is no need to worry about the vestiges of a footprint remaining in iPSCs reprogrammed by mRNA, but the method is cumbersome and as of now only appears to work with fibroblasts. PiggyBac may be an attractive alternative but published studies in human cells are limited and excision of the piggyBac insertion has not been reported in human iPSCs.

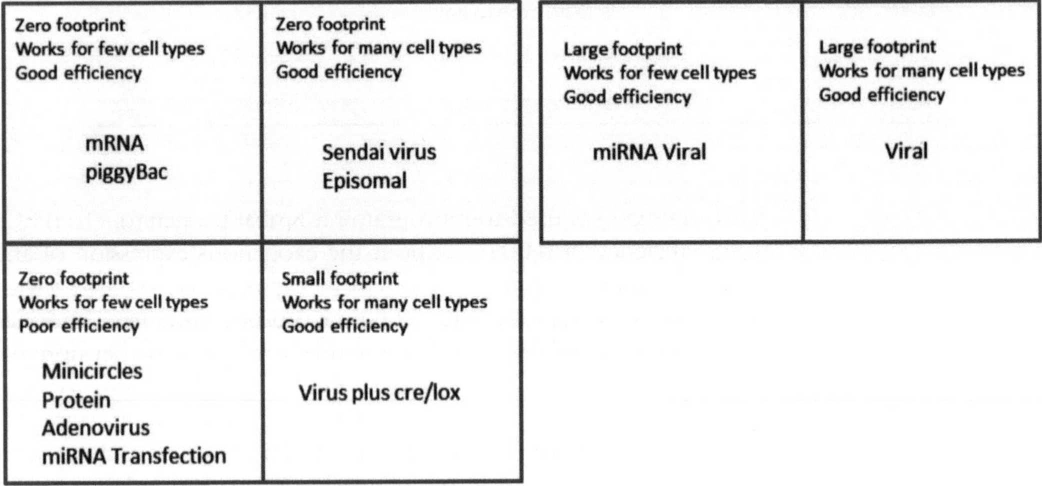

Fig. 1 Classification of reprogramming methods by footprint, validation that multiple cell types can be reprogrammed, and efficiency. Further details are found in the text

The remaining methods that have been discussed all either have severe limitations or have not been validated with the stringency required to express full confidence in their capabilities. Although there is no universal method that can handle every situation at least one of the methods described in this chapter should be able to cover nearly every need for a researcher attempting to produce iPSCs.

Acknowledgment

We thank Anastasia Efthymiou for comments on the manuscript.

References

1. Takahashi K, Yamanaka S (2006) Induction of pluripotent stem cells from mouse embryonic and adult fibroblast cultures by defined factors. Cell 126:663–676
2. Takahashi K, Tanabe K, Ohnuki M, Narita M, Ichisaka T, Tomoda K, Yamanaka S (2007) Induction of pluripotent stem cells from adult human fibroblasts by defined factors. Cell 131:861–872
3. Yu J, Vodyanik MA, Smuga-Otto K, Antosiewicz-Bourget J, Frane JL, Tian S, Nie J, Jonsdottir GA, Ruotti V, Stewart R, Slukvin II, Thomson JA (2007) Induced pluripotent stem cell lines derived from human somatic cells. Science 318:1917–1920
4. Park IH, Zhao R, West JA, Yabuuchi A, Huo H, Ince TA, Lerou PH, Lensch MW, Daley GQ (2008) Reprogramming of human somatic cells to pluripotency with defined factors. Nature 451:141–146
5. Wernig M, Meissner A, Foreman R, Brambrink T, Ku M, Hochedlinger K, Bernstein BE, Jaenisch R (2007) In vitro reprogramming of fibroblasts into a pluripotent ES-cell-like state. Nature 448:318–324
6. Papapetrou EP, Tomishima MJ, Chambers SM, Mica Y, Reed E, Menon J, Tabar V, Mo Q, Studer L, Sadelain M (2009) Stoichiometric and temporal requirements of Oct4, Sox2, Klf4, and c-Myc expression for efficient human iPSC induction and differentiation. Proc Natl Acad Sci USA 106:12759–12764
7. Carey BW, Markoulaki S, Hanna J, Saha K, Gao Q, Mitalipova M, Jaenisch R (2009) Reprogramming of murine and human somatic cells using a single polycistronic vector. Proc Natl Acad Sci USA 106:157–162
8. Chang CW, Lai YS, Pawlik KM, Liu K, Sun CW, Li C, Schoeb TR, Townes TM (2009) Polycistronic lentiviral vector for "hit and run" reprogramming of adult skin fibroblasts to induced pluripotent stem cells. Stem Cells 27:1042–1049
9. Sommer CA, Stadtfeld M, Murphy GJ, Hochedlinger K, Kotton DN, Mostoslavsky G (2009) Induced pluripotent stem cell generation using a single lentiviral stem cell cassette. Stem Cells 27:543–549
10. Soldner F, Hockemeyer D, Beard C, Gao Q, Bell GW, Cook EG, Hargus G, Blak A, Cooper O, Mitalipova M et al (2009) Parkinson's disease patient-derived induced pluripotent stem cells free of viral reprogramming factors. Cell 136:964–977
11. Somers A, Jean JC, Sommer CA, Omari A, Ford CC, Mills JA, Ying L, Sommer AG, Jean JM, Smith BW et al (2010) Generation of transgene-free lung disease-specific human induced pluripotent stem cells using a single excisable lentiviral stem cell cassette. Stem Cells 28:1728–1740
12. Stadtfeld M, Nagaya M, Utikal J, Weir G, Hochedlinger K (2008) Induced pluripotent stem cells generated without viral integration. Science 322:945–949
13. Zhou W, Freed CR (2009) Adenoviral gene delivery can reprogram human fibroblasts to induced pluripotent stem cells. Stem Cells 27:2667–2674
14. Fusaki N, Ban H, Nishiyama A, Saeki K, Hasegawa M (2009) Efficient induction of transgene-free human pluripotent stem cells using a vector based on Sendai virus, an RNA virus that does not integrate into the host genome. Proc Jpn Acad Ser B Phys Biol Sci 85:348–362
15. Seki T, Yuasa S, Oda M, Egashira T, Yae K, Kusumoto D, Nakata H, Tohyama S, Hashimoto H, Kodaira M et al (2010) Generation of induced pluripotent stem cells from human terminally differentiated circulating T cells. Cell Stem Cell 7:11–14
16. Ban H, Nishishita N, Fusaki N, Tabata T, Saeki K, Shikamura M, Takada N, Inoue M, Hasegawa M, Kawamata S, Nishikawa SI (2011) Proc Natl Acad Sci USA 108:14234–14239

17. Zhou H, Wu S, Joo JY, Zhu S, Han DW, Lin T, Trauger S, Bien G, Yao S, Zhu Y et al (2009) Generation of induced pluripotent stem cells using recombinant proteins. Cell Stem Cell 4:381–384
18. Kim D, Kim CH, Moon JI, Chung YG, Chang MY, Han BS, Ko S, Yang E, Cha KY, Lanza R et al (2009) Generation of human induced pluripotent stem cells by direct delivery of reprogramming proteins. Cell Stem Cell 4:472–476
19. Warren L, Manos PD, Ahfeldt T, Loh YH, Li H, Lau F, Ebina W, Mandal PK, Smith ZD, Meissner A et al (2010) Highly efficient reprogramming to pluripotency and directed differentiation of human cells with synthetic modified mRNA. Cell Stem Cell 7:618–630
20. Subramanyam D, Lamouille S, Judson RL, Liu JY, Bucay N, Derynck R, Blelloch R (2011) Multiple targets of miR-302 and miR-372 promote reprogramming of human fibroblasts to induced pluripotent stem cells. Nature Biotechnol 29:443–448
21. Anokye-Danso F, Trivedi CM, Juhr D, Gupta M, Cui Z, Tian Y, Zhang Y, Yang W, Gruber PJ, Epstein JA et al (2011) Highly efficient miRNA-mediated reprogramming of mouse and human somatic cells to pluripotency. Cell Stem Cell 8:376–388
22. Miyoshi N, Ishii H, Nagano H, Haraguchi N, Dewi DL, Kano Y, Nishikawa S, Tanemura M, Mimori K, Tanaka F et al (2011) Reprogramming of mouse and human cells to pluripotency using mature microRNAs. Cell Stem Cell 8:633–638
23. Kaji K, Norrby K, Paca A, Mileikovsky M, Mohseni P, Woltjen K (2009) Virus-free induction of pluripotency and subsequent excision of reprogramming factors. Nature 458:771–775
24. Woltjen K, Michael IP, Mohseni P, Desai R, Mileikovsky M, Hamalainen R, Cowling R, Wang W, Liu P, Gertsenstein M et al (2009) piggyBac transposition reprograms fibroblasts to induced pluripotent stem cells. Nature 458:766–770
25. Mali P, Chou BK, Yen J, Ye Z, Zou J, Dowey S, Brodsky RA, Ohm JE, Yu W, Baylin SB et al (2010) Butyrate greatly enhances derivation of human induced pluripotent stem cells by promoting epigenetic remodeling and the expression of pluripotency-associated genes. Stem Cells 28:713–720
26. Narsinh KH, Jia F, Robbins RC, Kay MA, Longaker MT, Wu JC (2011) Generation of adult human induced pluripotent stem cells using nonviral minicircle DNA vectors. Nature Protoc 6:78–88
27. Okita K, Nakagawa M, Hyenjong H, Ichisaka T, Yamanaka S (2008) Generation of mouse induced pluripotent stem cells without viral vectors. Science 322:949–953
28. Yu J, Vodyanik MA, Smuga-Otto K, Antosiewicz-Bourget J, Frane JL, Tian S, Nie J, Jonsdottir GA, Ruotti V, Stewart R et al (2007) Induced pluripotent stem cell lines derived from human somatic cells. Science 318:1917–1920
29. Hu K, Yu J, Suknuntha K, Tian S, Montgomery K, Choi KD, Stewart R, Thomson JA, Slukvin II (2011) Efficient generation of transgene-free induced pluripotent stem cells from normal and neoplastic bone marrow and cord blood mononuclear cells. Blood 117:e109–e119
30. Chen G, Gulbranson DR, Hou Z, Bolin JM, Ruotti V, Probasco MD, Smuga-Otto K, Howden SE, Diol NR, Propson NE et al (2011) Chemically defined conditions for human iPSC derivation and culture. Nat Methods 8:24–429
31. Chou BK, Mali P, Huang X, Ye Z, Dowey SN, Resar LM, Zou C, Zhang YA, Tong J, Cheng L (2011) Efficient human iPS cell derivation by a non-integrating plasmid from blood cells with unique epigenetic and gene expression signatures. Cell Res 21:518–529
32. Cheng L, Hansen NF, Zhao L, Du Y, Zou C, Donovan FX, Chou BK, Zhou G, Li S, Dowey SN et al (2012) Low incidence of DNA sequence variation in human induced pluripotent stem cells generated by nonintegrating plasmid expression. Cell Stem Cell 10:337–344
33. Wakayama T, Perry AC, Zuccotti M, Johnson KR, Yanagimachi R (1998) Full-term development of mice from enucleated oocytes injected with cumulus. Nature 394:369–374
34. Noggle S, Fung H-L, Gore A, Martinez H, Satriani KS, Prosser R, Oum K, Paull D, Druckenmiller S, Freeby M et al (2011) Human oocytes reprogram somatic cells to a pluripotent state. Nature 478:70–75
35. Huangfu D, Osafune K, Maehr R, Guo W, Eijkelenboom A, Chen S, Melton DA (2008) Induction of pluripotent stem cells from primary human fibroblasts with only *Oct4* and *Sox2*. Nat Biotechnol 26:1269–1275
36. Lin T, Ambasudhan R, Yuan X, Li W, Hilcove S, Abujarour R, Lin X, Hahm HS, Hao E, Hayek A, Ding S (2009) A chemical platform for improved induction of human iPSCs. Nat Methods 6:805–808
37. Ichida JK, Blanchard J, Lam K, Son EY, Chung JE, Egli D, Loh KM, Carter AC, DiGiorgio FP, Koszka K et al (2009) A small-molecule inhibitor of TGF-β signaling replaces *Sox2* in reprogramming by inducing *nanog*. Cell Stem Cell 5:491–503
38. Esteban MA, Wang T, Qin B, Yang J, Qin D, Cai J, Li W, Weng Z, Chen J, Ni S et al (2010) Vitamin C enhances the generation of mouse and human induced pluripotent stem cells. Cell Stem Cell 6:71–79

39. Zhu S, Li W, Zhou H, Wei W, Ambasudhan R, Lin T, Kim J, Zhang K, Ding S (2010) Reprogramming of human primary somatic cells by OCT4 and chemical compounds. Cell Stem Cell 7:651–655
40. Yoshida Y, Takahashi K, Okita K, Ichisaka T, Yamanaka S (2009) Hypoxia enhances the generation of induced pluripotent stem cells. Cell Stem Cell 5:237–241
41. Boyer LA, Lee TI, Cole MF, Johnstone SE, Levine SS, Zucker JP, Guenther MG, Kumar RM, Murray HL, Jenner RG et al (2005) Core transcriptional regulatory circuitry in human embryonic stem cells. Cell 122:947–956
42. Loh YH, Wu Q, Chew LW, Vega VB, Zhang W, Chen X, Bourque G, George J, Leong B, Liu J et al (2006) The Oct4 and Nanog transcription network regulates pluripotency in mouse embryonic stem cells. Nat Genet 38:431–440
43. Kim J, Chu J, Shen X, Wang J, Orkin SH (2008) An extended transcriptional network of pluripotency of embryonic stem cells. Cell 132:1049–1061
44. Balasubramanian S, Babai N, Chaudhuri A, Qiu F, Bhattacharya S, Dave BJ, Parameswaran S, Carson SD, Thoreson WB, Sharp JG et al (2009) Non cell-autonomous reprogramming of adult ocular progenitors: generation of pluripotent stem cells without exogenous transcription factors. Stem Cells 27:3053–3062

Chapter 4

Generation of Human-Induced Pluripotent Stem Cells by Lentiviral Transduction

Jennifer C. Moore

Abstract

Human somatic cells can be reprogrammed to the pluripotent state to become human-induced pluripotent stem cells (hiPSC). This reprogramming is achieved by activating signaling pathways that are expressed during early development. These pathways can be induced by ectopic expression of four transcription factors—Oct4, Sox2, Klf4, and c-Myc. Although there are many ways to deliver these transcription factors into the somatic cells, this chapter will provide protocols that can be used to generate hiPSC from lentiviruses.

Key words iPSC, Lentivirus, Reprogramming, Viral method

1 Introduction

Although the generation of the first human-induced pluripotent stem cell lines (hiPSC) was first published in 2006, the groundwork for this breakthrough had been laid down over the past 60 years (reviewed in ref. 1). In 1967 Stevens and Pierce independently published review papers summarizing the then current status of embryocarcinoma cells and in these papers they established the idea that developing embryos contain a population of cells capable of self-renewing and of differentiating into specific lineages (2, 3). These ideas became the foundation for the discovery and generation of the first mouse embryonic stem cell lines (mESC) in 1981 by both Evans and Martin (4, 5). Despite a number of technological advances in the field of mESC biology (the ability to generate chimeric mice and the ability to genetically modify mESC) the generation of human embryonic stem cells took another 17 years (1).

In 1998 Thomson and colleagues published the first paper detailing the isolation and characterization of human embryonic stem cells (6). Although initial publications showed that derivatives

Uma Lakshmipathy and Mohan C. Vemuri (eds.), *Pluripotent Stem Cells: Methods and Protocols*, Methods in Molecular Biology, vol. 997, DOI 10.1007/978-1-62703-348-0_4, © Springer Science+Business Media New York 2013

of all three germ layers could be generated from embryonic stem cells, the methods used to generate these cells types generally relied on spontaneous differentiation rather than on directed differentiation. Soon however, many different protocols to direct differentiation into particular lineages were developed (reviewed in ref. 7), and this allowed the cells to be commercially useful for drug screening and potentially useful for cellular transplantation therapies (reviewed in refs. 8 and 9). However, despite all the progress that has been made and the amount of effort put into helping these cells reach their full potential, several hurdles remain to their widespread therapeutic and commercial use. The first is the ethical concern arising from the destruction of human embryos. A second is the need to have cells that represent a population that is genetically diverse, and that consists of normal and disease states, particularly diseases that cannot be identified until adulthood. Finally, in order for proposed cellular transplantations to become a reality, cells that are a genetic match to the host must be readily obtainable.

Before 2006 it was generally accepted that once a cell underwent differentiation to generate a functional specialized cell it was forever locked out of the pluripotent state. The discovery that a group of transcription factors (Oct4, Sox2, Klf4, and c-Myc) could reprogram adult fibroblasts back into a pluripotent state in 2006 created completely new paradigms for understanding and recapitulating pluripotency (10). By 2007 two groups had successfully generated human-induced pluripotent stem cells using the same strategy (reviewed in ref. 11). In the following years additional cell types such as hepatocytes, blood cells, and keratinocytes and many others have all been used to generate hiPSC (11). An immediate advantage of this technique is the ability to obtain somatic cells from a diseased subject and to make iPSC capable of modeling the genotype, development, and treatment of the disease. Thus far, many different disease models have been generated (reviewed in ref. 12) and many more are in progress. Several have resulted in endpoint cells that display the affected phenotype such as spinal muscular atrophy (13) and long-QT syndrome (14), while a few have even showed improvement when treated with current or exploratory drug therapies (12).

To date, hiPSC have been successfully generated by many methods and by using many different combinations of transcription factors, small molecules, and microRNAs (reviewed in ref. 15). The earliest and most common method has been the delivery of Oct4, Sox2, Klf4, and c-Myc by retroviral vectors. The drawback of this method is the possibility of the retroviral integration causing insertional mutagenesis. To eliminate this concern, methods have been developed that utilized virally based vectors that can be excised using the CRE-recombinase-based system via LoxP sites (16, 17) or using the PiggyBac system, a method that completely removes all integrated viral DNA via the transposase enzyme (18, 19).

Other nonviral strategies have also been developed and these include transcription factor delivery by episomal plasmids (20), nonintegrating viruses (21), and small molecules (22). In this chapter we will present protocols to reprogram somatic cells to hiPSC using lentiviral vectors.

2 Materials

2.1 Equipment

1. 37°C water bath.
2. 5% CO_2 humidified incubator.
3. Benchtop centrifuge capable of centrifuging 15 mL screw-capped tubes.
4. Hemocytometer.
5. Mr. Frosty polypropylene cell-freezing container to freeze cells at a rate of 1°C/min.

2.2 Disposables

1. 15 mL screw-capped tubes.
2. 0.45 μM filter.
3. Pasteur pipets.
4. 6 well plates.
5. T25 flasks.
6. 10 cm dishes.
7. Ultracentrifuge tubes.

2.3 Medium

2.3.1 293FT Growth Medium

1. 435 mL high-glucose DMEM/F12.
2. 50 mL FBS.
3. 5 mL L-glutamine.
4. 5 mL NEAA.
5. 5 mL 100× pen/strep.

2.3.2 293 Cell-Freezing Medium

1. 10 mL growth medium.
2. 80 mL FBS.
3. 10 mL DMSO.

2.3.3 Fibroblast Growth Medium (hFib Medium)

1. 450 mL DMEM/F12.
2. 75 mL FBS.
3. 5 mL Nonessential amino acids.
4. 5 mL 100× pen/strep.

2.3.4 KOSR Medium

1. 400 mL DMEM/F12.
2. 100 mL Knockout Serum Replacer.

3. 5 mL Nonessential amino acids.
4. 5 mL glutamine.
5. 20 ng/mL bFGF.
6. 3.5 μL 14.3 M β-Mercaptoethanol.

2.3.5 mTeSR Medium

1. mTeSR(tm)1 (Stemcell Technologies, Vancouver, Canada).

2.4 Transfection Materials

1. Lipofectamine.
2. Opti-MEM media.
3. Endo-free lentiviral vectors. Use of endo-free viral vector preps can greatly increase your packaging efficiency leading to higher titer virus preps.

2.5 Chemicals

1. Ca^{2+}/Mg^{2+}-free phospho-buffered saline.
2. 0.01% gelatin.
3. G418.
4. Trypsin.
5. Accutase.
6. mTeSR.
7. hESC-qualified Matrigel.
8. Polybrene.
9. Rho-Kinase inhibitor (Y27632).

2.6 Cells

1. 293FT cells.
2. Fibroblasts.
3. mEFs.

3 Methods

3.1 293FT Culture

3.1.1 Gelatin Coating (See Note 1)

1. Coat 10 cm dishes with enough gelatin (approximately 4 mL) to cover the surface of the dish.
2. Incubate at 37°C for 30 min.
3. Unused dishes can be wrapped in Parafilm and stored at 4°C for up to 2 weeks.

3.1.2 Thawing

1. Place vial of frozen cells in 37°C water bath until cells are thawed (see Note 2). Transfer thawed cell suspension to 9 mL of 293FT growth medium in a 15 mL screw-capped tube.
2. Pellet at $200 \times g$ for 5 min.
3. Aspirate media and resuspend in 10 mL of 293FT growth medium.
4. Plate ~2.5×10^4 cells/cm^2 on pre-warmed gelatin-coated dish.

5. Add 500 μg/mL G418.
6. Feed every other day with 10 mL 293FT growth medium.

3.1.3 Passaging

1. Rinse plate twice with Ca^{2+}/Mg^{2+}-free PBS.
2. Add enough trypsin to cover cells (~3 mL).
3. Incubate at 37°C for 5 min.
4. Examine dishes under a microscope to ensure nearly all cells are detached, if necessary sharply tap plate or flask to detach remaining cells.
5. Transfer cell suspension to a 15 mL screw-capped tube containing ~7 mL of 293 growth medium.
6. Centrifuge at 200 × *g* for 5 min and remove the supernatant.
7. Resuspend in 10 mL of 293 growth medium.
8. Add 500 μg/mL Geneticin.
9. Plate on gelatin-coated dish at ~2.5e4 cells/cm^2 and transfer to a humidified 37°C incubator.

3.1.4 Freezing

1. Trypsinize cells as described in the above section.
2. Count using a hemocytometer.
3. Resuspend 1–2 million cells in 1 mL ice-cold freezing medium.
4. Freeze in Mr. Frosty cell freezer at −80°C overnight.
5. After 24 h at −80°C, transfer to liquid nitrogen for long-term storage.

3.2 Lentivirus

3.2.1 Lentivirus Generation

1. On a gelatin-coated dish, plate 293FT cells at 8×10^4 cells/cm^2 in 293FT growth medium + 500 μg/mL G418 and grow overnight in a humidified 37°C incubator.
2. The next morning, replace media on 293FT cells with 293FT growth medium without antibiotics (see Note 3).
3. Mix 45 μL of Lipofectamine and 1.5 mL of Opti-MEM medium and incubate at room temperature for 5 min (see Note 4).
4. Mix a total of 18 μg of iPSC reprogramming and helper plasmids in 1.5 mL of Opti-MEM and incubate at room temperature for 5 min (see Note 5). Lentiviral vectors and their packaging systems are reviewed in ref. (23).
5. Mix the Lipofectamine/Opti-MEM and DNA/Opti-MEM mixtures together and incubate at room temperature for 20 min.
6. Add the Lipofectamine/DNA/Opti-MEM mixture dropwise to the 293FT cells.
7. Incubate for 48 h at 37°C in a humidified incubator.
8. Collect supernatant and filter with 0.45 μm filter.

3.2.2 Lentiviral Concentration

1. Concentrate lentivirus by centrifuging the filtered medium for 1.5 h at 82,000 × *g* at 4°C.
2. Discard the supernatant.
3. Resuspend the virus in PBS at 100–200× (see Note 6).
4. Aliquot into single-use amounts and store at −80°C.

3.3 Fibroblast Culture (See Note 7)

3.3.1 Thawing

1. Place vial of frozen fibroblasts in 37°C water bath until cells are thawed (see Note 2).
2. Transfer to a 15 mL screw-capped tube containing 9 mL of hFib media.
3. Centrifuge at 200 × *g* for 5 min and remove the supernatant.
4. Resuspend in 5 mL of hFib media.
5. Plate cell suspension in T25 flask at ~1.25e4 cells/cm^2.
6. Feed every other day with hFib media.

3.3.2 Passaging

1. Rinse flask twice with Ca^{2+}/Mg^{2+}-free PBS.
2. Add enough trypsin to cover cells (~3 mL).
3. Incubate flask at 37°C for 5 min.
4. Examine dishes under a microscope to ensure nearly all cells are detached, if necessary sharply tap plate or flask to detach remaining cells.
5. Transfer to 15 mL screw-capped tube containing ~7 mL of hFib media.
6. Centrifuge at 200 × *g* for 5 min and remove the supernatant.
7. Resuspend in 5 mL of hFib media.
8. Plate in T25 flask at ~1.25e4 cells/cm^2 and grow in a humidified incubator at 37°C.
9. Feed every other day with hFib media.

3.4 iPSC Generation (See Fig. 1 for a Timeline for Reprogramming) (See Note 8)

3.4.1 Day 1: Plating Fibroblasts

1. Plate fibroblasts at 9×10^3 cells/cm^2 in hFib medium.
2. Incubate overnight at 37°C in a humidified incubator.

3.4.2 Day 2: Infecting Fibroblasts

1. Refresh hFib medium and add 8 μg/mL polybrene.
2. Infect with lentivirus at an MOI of 5.
3. Incubate overnight at 37°C in a humidified incubator.

3.4.3 Days 3–7: Growth in hFib Medium

1. Replace hFib medium every other day.
2. Prepare 10 cm mEF plates on day 6 as described in Lin et al. (24).

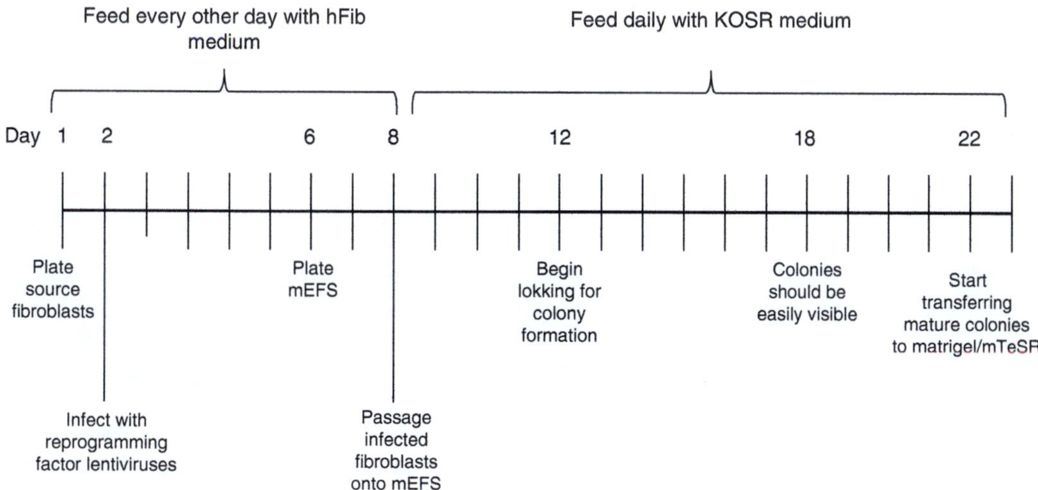

Fig. 1 Timeline for iPSC generation

3.4.4 Day 8: Passage onto mEFs

1. Rinse flask twice with Ca^{2+}/Mg^{2+}-free PBS.
2. Add enough trypsin to cover cells (~3 mL).
3. Incubate at 37°C for 5 min.
4. Examine dishes under a microscope to ensure nearly all cells are detached, if necessary sharply tap plate or flask to detach remaining cells.
5. Transfer cell suspension to 15 mL screw-capped tube containing ~7 mL of hFib media.
   (a) Centrifuge at 200×*g* for 5 min and remove the supernatant.
   (b) Resuspend in 10 mL of hFib media.
   (c) Plate 5×10^5 and 2.5×10^5 cells on mEF dishes and grow in a humidified incubator at 37°C.

3.4.5 Days 9–18: Colony Formation

1. Replace medium daily with KOSR.
2. Examine dishes under the microscope for colony formation (see Note 9).

3.4.6 Days 19–24: Matrigel Plate Preparation and Colony Picking

1. Reprogrammed colonies should now be visible as slightly raised colonies with sharply defined edges (see Fig. 2).
2. Prepare Matrigel dishes according to the protocol supplied with the Matrigel.
3. Using a pulled Pasteur pipet, gently break up and scrape off reprogrammed colonies.
4. Using a 100 μL pipet, pick up colony pieces and transfer to Matrigel plates containing mTeSR with 5 μM Rho-Kinase Inhibitor (Y27632).
5. Feed daily with mTeSR.

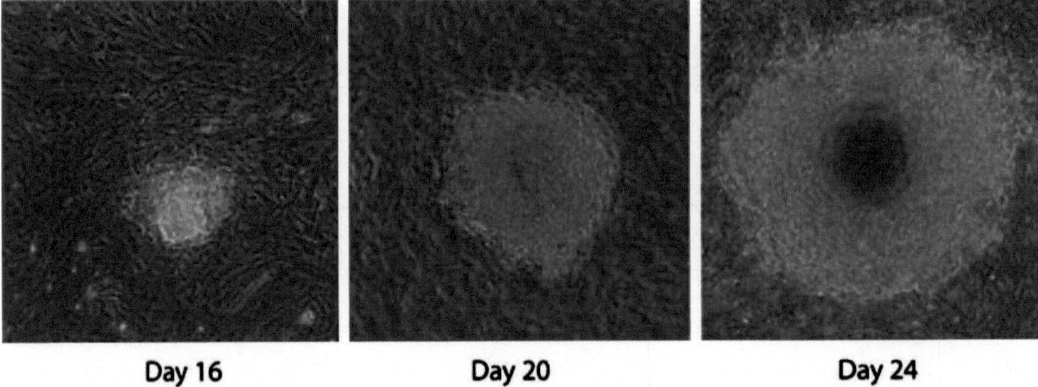

Fig. 2 Reprogrammed fibroblast colonies at days 16, 20, and 24 after initial reprogramming. Colonies initially appear as small aggregates of cells that develop into more tightly packed colonies with sharply defined edges. Colonies are ready for transfer to Matrigel when they reach 0.1–0.5 mm

4 Notes

1. This protocol is for growing 293FT cells in 10 cm dishes, but any size dish or flask can be used and the reagent amounts listed can be scaled accordingly.
2. Immediately remove cell vial from water bath upon thawing.
3. The presence of antibiotics can reduce lentiviral packaging efficiency.
4. Lipofectamine should be added directly to the Opti-MEM without touching the sides of the tube to increase transfection efficiency.
5. The helper plasmids and amounts of each will depend on whether you are using second- or third-generation lentivirus system.
6. If you started with 100 mL of filtered viral-containing medium from the 293FT cells you would resuspend in 100 μL for 100× concentration or 50 μL for 200× concentration.
7. This protocol is for growing fibroblasts in T25 flasks, but any size dish or flask can be used and the reagent amounts listed can be scaled accordingly.
8. This protocol is for generating iPSC in a 10 cm dish, but any size dish or flask can be used and the reagent amounts listed can be scaled accordingly.
9. Colonies should begin forming around day 14 and appear first as groups of cells that have large nuclei and are much smaller than the surrounding mEFs.

Acknowledgment

Supported by NIH U24 MH068457, RC1 CA147187, and R21 DA032984-01.

References

1. Evans M (2011) Discovering pluripotency: 30 years of mouse embryonic stem cells. Nat Rev Mol Cell Biol 12:680–686
2. Pierce GB (1967) Teratocarcinoma: model for a developmental concept of cancer. Curr Top Dev Biol 2:223–246
3. Stevens LC (1967) The biology of teratomas. Adv Morphog 6:1–31
4. Evans MJ, Kaufman MH (1981) Establishment in culture of pluripotential cells from mouse embryos. Nature 292:154–156
5. Martin GR (1981) Isolation of a pluripotent cell line from early mouse embryos cultured in medium conditioned by teratocarcinoma stem cells. Proc Natl Acad Sci USA 78:7634–7638
6. Thomson JA, Itskovitz-Eldor J, Shapiro SS, Waknitz MA, Swiergiel JJ, Marshall VS, Jones JM (1998) Embryonic stem cell lines derived from human blastocysts. Science 282:1145–1147
7. Yabut O, Bernstein HS (2011) The promise of human embryonic stem cells in aging-associated diseases. Aging 3:494–508
8. Jensen J, Hyllner J, Bjorquist P (2009) Human embryonic stem cell technologies and drug discovery. J Cell Physiol 219:513–519
9. Leeb C, Jurga M, McGuckin C, Forraz N, Thallinger C, Moriggl R, Kenner L (2011) New perspectives in stem cell research: beyond embryonic stem cells. Cell Prolif 44(Suppl 1):9–14
10. Takahashi K, Yamanaka S (2006) Induction of pluripotent stem cells from mouse embryonic and adult fibroblast cultures by defined factors. Cell 126:663–676
11. Vitale AM, Wolvetang E, Mackay-Sim A (2011) Induced pluripotent stem cells: a new technology to study human diseases. Int J Biochem Cell Biol 43:843–846
12. Wu SM, Hochedlinger K (2011) Harnessing the potential of induced pluripotent stem cells for regenerative medicine. Nat Cell Biol 13:497–505
13. Ebert AD, Yu J, Rose FF Jr, Mattis VB, Lorson CL, Thomson JA, Svendsen CN (2009) Induced pluripotent stem cells from a spinal muscular atrophy patient. Nature 457:277–280
14. Moretti A, Bellin M, Welling A, Jung CB, Lam JT, Bott-Flugel L, Dorn T, Goedel A, Hohnke C, Hofmann F, Seyfarth M, Sinnecker D, Schomig A, Laugwitz KL (2010) Patient-specific induced pluripotent stem-cell models for long-QT syndrome. N Engl J Med 363:1397–1409
15. Hussein SM, Nagy K, Nagy A (2011) Human induced pluripotent stem cells: the past, present, and future. Clin Pharmacol Ther 89: 741–745
16. Soldner F, Hockemeyer D, Beard C, Gao Q, Bell GW, Cook EG, Hargus G, Blak A, Cooper O, Mitalipova M, Isacson O, Jaenisch R (2009) Parkinson's disease patient-derived induced pluripotent stem cells free of viral reprogramming factors. Cell 136:964–977
17. Sommer CA, Sommer AG, Longmire TA, Christodoulou C, Thomas DD, Gostissa M, Alt FW, Murphy GJ, Kotton DN, Mostoslavsky G (2010) Excision of reprogramming transgenes improves the differentiation potential of iPS cells generated with a single excisable vector. Stem Cells 28:64–74
18. Kaji K, Norrby K, Paca A, Mileikovsky M, Mohseni P, Woltjen K (2009) Virus-free induction of pluripotency and subsequent excision of reprogramming factors. Nature 458:771–775
19. Woltjen K, Michael IP, Mohseni P, Desai R, Mileikovsky M, Hamalainen R, Cowling R, Wang W, Liu P, Gertsenstein M, Kaji K, Sung HK, Nagy A (2009) piggyBac transposition reprograms fibroblasts to induced pluripotent stem cells. Nature 458:766–770
20. Si-Tayeb K, Noto FK, Sepac A, Sedlic F, Bosnjak ZJ, Lough JW, Duncan SA (2010) Generation of human induced pluripotent stem cells by simple transient transfection of plasmid DNA encoding reprogramming factors. BMC Dev Biol 10:81
21. Seki T, Yuasa S, Oda M, Egashira T, Yae K, Kusumoto D, Nakata H, Tohyama S, Hashimoto H, Kodaira M, Okada Y, Seimiya H, Fusaki N, Hasegawa M, Fukuda K (2010) Generation of induced pluripotent stem cells from human terminally differentiated circulating T cells. Cell Stem Cell 7:11–14
22. Miyoshi N, Ishii H, Nagano H, Haraguchi N, Dewi DL, Kano Y, Nishikawa S, Tanemura M, Mimori K, Tanaka F, Saito T, Nishimura J, Takemasa I, Mizushima T, Ikeda M, Yamamoto H, Sekimoto M, Doki Y, Mori M (2011) Reprogramming of mouse and human cells to pluripotency using mature microRNAs. Cell Stem Cell 8:633–638
23. Delenda C (2004) Lentiviral vectors: optimization of packaging, transduction and gene expression. J Gene Med 6(Suppl 1):S125–S138
24. Lin S, Talbot P (2011) Methods for culturing mouse and human embryonic stem cells. Methods Mol Biol 690:31–56

Chapter 5

Generation of Induced Pluripotent Stem Cells with CytoTune, a Non-Integrating Sendai Virus

Pauline T. Lieu, Andrew Fontes, Mohan C. Vemuri, and Chad C. MacArthur

Abstract

One of the major obstacles in generating induced pluripotent stem cells for research or downstream applications is the potential modifications of the cellular genome as a result of using integrating viruses during reprogramming. Another major disadvantage of reprogramming cells with integrating vectors is that silencing and activation of transgenes is unpredictable, which may affect terminal differentiation potential and increase the risk of using iPSC-derived cells. Here we describe a protocol for the generation of induced pluripotent stem cells using a non-integrating RNA virus, Sendai virus, to efficiently generate transgene-free iPSCs starting with different cell types as well as in feeder-free conditions.

Key words Induced pluripotent stem cells, Non-integrating, Sendai virus, High efficiency, Human cells, Reprogramming, Feeder free

1 Introduction

Generation of induced pluripotent stem cells (iPSCs) from somatic cells by transducing four transcription factors using retroviral vectors (1–4) has opened a new area of biology for personalized medicine using patient-derived iPSCs (4–7). However, the major limitation for potential clinical application is the integration of viral transgenes into the host genome that can result in multiple insertions and risk of tumorigenicity (8, 9). Although the process of obtaining iPSC lines is technically simple, reprogramming is a slow and inefficient process, and multiple methods have been developed to address these issues (10). However, the majority of these methods has one or more limitations, such as low reprogramming efficiency or requiring multiple rounds of transfections or is effective only with specific cell types, such as skin fibroblasts. Here we report a robust and an efficient system to generate transgene-free iPSCs in different conditions, and its ease of use can be applied to

Uma Lakshmipathy and Mohan C. Vemuri (eds.), *Pluripotent Stem Cells: Methods and Protocols*, Methods in Molecular Biology, vol. 997, DOI 10.1007/978-1-62703-348-0_5, © Springer Science+Business Media New York 2013

a wide range of different cell types. Sendai virus is a negative-strand RNA virus that belongs to the *Paramyxoviridae* family (11–13). Unlike other RNA viruses, it replicates in the cytoplasm of infected cells and does not go through a DNA phase that can integrate into the host genome (11). In addition, Sendai virus can infect a broad host range and is nonpathogenic to humans. Recent papers also demonstrated that Sendai virus can reprogram somatic cells with higher efficiency than other reprogramming methods (14). Thus, the nature of Sendai virus makes it an ideal tool for cell reprogramming and stem cell research. Generation of transgene-free iPSCs under these conditions will be important to facilitate downstream applications of iPSC-based therapies.

2 Materials

2.1 Cells and Vectors

1. CytoTune®-iPS Reprogramming Kit, containing the four reprogramming virus particles (Life Technologies).
2. Mammalian somatic cells such as fibroblasts or CD34+ cells.
3. Optional: Human neonatal foreskin fibroblast cells (strain BJ; ATCC no. CRL2522) as a positive reprogramming control.
4. Mouse embryonic fibroblasts (irradiated).

2.2 Media Components and Reagents

All reagents are from Life Technologies unless otherwise specified.

2.2.1 For Fibroblasts

1. D-MEM with GlutaMAX™-I (high glucose).
2. KnockOut™ D-MEM/F-12.
3. Fetal Bovine Serum (FBS), ES Cell-Qualified.
4. KnockOut™ serum replacement (KSR).
5. MEM nonessential Amino Acids (NEAA).
6. GlutaMAX™-I Supplement.
7. Basic FGF, recombinant human.
8. β-Mercaptoethanol Penicillin-Streptomycin, liquid.
9. Attachment factor.
10. TrypLE™ select cell dissociation reagent or 0.05% Trypsin/ EDTA.
11. D-PBS without Ca^{2+} or Mg^{2+}.

2.2.2 Feeder-Free Reagents

1. D-MEM with GlutaMAX™-I (high glucose).
2. StemPro® hESC SFM.
3. Geltrex® hESC-qualified Reduced Growth Factor Basement Membrane Matrix.

4. β-Mercaptoethanol.
5. Basic FGF, recombinant human.

2.2.3 For CD34+ cells

1. StemPro®-34 SFM complete medium.
2. StemPro®-34 nutrient supplement.
3. L-Glutamine.
4. GM-CSF recombinant human.
5. IL3 recombinant human.
6. SCF recombinant human.

2.3 PCR Reagents

1. TRIzol® reagent.
2. SuperScript® VILO™ cDNA Synthesis Kit (Cat. no. 11754-050).
3. AccuPrime™ SuperMix I (Cat. no. 12342-010).

2.4 Equipment

1. Sterile cell culture hood (i.e., biosafety cabinet) equipped with a stereomicroscope.
2. Inverted microscope.
3. Incubator set at 37°C, 5% CO_2.
4. Water bath set at 37°C.
5. Sterile serological pipettes (5-mL, 10-mL).
6. Centrifuge.
7. 15-mL centrifuge tubes.
8. 60-mm and 100-mm tissue culture-treated dishes.
9. 6-well tissue culture-treated plates.
10. 25-gauge 1 in. needle.

2.5 Solutions

2.5.1 Complete Fibroblast Medium

To prepare 100 mL of complete MEF/fibroblast medium, add the following:

D-MEM with GlutaMAX™-I—1×	89 mL
FBS, ESC qualified—10%	10 mL
MEM nonessential amino acids solution 10 mM	1 mL

Complete MEF/fibroblast medium can be stored at 2–8°C for up to 1 week.

2.5.2 Basic FGF Stock Solution

To prepare 1,000 μL of 10-μg/mL Basic FGF solution, make the following:

Basic FGF	10 μg
D-PBS without Ca^{2+} and Mg^{2+}	980 μL
10% BSA	10 μL

Aliquot and store the Basic FGF solution at –20°C for up to 6 months.

2.5.3 Human iPSC Medium

To prepare 100 mL of human iPSC medium, add the following:

KnockOut™ D-MEM/F-12	89 mL
KnockOut™ serum replacement—20%	20 mL
MEM nonessential amino acids solution 10 mM	1 mL
GlutaMAX™-I supplement 100×	1 mL
β-Mercaptoethanol 55 mM	100 μL
Penicillin–Streptomycin (*optional*) 100×	1 mL
Basic FGF* 10 μg/mL	40 μL

*Prepare the iPSC medium without bFGF, and then supplement with fresh bFGF when the medium is used. Human iPSC medium can be stored at 2–8°C for up to 1 week.

2.5.4 Feeder-Free iPSC Medium: StemPro® hESC SFM

To prepare 100 mL of human iPSC medium, aseptically mix the components listed in the table below. Human iPSC medium can be stored at 2–8°C for up to 1 week.

D-MEM with GlutaMAX™-I 1×	90.8 mL
StemPro® hESC SFM growth 50× supplement	2.0 mL
BSA 25%	7.2 mL
β-Mercaptoethanol 55 mM	182 μL
Basic FGF* 10 μg/mL	80 μL

*Prepare the iPSC medium without bFGF, and then supplement with fresh bFGF when the medium is used.

2.5.5 StemPro®-34 Medium for Culturing CD34+

Thaw the frozen StemPro®-34 nutrient supplement in a 37°C water bath, mix well by gently vortexing, and aseptically transfer the complete contents of the bottle to the StemPro®-34 SFM. Swirl to obtain a homogenous complete medium. Aseptically add l-Glutamine (Cat. no. 25030) to a final concentration of 2 mM, by adding 5 mL 200 mMl-Glutamine to 500 mL medium. StemPro®-34 SFM complete medium should be stored at 2–8°C, in the dark. The complete medium has a shelf life of 30 days when stored as recommended.

StemPro®-34 + Cytokines

To prepare 10 mL of StemPro®-34 + cytokines, add the following:

GM-CSF recombinant human	0.1 mg/mL	2.5 μL
IL3 recombinant human	0.1 mg/mL	5.0 μL
SCF recombinant human	0.1 mg/mL	10.0 μL

StemPro®-34 + cytokines should be made fresh every other day.

2.5.6 Collagenase IV Solution

To prepare 50 mL of a 1-mg/mL Collagenase IV solution, add the following:

Collagenase IV	50 mg
KnockOut™ D-MEM/F-12	50 mL
Sterilize the Collagenase IV solution through a 0.2 μm filter.	
Aliquot and store the Collagenase IV solution at -20°C for up to 6 months.	

2.6 Preparation of Coated Dishes

2.6.1 Gelatin-Coating Culture Vessels

1. Cover the whole surface of each culture vessel with Attachment Factor (AF, Life Technologies) solution and incubate the vessels for 30 min at 37°C or for 2 h at room temperature.
2. Using sterile technique in a laminar flow culture hood, completely remove the AF solution from the culture vessel by aspiration.
3. Coated vessels may be used immediately or wrapped in Parafilm® sealing film and stored at room temperature for up to 24 h.

2.6.2 Geltrex™ Matrix-Coated Dishes

1. Geltrex™ matrix gels rapidly at temperatures above 2–8°C. When working with Geltrex™ matrix, keep solutions on ice at all times.
2. Thaw one tube of Geltrex™ matrix (1 mL) slowly at 2–8°C.
3. Dilute it 1:100 in 99 mL of D-MEM with GlutaMAX™-I. Mix the solution gently.
4. Cover the whole surface of each culture dish with the Geltrex™ matrix solution (e.g., 1.5 mL for a 35-mm dish, 3 mL for a 60-mm dish, 9 mL for a 100-mm dish).
5. Incubate the coated dishes for 1 h at 37°C. At this point you may store the Geltrex™ matrix-coated culture dishes at 2–8°C for up to 1 month. Seal each dish with Parafilm® sealing film to prevent the Geltrex™ matrix from drying out.
6. If plates have been stored at 2–8°C, transfer the Geltrex™ matrix-coated dishes to a laminar flow hood and allow them to equilibrate to room temperature (about 1 h) prior to using. Aspirate Geltrex™ solution immediately prior to use (do not allow plates to dry out).

3 Methods

3.1 Reprogramming of Fibroblasts

3.1.1 Transduction Human Fibroblasts with CytoTune®-iPS Reprogramming Kit

1. Two days before transduction, plate human fibroblast cells onto two 6-well plate at the appropriate density of 100,000 cells per well.
2. Culture the cells for 2 more days in human fibroblast medium, ensuring the cells have fully adhered and are healthy (see Note 10).

3. On the day of transduction, determine the total cell number on one of the 6-well plates. Add 0.5 mL of TrypLE™ Select reagent or 0.05% trypsin/EDTA to treat one well of the 6-well plates. When the cells have rounded up (1–3 min later), add 2 mL of fibroblast medium into each well, and count cells using the desired method.
4. Add the Sendai virus, OCT3/4, SOX2, KLF4, and c-Myc, to 1 mL of MEF/fibroblast medium, at the MOI of three based on your cell count (see Note 8).
5. Aspirate the fibroblast medium from the other well of the 6-well plate, and the virus solution. Place the cells in a 37°C, 5% CO_2 incubator and incubate for 24 h.
6. 24 h after transduction, remove the virus and replace with fresh fibroblast medium (see Note 4).
7. Culture the cells for 6 more days, changing the spent medium with fresh fibroblast medium every other day (see Note 5).

3.1.2 Replate Cells on MEF Feeder Cells

1. Six days after transduction, prepare gelatin-coated vessel. Plate inactivated mouse embryonic fibroblasts at the density of 2.5×10^4 cells/cm^2 (Gibco mouse embryonic fibroblasts) onto 100 mm plate.
2. On day 7 post transduction, fibroblast cells are ready to be harvested and plated on MEF culture dishes. Remove the medium from the fibroblasts, and wash cells once with D-PBS.
3. To remove the cells from the 6-well plate, use 0.5 mL of TrypLE™ Select reagent or 0.05% trypsin/EDTA for 1–3 min at room temperature. Add 2 mL of fibroblast medium into each well, and collect the cells in a 15-mL conical centrifuge tube (see Note 1). Centrifuge the cells at $200 \times g$ for 4 min, aspirate the medium, and resuspend the cells in an appropriate amount of fibroblast medium.
4. Count the cells using the desired method and seed approximately 5×10^4 to 2×10^5 cells per onto MEF culture dishes. Incubate at 37°C, 5% CO_2 incubator overnight (see Note 2).
5. 24 h later, change the medium to human iPSC medium Subheading 2.5.3 and replace the spent medium everyday thereafter.
6. Starting on day 11 (8 days after transduction), observe the plates every other day under a microscope for the emergence of cell clumps indicative of transformed cells.
7. Three to four weeks after transduction, colonies should have grown to an appropriate size for transfer (see Note 6). The day before transferring the colonies, prepare MEF culture plates using Attachment Factor-coated 12- or 24-well plates. When colonies are ready for transfer, perform live staining using Tra1-60 or Tra1-81 for selecting reprogrammed colonies (see Note 7).

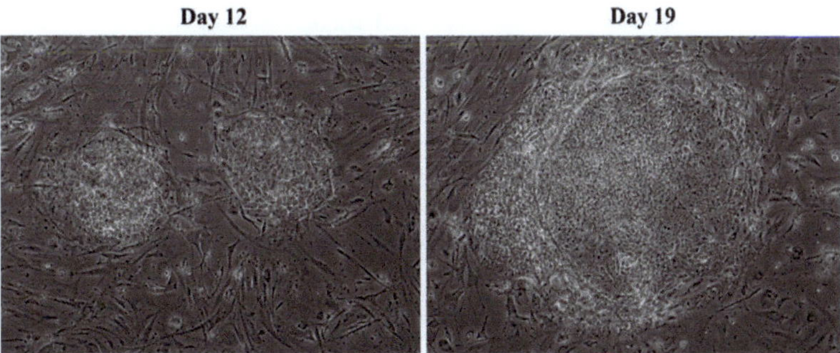

Fig. 1 BJ fibroblasts were transduced with the CytoTune™-iPS Reprogramming Kit. Seven days post transduction, cells were transferred to iMEFs, and colony formation appears on day 12 and day 19 post transduction

3.1.3 Replate Cells on Feeder-Free Condition

1. Six days after transduction, prepare Geltrex™ matrix-coated dishes Subheading 2.6.1.
2. If plates have been stored at 2–8°C, transfer the Geltrex™ matrix-coated dishes to a laminar flow hood and allow them to equilibrate to room temperature (about 1 h) prior to using. Aspirate Geltrex™ solution immediately prior to use (do not allow plates to dry out).
3. Six to seven days after transduction, the fibroblast cells are ready to be harvested and plated on Geltrex™ matrix-coated dishes. Remove the medium from the fibroblasts, and wash cells once with D-PBS (without Ca^{2+} and Mg^{2+}).
4. To remove the cells from the 6-well plate, use 0.5 mL of TrypLE™ Select reagent or 0.05% trypsin/EDTA following the procedure recommended by the manufacturer and incubate at room temperature.
5. When the cells have rounded up (1–3 min later), add 2 mL of fibroblast medium into each well and collect the cells in a 15-mL conical centrifuge tube (see Note 1).
6. Centrifuge the cells at $200 \times g$ for 4 min, aspirate the medium, and resuspend the cells in an appropriate amount of fibroblast medium.
7. Count the cells using the desired method (e.g., Countess® Automated Cell Counter), and seed the Geltrex™ culture dishes with 500,000 cells per 100-mm dish and incubate at 37°C, 5% CO_2 incubator overnight (see Note 4).
8. 24 h later, change the medium to StemPro® hESC SFM medium and replace the spent medium everyday thereafter.
9. Starting on day 11 (8 days after transduction), observe the plates every other day under a microscope for the emergence of cell clumps indicative of transformed cells (Fig. 1).
10. Three to four weeks after transduction, colonies should have grown to an appropriate size for transfer.

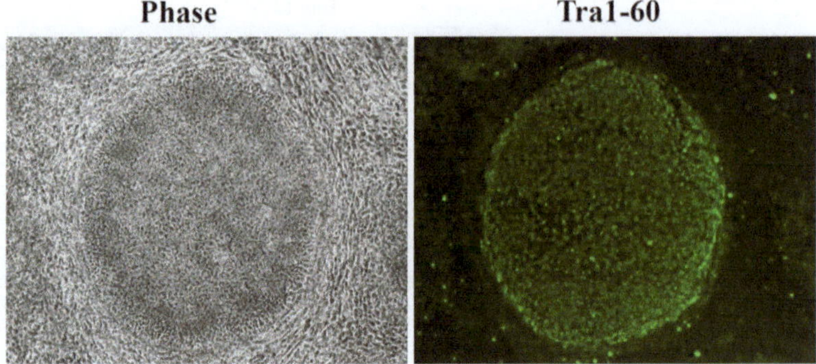

Fig. 2 Characterization of iPS colonies. iPS colony was stained with Tra1-60 antibody 21 days post transduction

11. When colonies are ready for transfer, perform live staining using Tra1-60 or Tra1-81 for selecting reprogrammed colonies (Fig. 2).

3.2 Reprogramming Blood Cells

3.2.1 Transduction CD34+ Cells with Sendai Virus

1. Three days before transduction, thaw 500,000 CD34+ cells into 1 mL StemPro®-34 in one well of a 24-well dish. Spin down cells and media at 200 × *g* for 10 min. Resuspend in StemPro®-34+cytokines Subheading 2.5.5 in one well of a 24-well dish at the density of 0.5×10^6 cells/mL.
2. Next day, count cells to ensure good viability and resuspend cells in 2 mL of StemPro®-34+cytokines.
3. Next day, remove 1 mL of medium without disturbing the cells and add 1 mL of StemPro®-34 + cytokines to cells.
4. Next day, place 250 K cells in each 24 well for transduction (in 200 μL volume)
5. Add the Sendai virus, OCT3/4, SOX2, KLF4, and c-Myc, and 4 μg/mL of polybrene to 300 μL of StemPro®-34 medium, at the MOI between 2 and 5.
6. Incubate at 37°C, 5% CO_2 incubator overnight.
7. Next day, remove virus by centrifugation at 400 × *g* for 10 min, and resuspend cells in 0.5 ml of StemPro®-34 + cytokines. Incubate at 37°C, 5% CO_2 incubator overnight.

3.2.2 Replate Cells on MEF Feeder Cells

1. Two days later, plate 50,000–200,000 of transduced cells onto 100 mm well plate of inactivated MEFs in 10 ml of StemPro®-34 + cytokines medium.
2. Every other day, gently remove 5 mL of StemPro®-34 + cytokines and replace with 1 mL fresh medium.
3. Continue replacing medium for the next 6 days.
4. On day 7, remove 5 mL of medium and 5 mL of human iPSC medium to transition the cells to new culture media.

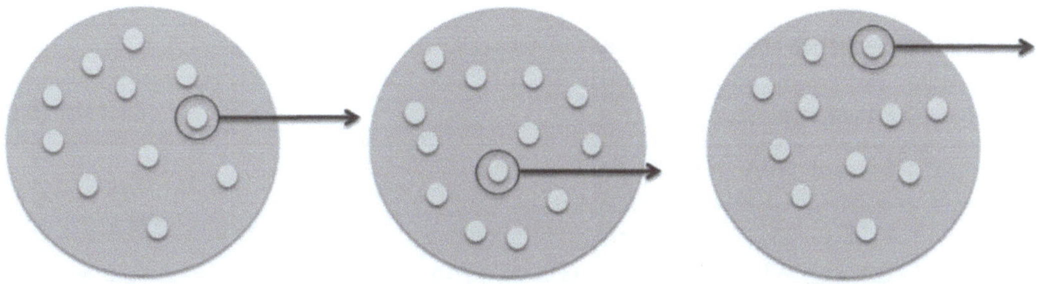

Fig. 3 Single-colony sub-cloning method. Pick and expand a single individual colony onto one 6-well plate. Continue single-colony sub-cloning for at least three passages and check for the presence of the virus

5. On day 8, completely remove media and add 2 mL of human iPSC medium. Replace the spent medium everyday thereafter.
6. Starting on day 11 (8 days after transduction), observe the plates every other day under a microscope for the emergence of cell clumps indicative of transformed cells.
7. Three to four weeks after transduction, colonies should have grown to an appropriate size for transfer. The day before transferring the colonies, prepare MEF culture plates using Attachment Factor-coated 12- or 24-well plates.
8. When colonies are ready for transfer, perform live staining using Tra1-60 or Tra1-81 for selecting reprogrammed colonies (Fig. 2).

3.3 Protocol for Generating and Identifying Vector-Free iPSCs

1. To obtain virus-free clones faster, we recommend performing single-colony sub-cloning for first few passages (minimum 5) versus bulk or pooled-clone passaging (see Note 9).
2. Pick from a single colony to transfer to a 6-well plate (passage 1).
3. From passage 1, pick a single colony and transfer to another 6-well plate (passage 2) and repeat sub-cloning for the next few passages (Fig. 3).
4. We recommend sub-cloning for three to five passages and test for virus-free iPSCs.
5. However, the time needed to derive vector-free iPSCs may vary depending on culture and passage conditions. In some cases, it may take longer.

3.4 Perform Immunostaining Using Anti-SeV Antibodies

To determine if iPS colonies still contain the virus, one can choose to perform immunostaining on one plate using anti-SeV antibodies or RT-PCR method.

1. Wash cells once with D-PBS.
2. Fix the cells in 4% paraformaldehyde for 5 min at room temperature.

Table 1
RT-PCR primer sequence and product sizes

Target	Primer sets	Product size (bp)
SeV	Forward: GGA TCA CTA GGT GAT ATC GAG C Reverse: ACC AGA CAA GAG TTT AAG AGA TAT GTA TC	181
Sox2	Forward: ATG CAC CGC TAC GAC GTG AGC GC Reverse: AAT GTA TCG AAG GTG CTC AA	451
Klf4	Forward: TTC CTG CAT GCC AGA GGA GCC C Reverse: AAT GTA TCG AAG GTG CTC AA	410
c-Myc	Forward: TAA CTG ACT AGC AGG CTT GTC G Reverse: TCC ACA TAC AGT CCT GGA TGA TGA TG	532
Oct3/4	Forward: CCC GAA AGA GAA AGC GAA CCA G Reverse: AAT GTA TCG AAG GTG CTC AA	483

RT-PCR primers for transgenes, SeV sequence, and the expected product sizes

3. Wash cells twice with D-PBS.
4. Add the anti-SeV antibody (MBL, Cat. no. PD029) diluted in 0.1% Triton®X-100 in D-PBS to the cells and incubate for 1 h at 37°C.
5. Remove the antibody solution. Wash the cells three times with D-PBS.
6. Add the secondary antibody diluted in 0.1% Triton® X-100 in D-PBS to the cells and incubate for 1 h at 37°C.
7. Remove the secondary antibody solution from the dish. Wash the cells three times with D-PBS.
8. Visualize the cells under a fluorescence microscope. If any colonies stain positive, repeat cell sub-cloning method for the next few passages Section 3.3. If all colonies are negative for anti-SeV antibodies, passage the cells and confirm the absence of the CytoTune™ Sendai-iPS Reprogramming vectors by RT-PCR.

3.5 RT-PCR for Detecting the SeV Genome and Transgenes

1. Extract the total RNA from 2 to 5×10^6 iPSCs using the TRIzol® Reagent following the instructions provided with the reagent. As a positive control, use cells set aside during the reprogramming procedure.
2. Carry out a reverse transcription reaction using 1 μg of RNA (from step 1) and the SuperScript® VILO™ cDNA Synthesis Kit following the instructions provided with the kit (see Note 2).
3. Carry out the PCR using 10 μL of cDNA from the reverse transcription reaction (step 2, above) and AccuPrime™ SuperMix I with the parameters below. For the RT-PCR primer sequences and the expected product size, see Table 1.

4. Step temperature time cycles: denaturation 95°C 30 s, annealing 55°C 30 s, 30–35 cycle, and elongation 72°C 30 s.
5. Analyze the PCR products using 2% agarose gel electrophoresis.

4 Notes

1. Because the cells can be very sensitive to trypsin at this point, minimize trypsin exposure time and incubate the cells at room temperature.
2. Set aside any remaining cells for RNA extraction to be used as a positive control in the RT-PCR detection of the SeV genome.
3. Because the CytoTune™ Sendai-iPS Reprogramming Kit is based on Sendai virus, which is an RNA virus, reverse transcription is required for detecting the presence of the Sendai genome in your reprogrammed cells.
4. For certain cell types, you may observe high cytotoxicity (can be greater than 50%) after viral transduction. Continue culturing cells and allow them to recover before transferring to iMEFs. In addition, you may want to reduce the MOI to less than what you started or increase the density of your starting cell to achieve 85–90% confluency.
5. Depending on the growth of your cell type, you may observe high cell density before day 5. We don't recommend passaging your cells after viral transduction onto MEF culture dishes before 6–7 days post transduction.
6. Depending on your cell type, you may observe colony formation between day 12 and 15. However, some may require longer culture for up to 4 weeks before seeing colonies.
7. If you do not obtain iPS colony in culture dish after 4 weeks, you may need to increase the MOI of the virus to two- to threefolds from the starting MOI. Normally, we observe drastic morphology change after 3–4 days post transduction which indicates good uptake of the virus.
8. We do not recommend re-free-thaw the virus since the titers will not maintain upon thawed.
9. In order to obtain faster transgene iPSCs, we recommend performing sub-cloning from one individual colony for a couple of passages (minimum 5) rather than bulk or pooled passaging.
10. Depending on your cell type, efficiency of reprogramming may vary between 0.01 and 0.1%. The efficiency of reprogramming depends on the "health" of your cells. It's important to start with low passage culture and maintain in conditions that allow proper proliferation.

Acknowledgments

This work was supported by Life Technologies Corporation. The products within this publication are for Research Use Only, not intended for animal or human therapeutic or diagnostic use.

References

1. Takeda A, Igarashi H, Kawada M, Tsukamoto T, Yamamoto H, Inoue M, Iida A, Shu T, Hasegawa M, Matano T (2008) Evaluation of the immunogenicity of replication-competent V-knocked-out and replication-defective F-deleted Sendai virus vector-based vaccines in macaques. Vaccine 26:6839–6843
2. Wernig M, Meissner A, Foreman R, Brambrink T, Ku M, Hochedlinger K, Bernstein BE, Jaenisch R (2007) In vitro reprogramming of fibroblasts into a pluripotent ES-cell-like state. Nature 448:318–324
3. Yu J, Vodyanik MA, Smuga-Otto K, Antosiewicz-Bourget J, Frane JL, Tian S, Nie J, Jonsdottir GA, Ruotti V, Stewart R, Slukvin II, Thomson JA (2007) Induced pluripotent stem cell lines derived from human somatic cells. Science 318:1917–1920
4. Park IH, Lerou PH, Zhao R, Huo H, Daley GQ (2008) Generation of human-induced pluripotent stem cells. Nat Protoc 3: 1180–1186
5. Dimos JT, Rodolfa KT, Niakan KK, Weisenthal LM, Mitsumoto H, Chung W, Croft GF, Saphier G, Leibel R, Goland R, Wichterle H, Henderson CE, Eggan K (2008) Induced pluripotent stem cells generated from patients with ALS can be differentiated into motor neurons. Science 321:1218–1221
6. Ebert AD, Yu J, Rose FF Jr, Mattis VB, Lorson CL, Thomson JA, Svendsen CN (2009) Induced pluripotent stem cells from a spinal muscular atrophy patient. Nature 457: 277–280
7. Soldner F, Hockemeyer D, Beard C, Gao Q, Bell GW, Cook EG, Hargus G, Blak A, Cooper O, Mitalipova M, Isacson O, Jaenisch R (2009) Parkinson's disease patient-derived induced pluripotent stem cells free of viral reprogramming factors. Cell 136:964–977
8. Okita K, Ichisaka T, Yamanaka S (2007) Generation of germline-competent induced pluripotent stem cells. Nature 448:313–317
9. Okita K, Matsumura Y, Sato Y, Okada A, Morizane A, Okamoto S, Hong H, Nakagawa M, Tanabe K, Tezuka K, Shibata T, Kunisada T, Takahashi M, Takahashi J, Saji H, Yamanaka S (2011) A more efficient method to generate integration-free human iPS cells. Nat Methods 8:409–412
10. Muller LU, Daley GQ, Williams DA (2009) Upping the ante: recent advances in direct reprogramming. Mol Ther 17:947–953
11. Li HO, Zhu YF, Asakawa M, Kuma H, Hirata T, Ueda Y, Lee YS, Fukumura M, Iida A, Kato A, Nagai Y, Hasegawa M (2000) A cytoplasmic RNA vector derived from nontransmissible Sendai virus with efficient gene transfer and expression. J Virol 74:6564–6569
12. Ikeda Y, Yonemitsu Y, Sakamoto T, Ishibashi T, Ueno H, Kato A, Nagai Y, Fukumura M, Inomata H, Hasegawa M, Sueishi K (2002) Recombinant Sendai virus-mediated gene transfer into adult rat retinal tissue: efficient gene transfer by brief exposure. Exp Eye Res 75:39–48
13. Masaki I, Yonemitsu Y, Komori K, Ueno H, Nakashima Y, Nakagawa K, Fukumura M, Kato A, Hasan MK, Nagai Y, Sugimachi K, Hasegawa M, Sueishi K (2001) Recombinant Sendai virus-mediated gene transfer to vasculature: a new class of efficient gene transfer vector to the vascular system. FASEB J 15:1294–1296
14. Fusaki N, Ban H, Nishiyama A, Saeki K, Hasegawa M (2009) Efficient induction of transgene-free human pluripotent stem cells using a vector based on Sendai virus, an RNA virus that does not integrate into the host genome. Proc Jpn Acad 85:348–362

Chapter 6

Generation of Human-Induced Pluripotent Stem Cells (hiPSCs) Using Episomal Vectors on Defined Essential 8™ Medium Conditions

Andrew Fontes, Chad C. MacArthur, Pauline T. Lieu, and Mohan C. Vemuri

Abstract

Human-induced pluripotent stem cells (iPSCs) are an important potential source of cells for regenerative medicine due to their inherent ability to differentiate into all cell types of the three germ layers. Generation of iPSCs with a non-integrating reprogramming method and in culture conditions that are completely absent of animal proteins will be ideal for such regenerative and cell therapy applications. Here we describe a method to generate non-integrating iPSCs using the Episomal iPSC Reprogramming Vectors.

Key words iPSC, Episomal vectors, Essential 8 medium, Pluripotency

1 Introduction

The ability to generate induced pluripotent stem cells (iPSCs) has opened the opportunity to generate patient-specific pluripotent stem cells from somatic cells. During the past few years, iPSCs have been demonstrated to exhibit similar characteristics to embryonic stem cells (1). Thus, iPSCs are an invaluable source for drug discovery, cell therapy, and basic research.

There are many methods to generate iPSCs, including retrovirus-mediated gene transduction and chemical induction. While retroviral vectors require integration into host chromosomes to express reprogramming gene, DNA-based vectors and plasmid vectors exist episomally and do not require integration. Here we describe a procedure to generate non-integrating iPSCs using three episomal vectors. The episomal vectors have the oriP/EBNA-1 (Epstein-Barr nuclear antigen-1) backbone that delivers the reprogramming genes, Oct4, Sox2, Nanog, Lin28, L-Myc, Klf4, and SV40LT. This system has been demonstrated successfully to

Uma Lakshmipathy and Mohan C. Vemuri (eds.), *Pluripotent Stem Cells: Methods and Protocols*, Methods in Molecular Biology, vol. 997, DOI 10.1007/978-1-62703-348-0_6, © Springer Science+Business Media New York 2013

reprogram fibroblasts, CD34+ cells, and peripheral blood mononuclear Cells (PBMCs). High transfection efficiency due to oriP/EBNA-1-mediated nuclear import and retention of vector DNA allows iPSC derivation in a single transfection (2). In addition, the removal of episomal vectors from the iPSCs can be accomplished by cell culture without any additional manipulation, due to the silencing of the viral promoter driving EBNA-1 expression and the loss of the episomes at a rate of ~5% per cell cycle due to defects in vector synthesis and partitioning (3). In addition, it has been demonstrated that additional reagents, such as MEK inhibitor PD0325901, GSK3β inhibitor, CHIR99021, TGF-β/Activin/Nodal receptor inhibitor A-83-01, ROCk inhibitor HA-100, and human leukemia inhibitory factor (hLIF), enhance reprogramming efficiency (3). This system has been optimized with a defined media condition, Essential 8™ Medium, which is a serum-free, xeno-free medium, and minimizes variability while improving feeder-free culture conditions for iPSCs.

The purpose of this chapter is to provide a protocol for derivation of iPSCs using viral free non-integrating episomal vectors and further expansion of iPSC clones in defined culture conditions with Essential 8 Medium.

2 Materials

The protocol for generating integration-free iPSCs uses the Episomal iPSC Reprogramming Vectors. Unless otherwise specified, all materials are from Life Technologies.

2.1 Reagents

1. Episomal iPSC Reprogramming Vectors (50 μL, 1 μg/μL).
2. Dulbecco's Modified Eagle Medium (DMEM) with GlutaMAX™-I (high glucose).
3. KnockOut™ DMEM/F-12.
4. Fetal bovine serum (FBS), ESC-qualified, US origin.
5. MEM non-essential amino acids solution, 10 mM.
6. Basic fibroblast growth factor (bFGF).
7. HA-100 (ROCk inhibitor).
8. Bovine Albumin Fraction V Solution (BSA).
9. Essential 8™ Medium.
10. DMEM/F-12 with HEPES.
11. N-2 Supplement (100×).
12. B-27® Supplement (50×).
13. GlutaMAX™-I (100×).
14. β-Mercaptoethanol, 1,000×.

15. PD0325901(MEK Inhibitor).
16. CHIR99021 (GSK3b inhibitor).
17. A-83-01 (TGF-b/Activin/Nodal receptor inhibitor).
18. hLIF (human leukemia inhibitory factor).
19. Vitronectin or Geltrex® LDEV-Free hESC-qualified reduced growth factor basement membrane matrix.
20. Antibiotic-antimycotic (100×).
21. 0.05% Trypsin–EDTA (1×), Phenol Red.
22. UltraPure™ 0.5 M EDTA, pH 8.0.
23. Dulbecco's PBS (DPBS) without Calcium and Magnesium.

Characterization reagents (surface marker staining)

24. Mouse primary antibodies (one is required):
    (a) Mouse Anti-Tra1-60 antibody.
    (b) Mouse Anti-Tra1-81 antibody.
    (c) Mouse Anti-SSEA4 antibody.
25. Alexa Fluor® secondary antibodies (one is required):
    (a) Alexa Fluor® 488 Goat Anti-Mouse IgG (H+L) antibody.
    (b) Alexa Fluor® 594 Goat Anti-Mouse IgG (H+L) antibody.
    (c) Alexa Fluor® 488 Goat Anti-Rabbit IgG (H+L) antibody.
    (d) Alexa Fluor® 594 Goat Anti-Rabbit IgG (H+L) antibody.

Detection reagents (for detection of episomal vectors using PCR)

26. CellsDirect Resuspension & Lysis Buffers.
27. AccuPrime Taq High Fidelity.
28. Forward and Reverse primers for PCR (primer sequences are given in the PCR protocol).
29. Electroporation instrument (e.g., Neon® Transfection System).
30. 37°C water bath.
31. Appropriate tissue culture plates and supplies.

2.2 Equipment

1. Biosafety cabinet.
2. 37°C/5% CO_2 incubator.
3. 4°C refrigerator.
4. –20°C freezer.
5. –80°C freezer.
6. Centrifuge for 15 mL and 50 mL tubes.
7. Colony marker for microscope.
8. Light microscope.

9. Pipet aid.
10. Liquid nitrogen storage tank.
11. Mr Frosty (isopropanol freezing container).
12. Plastic Cryovial holders.
13. Water bath set at 37°C.
14. X-ray source (RUCDR).

2.3 Workflow

A typical reprogramming schedule using the Episomal iPSC Reprogramming Vectors is shown below:

Day –4 to –2: Plate human fibroblasts into a T75 flask in Fibroblast Medium so that they are 75–90% confluent on the day of transfection (Day 0).

Day 0: Transfect the cells using the Neon® Transfection System. Plate transfected cells onto vitronectin-coated culture dishes and incubate overnight in Supplemented Fibroblast Medium.

Day 1–14: Change the medium to N2B27 Medium supplemented with CHALP molecule cocktail and bFGF prior to use; replace the spent medium every other day.

Day 15: Change the medium to Essential 8™ Medium and monitor the culture vessels for the emergence of iPSC colonies.

Day 21: Pick and transfer undifferentiated iPSCs onto fresh vitronectin-coated culture dishes for expansion.

2.4 Media Preparation

2.4.1 10 μg/mL bFGF Solution (1,000 μL)

1. To prepare 1 mL of 10 μg/mL bFGF solution, aseptically mix the following components:

bFGF	10 μg
DPBS without Calcium and Magnesium	980 μL
BSA	10 μL

2. Aliquot and store at –20°C for up to 6 months.

2.4.2 Fibroblast Medium (for 100 mL of Complete Medium)

1. To prepare 100 mL of Fibroblast Medium, aseptically mix the following components:

DMEM	89 mL
FBS, ESC-qualified, US origin	10 mL
MEM non-essential amino acids solution, 10 mM	1 mL

2. Fibroblast Medium can be stored at 2–8°C for up to 2 weeks.

2.4.3 Supplemented Fibroblast Medium (for 100 mL of Complete Medium)

Note: You will need 30 mL of Supplemented Fibroblast Medium per transfection.

1. To prepare 100 mL of Supplemented Fibroblast Medium, add the following components to Fibroblast Medium freshly, prior to use:

HA-100 (ROCk inhibitor)	Varies (final concentration = 10 μM)
bFGF (10 μg/mL)	40 μL (final concentration = 4 ng/mL)

2. Supplemented Fibroblast Medium must be used once HA-100 and bFGF are added to the medium.

2.4.4 Essential 8™ Medium (500 mL of Complete Medium)

1. Thaw Essential 8™ Supplement (50×) at 2–8°C overnight.
2. To prepare 500 mL of complete Essential 8™ Medium, aseptically mix the following components:

DMEM/F-12 (HAM) 1:1 (A14625DJ)	490 mL
Essential 8™ Supplement (50×) (A14626SA)	10 mL

3. Complete Essential 8™ Medium can be stored at 2–8°C for up to 2 weeks (see Note 1).

2.4.5 N2B27 Medium (250 mL of Complete Medium)

1. To prepare 250 mL of N2B27 Medium, aseptically mix the following components:

DMEM/F-12 with HEPES	238.75 mL
N-2 Supplement (100×)	2.5 mL
B-27® Supplement (50×)	5.0 mL
MEM non-essential amino acids solution, 10 mM	2.5 mL
GlutaMAX™-I (100×)	1.25 mL
β-Mercaptoethanol, 1,000×	454.5 μL

2. To supplement N2B27 Medium with CHALP molecule cocktail and bFGF, add the following components to the desired volume of N2B27 Medium. *These must be added freshly, just prior to use.*

PD0325901	0.5 μM
CHIR99021	3 μM
A-83-01	0.5 μM
hLIF	10 ng/mL
HA-100	10 μM
bFGF (10 μg/mL)	2.5 μL (final concentration = 100 ng/mL)

(See Note 2)

3. N2B27 Medium (without CHALP molecules and bFGF) can be stored at 2–8°C for up to 1 week.

Table 1
Volume of diluted vitronectin required

Culture vessel	Surface area (cm²)	Volume of diluted substrate (mL)
6-well plate	10 cm²/well	1 mL/well
12-well plate	4 cm²/well	0.4 mL/well
24-well plate	2 cm²/well	0.2 mL/well
35-mm dish	10 cm²	1 mL
60-mm dish	20 cm²	2 mL
100-mm dish	60 cm²	6 mL

2.4.6 0.5 mM EDTA in DPBS (50 mL)

1. To prepare 50 mL of 0.5 mM EDTA in DPBS, aseptically mix the following components in a 50-mL conical tube in a biological safety cabinet:

DPBS without Calcium and Magnesium	50 mL
0.5 M EDTA	50 μL

2. Filter sterilize the solution. The solution can be stored at room temperature for up to 6 months.

2.4.7 Coating Culture Vessels with Vitronectin (VTN-N)

1. Remove a 1-mL vial of vitronectin from –80°C storage and thaw at 2–8°C overnight.
2. Prepare working aliquots by dispensing 60 μL of vitronectin into polypropylene tubes. The working aliquots can be frozen at –80°C or used immediately.
3. Calculation for working concentration: Prior to coating culture vessels, calculate the working concentration of vitronectin using the formula below and dilute the stock appropriately. Refer to Table 1 for culture surface area and volume required.

The optimal working concentration of VTN-N is cell line dependent. We recommend using a final coating concentration of 0.1–1.0 μg/cm² on the culture surface, depending on your cell line. We routinely use vitronectin at 0.5 μg/cm² for human PSC culture.

$$\text{Working conc.} = \text{Coating conc.} \times \frac{\text{Culture surface area}}{\text{Volume required for surface area}}$$

$$\text{Dilution factor} = \frac{\text{Stock concentration (0.5 mg / mL)}}{\text{Working concentration}}$$

Example: To coat a 6-well plate at a coating concentration of 0.5 μg/cm², you will need to prepare 6 mL of diluted vitronectin solution (10 cm²/well surface area and 1 mL of diluted vitronectin/well; see Table 1) at the following working concentration:

$$\text{Working conc.} = 0.5\ \text{g}/\text{cm}^2 \times \frac{10\ \text{cm}^2}{1\ \text{mL}} = 5\ \text{g}/\text{mL}$$

$$\text{Dilution factor} = \frac{0.5\ \text{mg}/\text{mL}}{5\ \text{g}/\text{mL}} = 100\times (\text{i.e., } 1:100 \text{ dilution})$$

4. To coat the wells of a 6-well plate, remove a 60-μL aliquot of vitronectin from −80°C storage and thaw at room temperature. You will need one 60-μL aliquot per 6-well plate.
5. Add 60 μL of thawed vitronectin into a 15-mL conical tube containing 6 mL of sterile DPBS without Calcium and Magnesium at room temperature. Gently resuspend by pipetting the vitronectin dilution up and down (see Note 3).
6. Aliquot 1 mL of diluted vitronectin solution to each well of a 6-well plate (refer to Table 1 for recommended volumes for other culture vessels) (see Note 4).
7. Incubate at room temperature for 1 h (see Note 5).
8. Aspirate the diluted vitronectin solution from the culture vessel and discard. It is not necessary to rinse off the culture vessel after removal of vitronectin. Cells can be passaged directly onto the VTN-N-coated culture dish (see Note 6).

2.5 Reprogramming Fibroblasts

The following protocol has been optimized for human neonatal foreskin fibroblast cells (strain BJ; ATCC no. CRL2522). Researchers need to optimize the protocol for the cell type used in their protocol for reprogramming.

Day −4 to −2: seed cells

1. Two to four days before transfection, plate human fibroblast cells in Fibroblast Medium into a T75 flask. Cells should be approximately 75–90% confluent on the day of transfection (Day 0) (see Note 7).

Day 0: Prepare the cells for transfection

Prepare cells for transfection using the Neon® Transfection System (see Note 8).

2. Prepare Supplemented Fibroblast Medium (30 mL per transfection).
3. Prepare a 15-mL conical tube by adding 6 mL of the Supplemented Fibroblast Medium to a 15-mL conical tube. Incubate tube at 37°C until needed (see step 25).
4. Aspirate medium from two 100-mm VTN-N-coated plates and replace with 12 mL fresh Supplemented Fibroblast Medium per plate. Place at 37°C until ready for use.
5. Aspirate the spent medium from fibroblasts in T75 flasks.
6. Wash the cells in DPBS without Calcium and Magnesium.
7. Add 2 mL 0.05% Trypsin/EDTA to each flask.

Table 2
Electroporation parameters for Neon® Transfection System

Parameter	Setting
Pulse voltage (V)	1,650
Pulse width (ms)	10
Pulse number	3
Cell density (cells/0.1 mL)	1×10^6
Tip type	100 μL

8. Incubate the flasks at 37°C for approximately 4 min.
9. Add 6 mL Supplemented Fibroblast Medium to each flask. Tap the plate against your hand to ensure cells have been dislodged from the plate, and carefully transfer cells into an empty 15-mL conical tube (see Note 9).
10. Remove a 20-μL sample to perform a viable cell count. You will need 1×10^6 cells for one transfection.
11. Transfer 1×10^6 cells into a new 15-mL conical tube.
12. Bring the volume to 10 mL in the new tube with Supplemented Fibroblast Medium and centrifuge cells at 1,000 rpm for 5 min at room temperature.
13. Carefully aspirate most of the supernatant, using a glass Pasteur pipette, leaving approximately 100–200 μL behind. Remove the remaining medium with a 200-μL pipette.

2.6 Transfection

1. Resuspend the cell pellet in Resuspension Buffer R (included with Neon® transfection kits) at a final concentration of 1.0×10^6 cells/0.1 mL.
2. Transfer the cells to a sterile 1.5-mL microcentrifuge tube.
3. Turn on the Neon® unit and enter the electroporation parameters in the input window (see Table 2).
4. Fill the Neon® tube with 3 mL electrolytic buffer (use buffer E2 for the 100 μL Neon® tip).
5. Insert the Neon® tube into the Neon® pipette station until you hear a click.
6. Transfer 8.5 μL Episomal Reprogramming Vectors to the tube containing cells and mix gently.
7. Insert a Neon® tip into the Neon® pipette.
8. Press the push-button on the Neon® pipette to the first stop and immerse the Neon® tip into the cell-DNA mixture. Slowly release the push-button on the pipette to aspirate the cell-DNA mixture into the Neon® tip (see Note 10).

9. Insert the Neon® pipette with the sample vertically into the Neon® tube placed in the Neon® pipette station until you hear a click.
10. Ensure that you have entered the appropriate electroporation parameters and press Start on the Neon® touch screen to deliver the electric pulse (see Note 11).
11. Remove the Neon® pipette from the Neon® pipette station and immediately transfer the samples from the Neon® tip into the 15-mL tube with pre-warmed Supplemented Fibroblast Medium (prepared in step 4).
12. Mix the transfected cells by gentle inversion and pipette 3 mL into the 100-mm VTN-N-coated plate (two plates per transfection). Evenly distribute cells across plate. Discard the Neon® tip into an appropriate biological hazardous waste container. Do not use Neon® tips more than twice.
13. Repeat the process for any additional samples. Use a new Neon® tip and Neon® tube for each new cell type.
14. Incubate the plates at 37°C in a humidified CO_2 incubator overnight.

Day 1: Switch to Supplemented N2B27 Medium

15. Aspirate the spent Supplemented Fibroblast Medium from the plates using a Pasteur pipette.
16. Add 10 mL N2B27 Medium supplemented with CHALP molecule cocktail and bFGF (added freshly prior to use) to each 100-mm plate.
17. Replace the spent medium every other day, up to day 15 post-transfection.

Day 15: Switch to Essential 8™ Medium

18. Aspirate the spent medium and replace with Essential 8™ Medium. Resume medium changes every other day.
19. Observe the plates every other day under a microscope for the emergence of cell clumps indicative of transformed cells (see Fig. 1). Within 15–21 days of transfection, the iPSC colonies will grow to an appropriate size for transfer

2.7 Identifying iPSC Colonies

1. By Day 21 post-transduction, the cell colonies on the VTN-N-coated plates are large and compact, covering the majority of the surface area of the culture vessel. However, only a fraction of these colonies will consist of iPSCs, which exhibit a hESC-like morphology characterized by a flatter cobblestone-like appearance with individual cells clearly demarcated from each other in the colonies (see Fig. 2). Therefore, we recommend that you perform live staining with Tra1-60 or Tra1-81 antibodies that recognize undifferentiated iPSCs (see Note 12).
2. Aspirate the medium from the reprogramming dish.
3. Wash the cells once with DMEM/F-12 (HAM) 1:1.

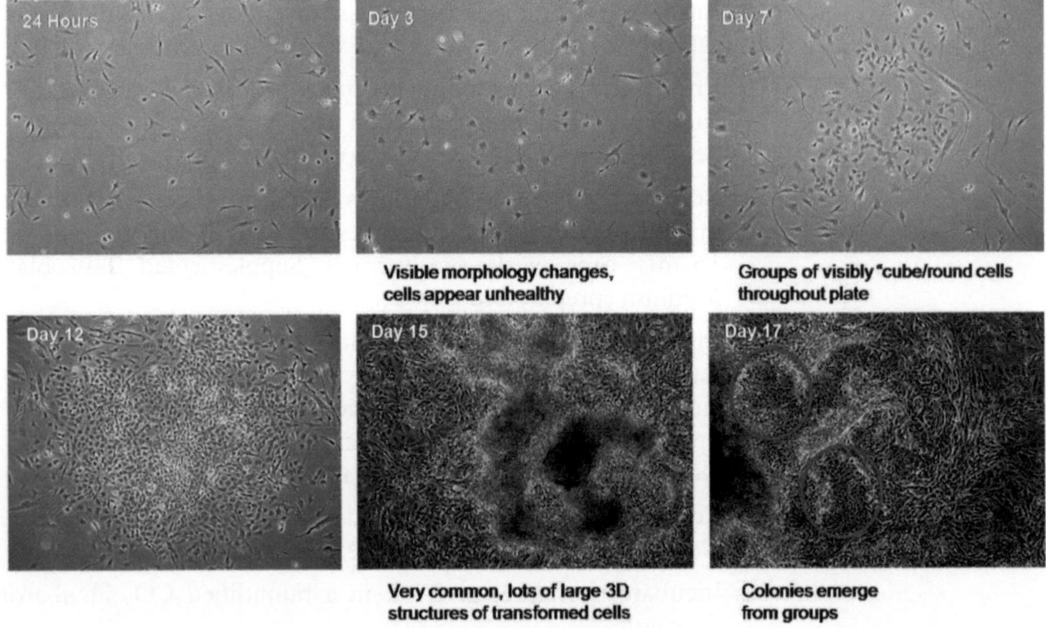

Fig. 1 Expected morphology of cells during the reprogramming experiment

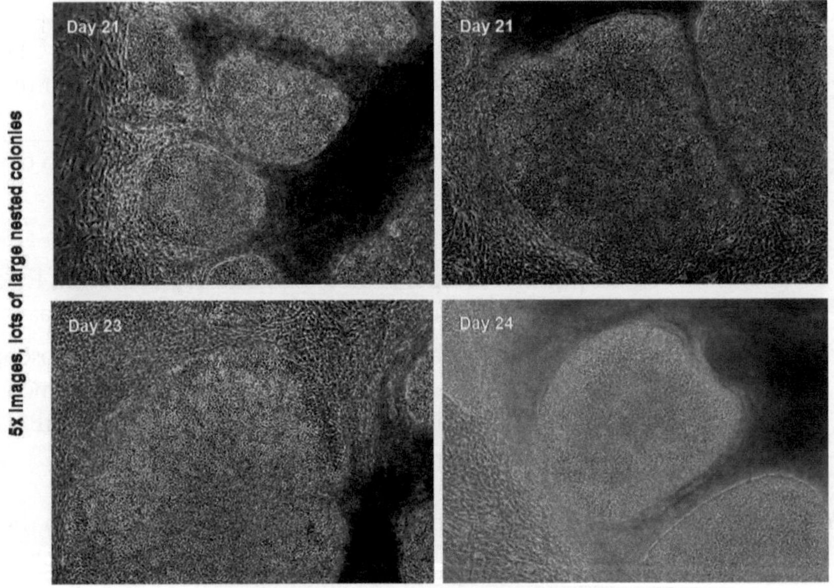

Fig. 2 Expected morphology of emerging iPSCs

4. Add the diluted primary antibody to the cells (6 mL per 100-mm dish) (see Note 13).
5. Incubate the primary antibody and the cells at 37°C for 60 min.
6. Remove the primary antibody solution from the dish (see Note 14).
7. Wash cells three times with DMEM/F-12 (HAM) 1:1.

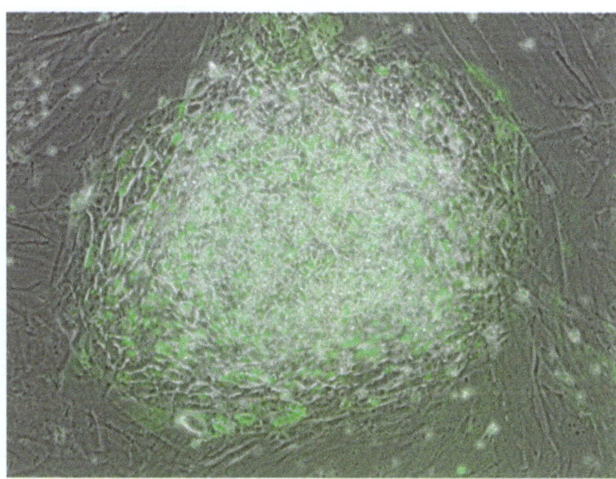

Fig. 3 Day 24 Tra 1-81 stain of an iPSC colony (10×)

8. Add the diluted secondary antibody to the cells (6 mL per 100-mm dish) (see Note 15).
9. Incubate the secondary antibody and the cells at 37°C for 60 min.
10. Remove the secondary antibody solution from the dish (see Note 16).
11. Wash cells three times with DMEM/F-12 (HAM) 1:1. Add fresh DMEM/F-12 (HAM) 1:1 to cover the surface of the cells (6 mL per 100-mm dish).
12. Visualize the cells under a standard fluorescent microscope and mark the successfully reprogrammed colonies for picking and expansion. Successful antibody staining can very specifically distinguish reprogrammed colonies from just plain transformed counterparts (see Fig. 3), and can be detected for up to 24–36 h. This is particularly useful because it helps identifying and tracking of candidate iPS colonies before picking and the day after they are transferred into a new culture dish for expansion.

2.8 Expected Results

2.8.1 Picking and Characterization of Episomal Reprogrammed iPSC Colonies

1. Examine the culture dish containing the reprogrammed cells under 10× magnification using an inverted microscope, and mark the colony to be picked on the bottom of the culture dish (see Note 17).
2. Transfer the culture dish to a sterile cell culture hood (i.e., biosafety cabinet) equipped with a stereomicroscope.
3. Cut the colony to be picked into five to six pieces in a grid-like pattern using a 25-gauge 1½ in. needle.
4. Using a 200-μL pipette, transfer the cut pieces to a freshly prepared 24-well VTN-N-coated plate containing Essential 8™ Medium.

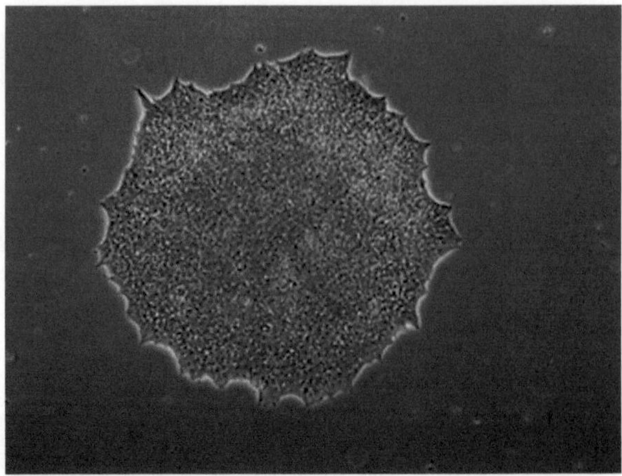

Fig. 4 Human neonatal foreskin fibroblast cells were transformed using Episomal iPSC Reprogramming Vectors and allowed to proliferate on VTN-N-coated plates in fibroblast medium, shown here at passage 5

5. Incubate the VTN-N-coated plate containing the picked colonies in a 37°C, 5% CO_2 incubator.
6. Allow the colonies to attach to the culture plate for 48 h before replacing the spent medium with fresh Essential 8™ Medium Fig. 4. After that, change the medium everyday.
7. Treat the reprogrammed colonies like normal human iPSC colonies; passage, expand, and maintain them using standard culture procedures until you have frozen cells from two 60-mm plates.

2.9 Detecting the Episomal Vectors Using PCR

1. Aspirate the medium from the dish containing iPSCs with a Pasteur pipette, and rinse the dish twice with Dulbecco's PBS (DPBS) without calcium and magnesium. Refer to Table 3 for the recommended volumes.
2. Add 0.5 mM EDTA in DPBS to the dish containing iPSCs. Adjust the volume of EDTA for various dish sizes (refer to Table 4). Swirl the dish to coat the entire cell surface.
3. Incubate the vessel at room temperature for 5–8 min or 37°C for 4–5 min. When the cells start to separate and round up, and the colonies will appear to have holes in them when viewed under a microscope, they are ready to be removed from the vessel (see Note 18).
4. Aspirate the EDTA solution with a Pasteur pipette.
5. Add pre-warmed complete Essential 8™ Medium to the dish according to Table 5.
6. Remove the cells by gently squirting the colonies from the well using a 5-mL glass pipette. Avoid creating bubbles. Collect cells in a 15-mL conical tube (see Note 19).
7. Centrifuge the cell suspension at 200 × *g* for 5 min to pellet cells.

Table 3
Volume of DPBS required

Culture vessel	Surface area (cm^2)	Volume (mL)
6-well plate	10 cm^2/well	2 mL/well
12-well plate	4 cm^2/well	1 mL/well
24-well plate	2 cm^2/well	0.5 mL/well
35-mm dish	10 cm^2	2 mL
60-mm dish	20 cm^2	4 mL
100-mm dish	60 cm^2	12 mL

Table 4
Volume of 0.5 mM EDTA in DPBS required

Culture vessel	Surface area (cm^2)	Volume (mL)
6-well plate	10 cm^2/well	1 mL/well
12-well plate	4 cm^2/well	0.4 mL/well
24-well plate	2 cm^2/well	0.2 mL/well
35-mm dish	10 cm^2	1 mL
60-mm dish	20 cm^2	2 mL
100-mm dish	60 cm^2	6 mL

Table 5
Volume of complete Essential 8™ Medium required

Culture vessel	Surface area (cm^2)	Volume (mL)
6-well plate	10 cm^2/well	2 mL/well
12-well plate	4 cm^2/well	1 mL/well
24-well plate	2 cm^2/well	0.5 mL/well
35-mm dish	10 cm^2	2 mL
60-mm dish	20 cm^2	4 mL
100-mm dish	60 cm^2	12 mL

8. Aspirate and discard the supernatant. Resuspend cell pellet in 500 μL DPBS and transfer resuspended cells to a thin-walled 0.5-mL PCR tube.
9. Centrifuge the cell suspension at $200 \times g$ for 5 min to pellet cells.

Table 6
Preparation of reactions for PCR

Component	Volume per reaction (μL)
10× PCR buffer II	5
Forward primer (10 μM stock)	1
Reverse primer (10 μM stock)	1
AccuPrime™ Taq (5 U/μL)	1
Cell lysate	3
Sterile distilled water	39

Table 7
Primers for standard PCR

Transgene	Primers	Sequence	Expected size (bp)
oriP	pEP4-SF1-oriP	5′-TTC CAC GAG GGT AGT GAA CC-3′	544
	pEP4-SR1-oriP	5′-TCG GGG GTG TTA GAG ACA AC-3′	
EBNA-1	pEP4-SF2-oriP	5′-ATC GTC AAA GCT GCA CAC AG-3′	666
	pEP4-SR2-oriP	5′-CCC AGG AGT CCC AGT AGT CA-3′	

10. Aspirate and discard the supernatant. Resuspend cell pellet in 20 μL of Resuspension Buffer with 2 μL of lysis solution added to the Resuspension Buffer.
11. Incubate the cells for 10 min in an incubator or thermal cycler that has been preheated to 75°C.
12. Spin the tube briefly to collect any condensation. Use 3 μL of the cell lysate in a 50-μL PCR (see below).

2.10 PCR Using AccuPrime™ High Fidelity Taq DNA Polymerase

1. Add the following components to a DNase/RNase-free, thin-walled PCR tube as directed in Table 6. Forward and reverse primers are shown in Table 7. For multiple reactions, prepare a master mix of common components to minimize reagent loss and enable accurate pipetting (see Note 20).

   Cap the tube, tap gently to mix, and centrifuge briefly to collect the contents.
2. Place the tube in the thermal cycler and use the PCR parameters shown in Table 8.
3. Analyze the PCR products using 2% agarose gel electrophoresis.

Table 8
PCR parameters

Step	Temperature (°C)	Time	Cycles
Initial denaturation	94	2 min	–
Denaturation	94	30 s	35–40
Annealing	55	30 s	
Elongation	72	1 min	
Final elongation	72	7 min	–

3 Notes

1. Before use, warm complete medium required for that day at room temperature until it is no longer cool to the touch. *Do not warm the medium at 37°C.*
2. CHALP molecule cocktail is an optimized mixture of small molecules (**CH**IR99021, **HA**-100, **A**-83-01, **L**IF, **P**D0325901) shown to greatly improve the episomal reprogramming efficiency.
3. This results in a working concentration of 5 μg/mL (i.e., a 1:100 dilution).
4. When used to coat a 6-well plate (10 cm^2/well) at 1 mL/well, the final concentration will be 0.5 μg/cm^2.
5. Dishes can now be used or stored at 2–8°C wrapped in laboratory film for up to a week. Do not allow the vessel to dry. Prior to use, pre-warm the culture vessel to room temperature for at least 1 h.
6. Geltrex® LDEV-Free hESC-qualified reduced growth factor basement membrane matrix may be substituted for vitronectin.
7. Growth rate is dependent on the cell line and culture conditions. Depending on the seeding density and culture conditions, the cells may take up to 5 days to reach 75–90% confluency. Since overconfluency results in decreased transfection efficiency, we recommend replating your cells to achieve 75–90% confluency if your cells have become overconfluent during culturing.
8. Gentle handling is essential prior to transfection.
9. Each T75 flask provides plenty of cells for transfection, so any residual cells still clinging to the flask after Trypsin/EDTA treatment may be left behind.

10. Avoid air bubbles during pipetting to avoid arcing during electroporation. If you notice air bubbles in the tip, discard the sample and carefully aspirate fresh sample into the tip again without any air bubbles.
11. The touch screen displays "Complete" to indicate that electroporation is complete.
12. Other methods of identifying iPSCs (such as alkaline phosphatase staining) are also acceptable.
13. Mouse Anti-Tra 1-60, Mouse Anti-Tra 1-81, or Mouse Anti-SSEA can be used.
14. The primary antibody solution can be stored at 4°C for 1 week and reused up to two times.
15. Any of the four Alexa Fluor® secondary antibodies listed in Subheading 2 can be used.
16. The secondary antibody solution can be stored at 4°C for 1 week and reused up to two times.
17. We recommend picking at least ten distinct colonies by the end of each reprogramming experiment and expanding them in separate 24-well VTN-N-coated plates.
18. In larger vessels or with certain cell lines, this may take longer than 5 min.
19. *Do not scrape the cells from the dish.* There may be obvious patches of cells that were not dislodged and left behind. Do not attempt to recover them through scraping. Depending upon the cell line, work with no more than one to three wells at a time, and work quickly to remove cells after adding Essential 8™ Medium to the well(s). The initial effect of the EDTA will be neutralized quickly by the medium. Some lines re-adhere very rapidly after medium addition, and must be removed one well at a time. Others are slower to reattach, and may be removed three wells at a time.
20. Assemble PCRs in a DNA-free environment. We recommend use of clean dedicated automatic pipettors and aerosol-resistant barrier tips.

References

1. Yu J, Chau KF, Vodyanik MA, Jiang J, Jiang Y (2011) Efficient feeder-free episomal reprogramming with small molecules. PLoS One 6:e17557
2. Takahashi K, Yamanaka S (2006) Induction of pluripotent stem cells from mouse embryonic and adult fibroblast cultures by defined factors. Cell 126(4):663–676
3. Nanbo A, Sugden A, Sugden B (2007) The coupling of synthesis and partitioning of EBV's plasmid replicon is revealed in live cells. EMBO J 26:4252–4262

Chapter 7

Feeder-Free Substrates for Pluripotent Stem Cell Culture

Alexandria Sams and Mark J. Powers

Abstract

A significant barrier to the therapeutic application of pluripotent stem cells (PSCs) is the risk associated with the presence of undefined, animal-derived elements that are routinely used to culture these cells. Originally, PSCs were derived on mouse feeder cells in media containing fetal calf serum. Such conditions could expose potential patients to animal pathogens or lead to immune rejection. Substantial efforts have been made to remove these components and successfully maintain these cells in a completely defined, xeno-free environment. In this chapter, we examine substrates consisting of animal-derived proteins, purified human proteins, recombinant human proteins, and synthetic polymers and their ability to maintain the undifferentiated growth of various pluripotent stem cell lines in a variety of supplemented media.

Key words Human, Pluripotent stem cells, Culture, Media, Matrix

1 Introduction

The generation of embryonic stem cells (ESCs) and induced pluripotent stem cells (iPSCs) has been well established (1–3). Due to their pluripotent nature, these cells have the potential to revolutionize cell-based therapeutic applications. A key step in realizing this potential is the development of PSC derivation and culture conditions that meet good manufacturing practice (GMP) standards. Many downstream therapeutic applications will require large-scale, uniform, and consistent populations of undifferentiated clinical-grade cells. However, the initial culture system for ESCs culture and expansion employed the use of basal medium supplemented with fetal calf serum and a mouse embryonic fibroblast (MEF) feeder layer, which contribute an undefined and complex mixture of extracellular matrix components, cytokines, and other factors (1). Concern over animal pathogens, potential immune rejection of animal proteins, and batch-to-batch variability led several groups to develop xeno-free media and human feeder cells for human ESC (hESC) growth (4). More recent efforts have

Uma Lakshmipathy and Mohan C. Vemuri (eds.), *Pluripotent Stem Cells: Methods and Protocols*, Methods in Molecular Biology, vol. 997, DOI 10.1007/978-1-62703-348-0_7, © Springer Science+Business Media New York 2013

been targeted at developing fully defined, xeno-free culture conditions for both the derivation and propagation of PSC lines. This chapter discusses an array of substrates and media options that have been developed for feeder-free culture. A thorough discussion of pluripotent stem cell culture systems has to take both the media and the substrate into consideration. As highlighted in a recent report that employed multiple media and substrates to culture the same pluripotent stem cell line, the maintenance of PSC pluripotency is a result of the synergy of inputs from both the substrate and the soluble factors in the medium (5).

Additional variables such as the tissue of origin, derivation conditions, passaging techniques, and passage numbers may also be contributing to emerging differences observed between PSC lines. Although outside the scope of this chapter there are studies that report comprehensive characterization of multiple hESC lines in parallel using standardized assay conditions and highlight differences that could be due to inherent genetic variation or environmental factors (6–8).

Greater understanding of the stem cell niche, the microenvironment that regulates the maintenance, self-renewal, activation, proliferation, and long-term regenerative capacity of stem cells via external signals, has generated significant improvements in PSC culture conditions. This chapter will review the media and feeder-free substrate alternatives including animal-derived, human-derived, and synthetic options that address some of the disadvantages of feeder-based culture systems.

2 Overview of Current Methods

2.1 Animal-Derived Feeder-Free Substrates

2.1.1 Engelbreth-Holm-Swarm (EHS) Tumor Basement Membrane Extract

In the first report of feeder-free hESC cultures the authors evaluated different matrices along with medium preconditioned on MEFs supplemented with basic fibroblast growth factor (bFGF) (9). The rationale behind this culture system was that most normal cells require adhesion to an extracellular matrix component for survival and growth and that soluble factors secreted by MEFs may be important for hESCs. Conditioning the medium requires incubating the hESC culture medium on a layer of MEF cells, which enables the secretion of important growth and attachment factors into the culture medium. Matrigel, a basement membrane extract (BME), was shown to support undifferentiated hESCs when coupled with MEF-conditioned media (MEF-CM) as evidenced by maintenance of a normal karyotype, a stable proliferation rate, high telomerase activity, undifferentiated marker expression, and teratoma formation capability (9). BME is a complex mixture of extracellular matrix proteins and growth factors. In normal tissues the basement membrane is typically present in small amounts and is difficult to isolate due to poor solubility. In contrast, the EHS tumor produces an abundance of basement membrane. The protocol to isolate basement membrane from an EHS tumor typically includes a 20% NaCl wash to remove cell- and serum-derived proteins, an extraction with 2 M urea to break up protein-protein interactions, and a dialysis step using chloroform to kill any bacteria and spores. The most abundant components common to all basement membranes include laminin, type IV collagen, heparan sulfate proteoglycan, and nidogen/entactin. Recently, a group identified more than 1,800 proteins present in BME that vary significantly from batch to batch (10).

It should be noted that the MEF-CM used in this study was supplemented with bFGF (also referred to as FGF-2) which is a widely used supplement for long-term PSC culture under both feeder and feeder-free conditions. Although bFGF is recognized as being necessary for maintaining undifferentiated hESCs its mechanism of action is not completely understood. Current evidence suggests numerous roles for bFGF that include inhibiting BMP signals, promoting secretion of hESC supportive factors (e.g., IGF-II, TGF-β1, Activin A), acting as a competence factor, and directly affecting the MEK/ERK signaling in hESC cultures (11).

Since the introduction of the first serum-free feeder-free system, additional technologies have been developed to obviate the need for conditioned medium. It has been reported that BME could support H7 and H9 hESCs in serum-free nonconditioned medium (KNOCKOUT™ DMEM+20% KNOCKOUT™ Serum Replacement) with 40 ng/mL bFGF (12). Another study noted that 40 ng/mL bFGF and 500 ng/mL noggin, a bone morphogenetic protein antagonist, were able to maintain undifferentiated proliferation of H1 hESCs on BME in the same nonconditioned medium (13). In this same study, noggin was able to be eliminated by using a higher bFGF concentration (100 ng/mL). Finally, another report described the use of serum-free nonconditioned medium containing 24–36 ng/mL bFGF on either BME or laminin to maintain H1 and H9 hESCs over multiple passages (14). The discrepancies surrounding the bFGF concentration needed could be due to various factors including differences in FGF receptor expression among different hESC lines or batch-to-batch variability in the BME that require different concentrations of bFGF added for hESC maintenance. While the process to prepare BME typically yields similar overall protein concentrations from one preparation to the next, the individual component concentrations can vary significantly. Such variations can have a dramatic impact on the behavior of a culture system. bFGF receptors, transmembrane tyrosine kinase FGF receptors and cell surface or extracellular matrix heparan sulfate proteoglycans are known to cooperate to generate a cellular response to bFGF (15, 16). Differences in the amount of heparan sulfate proteoglycans in the BME preparations used in previously mentioned reports could affect the concentration of bFGF that is necessary to add to the medium in order to maintain hESCs in an undifferentiated state. Additional studies have shown that exogenous addition of members of the TGF family (including Activin A, Nodal, and TGF-β1) can maintain hESC pluripotency as well (11). These factors and the cell signaling pathways they are involved in are reviewed elsewhere (11).

iPSCs represent another group of pluripotent stem cells that meet the same defining criteria for ESCs (i.e., normal karyotype, pluripotency marker expression, differentiation potential) except that these cells are not derived from embryos. Instead, iPSCs are generated by reprogramming somatic cells using various transcription factors (17). BME has been demonstrated to support iPSCs in MEF-CM (17, 18). It has been shown that iPS cell lines (iPS(IMR90)-3, iPS(IMR90)-4, and iPS(foreskin)-2) expressed similar levels of α and β integrin subunits but these levels were not identical to hESC expression levels which may be attributable to the different tissues of origin for the different cell lines (18). The β1 integrin subunit was shown to be required for both iPSC and hESC adhesion and proliferation on BME. However, the authors noted that the inhibition of the α6 integrin subunit did not lead to decreased cell adhesion which is surprising since laminin (which binds integrin α6β1) is a major component of BME. This could indicate that there are other α integrin subunits that the β1 integrin subunit can bind to in order to mediate hESC and iPSC attachment to BME.

Currently, BME and MEF-CM are the most common choices for feeder-free culture (Table 1). However, the risk of xenogeneic pathogen transmission limits potential medical application of PSCs cultured in BME and MEF-CM. The undefined nature of these materials is also undesirable. The complex interplay between the substrate and the soluble factors provided by the culture medium is more difficult to fully comprehend using such a non-reproducible system. The remaining portion of this chapter will focus on various substrates and media that have been developed to address the xenogeneic pathogen risk and/or the undefined nature of current systems.

2.1.2 Type I Collagen

As a substrate for feeder-free hESC culture, type I collagen has advantages that include being more defined, widely available, and FDA approved for various health and beauty care applications. Jones et al. evaluated the combination of type I collagen and media conditioned by SDEC (SDEC-CM), a human embryonic germ-cell-derived cell culture (19). H1 hESCs maintained in this system demonstrated similar growth kinetics, colony morphology, karyotypic stability, and teratoma formation containing differentiated tissues from all three germs layers compared to hESCs maintained on Matrigel in either SDEC-conditioned media or MEF-CM following a 1-month culture period. Transcriptional profiling of

Table 1
Feeder-free conditions used to culture PSCs

Substrate	Basal medium	Supplements	PSC lines	Length of culture	References
Matrigel	Knockout™ DMEM (KO-DMEM)	20% Knockout™ Serum Replacement (KSR), 40 ng/mL bFGF	H7, H9	15 passages	(12)
Matrigel	DMEM/F12	20% KSR, 40 ng/mL bFGF, 500 ng/mL noggin	H1, H9, H14	19–33 passages	(13)
Matrigel	MEF-CM	12 ng/mL bFGF	H9	> 35 passages	(14)
Matrigel	KO-DMEM	20%SR, 24–36 ng/mL bFGF	H1, H9	>8 passages, >31 passages	(14)
Matrigel	MEF-CM	8 ng/mL bFGF	H1, H7, H9, H14	170 days (~130 population doubling)	(8)
Matrigel	MEF-CM	8 ng/mL bFGF	H1, H7, H9	70–141 passages	(49)
Matrigel	X-VIVO™ 10	40–80 ng/mL FGF-2, 0.1 ng/mL TGF-β1	H9, CA1, 16	>30 passages	(50)
Matrigel	IMDM/F12	10 ng/mL Activin A, 12 ng/mL bFGF	H9	10 passages	(51)
Matrigel	DMEM/F12	20% KSR, 5 ng/mL Activin A	H1, 16	>20 passages	(52)
Matrigel	DMEM/F12	N2 or N2/B27, 20 ng/mL bFGF	H1, HSF6	>27 passages	(53)
Matrigel	DMEM/F12	20% KSR, 100 ng/mL bFGF	H1, H7, H9, H14	16–33 passages	(54)
Matrigel	KO-DMEM	20% KSR, 30 ng/mL IGF-II	H1, H9, HES3, CA1	>12 passages	(55)
Collagen type I	SDEC-CM	16 ng/mL bFGF	H1, H9	1 month	(20)
Collagen type I	ESF	10 ng/mL bFGF, 100 ng/mL heparin	Shef1, HUES-1	15 passages, 25 passages	(21)
Human serum	hES-dF-CM	12 ng/mL bFGF	H1, hES-NCL1	>27 passages	(22)

(continued)

Table 1 (continued)

Substrate	Basal medium	Supplements	PSC lines	Length of culture	References
Laminin	X-VIVO™ 10	80 ng/mL bFGF	H1	40 passages	(24)
Laminin	DMEM	KSR, 50 ng/mL Activin A, 50 ng/mL keratinocyte growth factor, 10 mM Nicotinamide	HSF6	>20 passages	(23)
Laminin	MEF-CM	8 ng/mL bFGF	H1	6 passages	(9)
Fibronectin		15% KSR, 0.12 ng/mL TGF-β1, 4 ng/mL bFGF, ±1,000 U/mL			
	KO-DMEM	LIF	H9, 1–6, 1–3	>30 passages	(25)
Fibronectin	DMEM/F12	20% KSR, 10 ng/mL bFGF, 100 μM Y27632, 0.95 μM CHIR99021, 0.4 μM PD0325901	H1, HSF1	25 passages	(27)
Vitronectin	DMEM/F12	mTeSR™1 Supplement	HUES1, HES2, HES-NL3	>7 passages	(30)
Vitronectin	DMEM/F12	100 ng/mL bFGF, 2 ng/mL TGF-β1 or 100 ng/mL Nodal	H1, H9, iPS foreskin	>20 passages	(31)
CELLstart™	DMEM/F12	StemPro® hESC Supplement, 8 ng/mL bFGF	I6	>25 passages	(32)
Laminin	MEF-CM	5 ng/mL bFGF	KhES-1, KhES-2, KhES-3	10 passages	(33)
E-cadherin	DMEM/F12	mTeSR™1 Supplement	H9	>35 passages	(37)

Poly(methyl vinyl ether-alt-maleic anhydride) (PMVE-alt-MA)	DMEM/F12	StemPro® hESC Supplement, 30 ng/mL bFGF	HUES1, HUES9	5 passages	(42)
Bone sialoprotein peptide-acrylate surface (BSP-PAS), vitronectin peptide-acrylate surface (VN-PAS)	X-VIVO™10	80 ng/mL bFGF, 0.5 ng/mL TGF-β1	H1, H7	>10 passages	(44)
Self-assembled monolayers (LTTAPKLPKVTR and TVKHRPDALHPQ)	DMEM/F12	mTeSR™1 Supplement + 5 μM Y-27632	H9	3 passages	(45)
Heparin-binding peptide GKKQRFRHRNRKG	DMEM/F12	m TeSR™1 Supplement + 5 μM Y-27632	H9, H13, H14, DF19-9 7T	>17 passages	(46)
Poly(2-(methacryloyloxy) ethyl dimethyl-(3-sulfo-propyl) ammonium hydroxide) (PMEDSAH)	DMEM/F12	StemPro® hESC Supplement, 8 ng/mL bFGF	H9	10 passages	(43)
Cell adhesion peptides (AG-10 + C-16 + AG-73)	X-VIVO™ 10	80 ng/mL bFGF, 0.5 ng/mL TGF-β1	H1	3 passages	(48)

In general, common supplements include L-glutamine, nonessential amino acids, and beta-mercaptoethanol. Refer to references for specific details

SDEC feeders and non-supportive feeders demonstrated numerous differentially expressed genes. Among those highlighted in the article the authors noted that non-supportive feeder cells exhibited high levels of sFRP-1, a secreted protein that is a known antagonist of the WNT/β-catenin signaling pathway. This pathway plays a key role in stem cell proliferation and maintenance of pluripotency via activation of various factors including Nanog and OCT4. The authors hypothesized that interference with this pathway could be the means through which the conditioned media of non-supportive feeder cells could negatively impact hES cells.

Several hESC lines were cultured on type I collagen in hESF8 medium, a variant of a medium originally developed for mouse ESCs that includes bFGF (10 ng/mL) and heparin (100 ng/mL) (20). It should be noted that the base medium used in this formulation differs significantly from DMEM and DMEM/F12, which are commonly used in other hESC formulations. It is hypothesized that the requirement for a relatively low concentration of bFGF is due to the presence of heparin, a known cofactor of bFGF, in the formulation which stabilizes endogenous bFGF. As discussed in the previous section multiple reports confirm that bFGF is necessary for maintenance of undifferentiated hESCs. In this study the authors reported that the addition of heparin in the absence of exogenous FGF results in the phosphorylation of FGF receptors in hES cells which could be involved in stabilizing endogenously produced FGF (20). Such a role could explain the fact that a rather low bFGF concentration was needed in the media compared to concentrations reported in other studies that do not report the use of heparin. The authors reported typical colony morphology, pluripotency maintenance, a normal karyotype, and differentiation capability following 15–25 passages as evidence that type I collagen and hESF9 media represent a defined system for long-term hESC culture.

2.2 Human-Derived Extracellular Matrix Proteins

2.2.1 Human Serum

Human serum (HS) derived from male clotted blood has been shown to support H1 hESC and hES-NCL1 cells using media conditioned by mitotically inactivated fibroblasts derived from differentiated hESCs (hES-dF-CM) (21). Both cell lines expressed surface and intracellular pluripotency markers (e.g., TRA-1-60, TRA-1-81, SSEA-4, AP, OCT-4), differentiation capability, and genomic stability after prolonged culture. HS contains multiple extracellular matrix (ECM) components including fibronectin, vitronectin, hyaluronic acid, and other factors. Earlier reports have described the use of HS as a serum supplement (instead of a coating substrate) to support hESCs cultured on human fibroblasts but after prolonged culture (ten passages) there was evidence of increased rates of differentiation (4).

2.2.2 Laminin

Basement membrane (BM) components are an important part of the in vivo environment that contributes to the differentiation, phenotype maintenance, and function of many types of somatic and stem cells. As mentioned earlier in this chapter the BME is a complex mixture of ECM proteins. Various research groups have studied the ability of individual ECM proteins to support hESC culture. Laminins (LN), the major glycoprotein component of basement membranes, are heterotrimeric proteins composed of α, β, and γ chains. Multiple types of these chains form 15 different combinations in human tissues. The naming convention is based upon the types of chains that are present in that particular laminin molecule. For example, LN-511 consists of α5, β1, and γ1 chains. The various isoforms demonstrate spatiotemporal expression patterns as well as tissue-specific locations and functions. LN-511 (LN-10) is the most common form found in BMs of the early embryo and most adult tissues. Importantly, it is found in the extracellular matrix between cells of the inner cell mass of blastocysts which is the natural origin of ES cells (22). Therefore, one could assume that laminin may be essential for maintaining undifferentiated hESC cells in vitro due to the fact that it is a natural component of the ESC niche in vivo.

H1 hES cultures maintained on laminin and MEF-CM exhibited compact colonies of undifferentiated cells (9). Flow cytometry analysis indicated that 97–99% of H1 hESCs were positive for α6 and β1 integrin subunits, which form a dimer with a high affinity for laminin (9). Another study showed that HSF6 hESCs grown on laminin in medium supplemented with Activin A maintained stem cell features for more than 20 passages (e.g., normal karyotype, undifferentiated cell markers, differentiation potential via teratoma formation) (23). Human laminin and a defined serum-free medium containing only human sourced and recombinant proteins were shown to provide the necessary elements for maintenance and growth of H1 hES cells for more than 240 days in culture (24). The medium was supplemented with a high concentration of human bFGF (80 ng/mL) alone or a combination of human bFGF (40 or 80 ng/mL) and flt3-L (15 ng/mL). The various mechanisms of action bFGF have been discussed earlier in this chapter. Although flt3 L demonstrates a beneficial effect as well the exact mechanism remains unknown.

2.2.3 Fibronectin

Fibronectin, a ubiquitous and abundant ECM protein, is often used to support cell adhesion to culture surfaces. Fibronectin dimers induce integrin clustering which produces a number of distinct intracellular signaling cascades that influence cell behaviors including proliferation, apoptosis, motility, gene expression, and differentiation. Amit et al. (25) described a feeder-free and serum-free system that includes a fibronectin substrate (isolated from human plasma or human fibroblasts) and a serum replacement media supplemented with a combination of growth factors (TFG-β1 + bFGF or TGF-β1 + bFGF + LIF). hESCs cultured in this system were shown to maintain multiple ES cell characteristics that include undifferentiated marker expression after prolonged passaging, stable karyotype, teratoma formation containing tissues representative of the three embryonic germ layers, and EB formation and differentiation into different cell types representative of the three germ layers. Another study leveraged a quantitative PCR method to measure the mRNA levels of multiple up- and down-regulated genes before and after hESC differentiation in order to calculate a collective expression index (26). These studies found that undifferentiated hESCs cultured on fibronectin in MEF-CM with serum replacement had a slightly lower expression index than undifferentiated hESCs maintained on MEFs, although the authors pointed out that it is not yet clear if this difference has any biological significance. More recently, a group identified three small molecule inhibitors against ROCK, GSK, and MEK that in combination with bFGF support long-term maintenance of hESCs using single-cell passaging on fibronectin-coated surfaces in serum-free medium supplemented with serum replacement (27).

2.2.4 Vitronectin

Vitronectin, an abundant multifunctional glycoprotein found in blood plasma and the ECM, contains binding sites for numerous factors (e.g., integrins, heparin, collagen, thrombin-antithrombin II complex, plasminogen, plasminogen activator inhibitor-1, urokinase receptor, epidermal growth factor, bFGF). As a matricellular protein it does not serve a structural role (like collagen or laminin) but does modulate cell-matrix interactions and cell functions including angiogenesis, hemostasis and thrombus formation, and wound repair (28). It has been reported that vitronectin is the primary adhesive protein in fetal bovine serum, the media component used in most animal cell culture. Comparison of multiple bovine serum batches found that cell attachment activity attributable to vitronectin was 8–16-fold greater than that of fibronectin (29).

Functional analysis of three independently derived hES cell lines (HES2, HUES1, and HESC-NL3) using integrin-blocking antibodies demonstrated that the αVβ5 integrin was responsible for binding to vitronectin (30). These three cell lines were chosen because of the differences used in culturing and passaging these lines HUES1 cells were cultured on MEFs in Plasmanate®/KNOCKOUT™ Serum Replacement-containing medium, HES2 cells were cultured on MEFs using fetal calf serum-containing medium, and HESC-NL3 cells were cultured on human foreskin feeders in KNOCKOUT™ Serum Replacement-containing medium. This analysis revealed that although these hES cell lines could attach and proliferate on major ECM components (i.e., laminin, collagen I, collage IV, fibronectin) in MEF-CM only vitronectin could support these cell lines as well as BME when cultured in mTeSR™1, a fully defined medium. The authors suggested that the differences observed between MEF-CM and mTeSR™1 could be due to additional factors such as laminin and fibronectin that are known to be secreted into the conditioned media by the MEFs that could potentially mediate hES cells binding to their ECM component substrate. Culture expansion on a given ECM component in the presence of blocking antibodies against α6β1 (laminin receptor) and α5β1 (fibronectin receptor) revealed decreased expansion on collagen IV and laminin substrates. However, no such decrease was observed for vitronectin.

iPS lines (iPS(IMR90)-3) were also shown to be successfully maintained on purified vitronectin for nine to ten passages using mTeSR™1 based on a similar morphology, normal karyotype, and differentiation potential compared to cells grown on BME in mTeSR™1 (18). In this study the authors noted that cell adhesion was higher on BME and hypothesized that it could be due to some adaptation that the cells underwent when they were transitioned from feeders to BME prior to being passaged on vitronectin. In fact, when the iPSCs were transitioned from feeders to vitronectin for two passages and then split onto either BME or vitronectin the level of cell adhesion was similar. These studies also confirmed that αVβ5 integrin heterodimer is required for iPSC cell adhesion on vitronectin. However, iPSC proliferation was significantly decreased only when both αVβ5 and β1 was blocked. The authors suggested that this could mean that there are other vitronectin-binding factors such as αVβ1 that are involved in iPSC proliferation or that there is an adhesive factor (either secreted by the iPSCs or available in the conditioned media) that is replacing the need for the vitronectin interactions.

More recently, it has been reported that vitronectin was able to support improved iPSC derivation efficiency in combination with E8, a simplified medium containing only eight components instead of 18 that are in the original TeSR™ medium formulation (31). This study reported that once β-mercaptoethanol was removed from the medium formulation bovine serum albumin (BSA) was no longer necessary for PSC culture (31). Once those two components were removed the analysis revealed that several other medium components (i.e., pipecolic acid, GABA, LiCl, chemically defined lipids, trace elements, glutathione, thiamine, l-glutamine) no longer had a positive effect on PSC culture. The E8-vitronectin system was shown to support higher reprogramming efficiencies for viral and episomal approaches using established fibroblasts and fresh biopsy samples (31).

2.2.5 CELLstart™

CELLstart™ is a commercially available xeno-free substrate that has been used to culture I6 hES cells in a defined media supplemented with FGF2 (8 ng/mL) for more than 25 passages (6 months) (32). This substrate can support long-term propagation of undifferentiated hES cells as assessed by morphology, pluripotency marker expression, and karyotypic stability (32). Using CELLstart™ these hESCs were used to derive neural stem cells which could be propagated for a prolonged period and still retain the ability to differentiate into neurons, oligodendrocytes, and astrocytes (32). Dopaminergic neurons generated under these defined, xeno-free conditions were functional as shown by their ability, following transplantation, to improve behavioral deficits in rodent Parkinson's disease models (32).

2.3 Recombinant Human Proteins

Recombinant human proteins, products of the manipulation and combination of DNA molecules from different sources, represent an appealing candidate for xeno-free substrates. Such proteins possess less batch-to-batch variability than proteins purified from natural sources (e.g., blood plasma).

2.3.1 Recombinant Human Laminin

One study analyzed multiple recombinant human laminin (rhLN)s (111, 211, 332, 411, and 511) based on the integrin expression profile for three different hES cell lines (KhES-1, KhES-2, and KhES-3) maintained in MEF-CM (33). $\alpha6\beta1$, the most abundantly expressed integrin on these hES cells, binds preferentially to rhLN-111, rhLN-332, and rhLN-511. $\alpha3\beta1$ and $\alpha7\beta1$ integrins, expressed at a lower level on these hES cells, show binding preferences for rhLN-211 and rhLN-411. rhLN-322 supported the highest growth rate while rhLN-111, rhLN-332, and rhLN-511 were all shown to support undifferentiated hES cultures for ten passages with normal karyotypes and in vitro differentiation capability similar to BME. This study also reported that hESCs predominantly express LN-511/521. The lack of LN-332 expression led the authors to hypothesize that hESCs can be cultured on any laminin isoform that can bind to integrin $\alpha6\beta1$ due to its high avidity.

Rodin et al. reported the culture of multiple hES cell lines (HS420, HS207, HS401) on rhLN-511 in both O3 media (a variant of mTeSR1 containing BSA) and H3 media (a variant of xeno-free TeSR1 medium containing human serum albumin) (34). Cell adhesion assays demonstrated that hESCs had an average contact area on rhLN-511 that was greater than that of cells plated on BME. Proliferation rates on rhLN-511 were similar to BME and karyotypes for hESCs on rhLN-511 were normal following 20 passages. Real-time RT-PCR and western blot analysis revealed that hESCs on rhLN-511 displayed expression levels of pluripotency markers (OCT4, SOX2, Nanog) that were higher than for those cells cultured on BME and similar to levels seen for cells maintained on feeders. In vivo differentiation capability was demonstrated for all three hES cell lines in both media. Additional hES and iPS cell lines were shown to be successfully maintained on rhLN-511. It was shown that $\alpha6\beta1$ integrins were critical for hES cell attachment on rhLN-511. This data, along with the increased contact area on rhLN-511 and the observation that cells formed monolayers within 24 h of being passaged as clumps, led the authors to postulate that LN-511 provides focal adhesion contacts for hESCs so that they can attach and migrate efficiently.

2.3.2 Recombinant E-Cadherin

Cadherins are cell surface, calcium-dependent glycoproteins that mediate cell-cell adhesions. E-cadherin is a transmembrane glycoprotein that is primarily expressed on epithelial cells and constitutes an essential component of adherens junctions via homophilic (binding to other E-cadherin molecules) interactions. ES cells grow in colonies that are maintained by various adhesion complexes that include adherens junctions. Reports have shown that upon differentiation HES4 hES cells demonstrate a loss of E-cadherin, a gain of N-cadherin, upregulation of E-cadherin repressor proteins, and other events that are all characteristic of an epithelial-mesenchymal transition (EMT), a crucial process involved in various stages of embryogenesis, tissue repair, and tumor invasion (35). Multiple studies have reported that the loss of E-cadherin affects ES cell architecture and the localization of a variety of cell surface molecules (36). It was also shown that the culture of HES4 and H1 hESCs in the presence of E-cadherin neutralizing antibody allows those cultures to be maintained without bFGF (on a minimal feeder layer in serum replacement media). Although the exact mechanism behind this phenomenon is unknown it does point to the fact that E-cadherin positively regulates pluripotent signaling pathways.

Nagaoka et al. first reported that mouse ES cells could be maintained on a surface coated with a fusion protein consisting of the E-cadherin extracellular domain and the IgG Fc domain (E-cad-Fc) (37). Although the mouse ES cells did not form characteristic colonies on this substratum, they still exhibited stem cell properties (e.g., pluripotency marker expression, differentiation capability). Additional studies using hESCs found similar results (38). H9 and H1 hESCs cultured on E-cad-Fc-coated surfaces in mTeSR™1 medium displayed evidence of pluripotency marker expression, normal karyotype, teratoma formation with tissue present from all three germ layers, and proliferative capacity equivalent to cells grown on BME. The plating efficiency was significantly less on E-cad-Fc in comparison to BME but that difference was attributed to the use of Accutase® during passaging which caused degradation of the cell surface E-cadherin. When non-proteolytic cell dissociation buffers were used, the plating efficiency on E-cad-Fc was equivalent to BME. In the same study several human iPSC lines were shown to be successfully maintained on E-cad-Fc as well. Cumulatively, this data effectively suggests that this recombinant fusion protein represents a defined, xeno-free substratum for the reproducible culture of human PSCs.

2.3.3 Recombinant Vitronectin

Given the success of purified vitronectin as a substrate, recent studies have described the development of full-length or truncated recombinant vitronectin to support hESC growth and self-renewal (30, 31). In developing a chemically defined culture system for PSC culture Chen et al. identified two vitronectin variants (VTN-NC and VTN-N), truncated at either the N-terminus and/or the C-terminus, that were able to support H1 and H9 hES cell attachment and survival better than wild type VTN in E8 medium (31).

Under long-term culture on recombinant vitronectin in mTeSR™1, HESC-NL3 hES cells were shown to be karyotypically stable, express pluripotency markers, and demonstrate differentiation potential (30). It should be noted that the recombinant vitronectin used in this study was produced in the mouse myeloma cell line NS0 and therefore could potentially contain immunogenic molecules like Neu5Gc, a nonhuman cell surface sialic acid.

2.4 Synthetic Polymers

Clinical applications for hESCs will likely require an extremely large number of cells per patient (39). In an effort to identify materials that could enhance hESC proliferation researchers are investigating the use of synthetic polymers. Although polymeric biomaterials have been utilized as substrates for the growth of a variety of cell types, their application as ESC substrates is relatively new (40, 41). Using an array-based high-throughput approach researchers identified multiple polymers including poly(methyl vinyl ether-alt-maleic anhydride) (PMVE-alt-MA) on which HUES1 and HUES9 hESCs demonstrated the characteristic morphology, proliferation, and pluripotency maintenance after a 5-day culture period (42). Following five passages in BSA-containing medium pluripotency marker expression (i.e., OCT4, SOX2, and Nanog), normal karyotype, and in vitro differentiation into all three lineages confirm that PMVE-alt-MA can successfully maintain these cell lines. These studies also found that hESCs on PMVE-alt-MA exhibited increased expression levels of ECM proteins (e.g., fibronectin, laminin, collagen V), integrin αV, and integrin α5 relative to cultures maintained on BME which suggests that while this synthetic polymer can support initial hESC attachment there are additional proteins being secreted by the hESCs themselves in order to produce a supportive microenvironment.

Another group used surface graft polymerization to synthesize multiple polymer coatings and tested their ability to support long-term hES cell attachment and proliferation in different media. They identified poly(2-(methacryloyloxy)ethyl dimethyl-(3-sulfopropyl)ammonium hydroxide) (PMEDSAH) as a polymer that was capable of supporting BG01 and H9 hES cells for 25 passages in MEF-CM with a normal karyotype, proliferation, pluripotency expression, and differentiation potential (43). Similar results were found when a xeno-free medium was used. Additional studies were conducted using serum-free media and while mTeSR™ could not support hES cells PMEDSAH StemPro® SFM medium was able to support H9 hES cells as evidenced by proliferation, pluripotency marker expression, a normal karyotype, and differentiation potential following ten passages.

Melkoumian et al. developed synthetic surfaces that incorporate extracellular matrix protein-derived peptides covalently linked to an acrylate coating (44). H1 and H7 hESCs were seeded onto several synthetic peptide-acrylate surfaces in a xeno-free, defined media containing recombinant bFGF and TGF-β1. Bone sialoprotein (BSP-PAS) and vitronectin (VN-PAS) were shown to support H1 and H7 hES cell attachment for more than ten serial passages as demonstrated by comparable morphology, karyotypic stability, pluripotency marker expression, and in vivo and in vitro differentiation potential in comparison to hESCs cultured on BME. Directed differentiation into functional cardiomyocytes on both VN-PAS and BSP-PAS confirmed that there was no difference in differentiation potential on these two substrates. This report also includes data that compares cell performance on these substrates in different commercially available media. X-VIVO™ 10 supported greater H7 hES cell proliferation on these surfaces compared to the other media tested which could be due to some adaptation that the cells have undergone having been routinely cultured on X-VIVO™ 10 prior to these experiments.

Derda et al. employed the use of phage display libraries to identify novel peptide ligands that can bind to hESCs (45). These peptides were then incorporated into substrates by attachment to alkanethiols to form self-assembled monolayers (SAM) on gold. Using arrays of these peptide-modified SAMs, the group reported six peptide sequences that were capable of mediating H9 hES cell adhesion. Further analysis demonstrated that two sequences (TVKHRPDALHPQ and LTTAPKLPKVTR) were able to maintain undifferentiated hESCs for three passages (20 days) in mTeSR medium. The cell receptors for these peptides are not obvious although the authors did demonstrate through EDTA and heparin addition that neither integrins nor proteoglycans are likely involved. More recently, this group reported using the peptide-modified SAM approach using known bioactive peptides reported to bind to a diverse number of cell surface receptors (46). Among the 500 unique surfaces that were screened, they identified a heparin-binding peptide (GKKQRFRHRNRKG) derived from vitronectin that was shown to be capable of supporting multiple hES and iPS cells in short-term culture (i.e., cell attachment, proliferation) and long-term culture (i.e., pluripotency maintenance, karyotypic stability, in vitro and in vivo differentiation) in mTeSR medium (46).

Saha et al. described a procedure in which polystyrene was treated with short-wavelength UV to create a surface chemistry that supports hES cells (BG01, WIBR1, and WIBR3) and iPSCs. The UV-treated surfaces supported threefold more cells per area and created at least twofold more colonies per cell seeded than untreated polystyrene (47). Long-term culture on these UV-treated surfaces was performed with serum-free mTeSR™1 medium. The authors also found that spatial patterning of the UV treatment could also increase the number of undifferentiated cells per growth area. It was also demonstrated that hESCs and iPSCs could be consistently passaged as single cells on these surfaces for more than ten passages if ROCK inhibitor was included for 8–12 h after dissociation as evidenced by a normal karyotype and pluripotency marker expression. It should be noted that before cell seeding the UV-treated surfaces were treated with recombinant human vitronectin, 20% human serum, or 20% fetal bovine serum. It is unclear if the carboxylic acid/ester and nitrogen-containing moieties generated on the polystyrene surface by the UV exposure are influencing the cells directly or affecting the presentation/binding of the attachment proteins (i.e., vitronectin, serum) coated on the surface.

Another group developed synthetic peptides that engaged the same integrins on H1 hESCs maintained on BME in serum-free media (48). Their analysis of the integrin expression profile found that αVβ3, α6, β1, and α2β1 integrins played a significant role in initial hESC adhesion to BME. Peptides were patterned after the integrin-engaging cell adhesion motifs of laminin-111, which is the major component of BME. A mixture of three of the peptides (60:16:24% molar ratio) maintained the hES cells in an undifferentiated state longer than individual peptide substrates. However, it was noted that an increasing amount of differentiation was seen following three passages on the blended peptide coating. The authors theorized that this increasing differentiation could be due to unligated α6β1 integrins that could cause integrin-mediated apoptosis and promotion of stromal cells derived from the ES cells.

References

1. Thomson JA, Itskovitz-Eldor J, Shapiro SS, Waknitz MA, Swiergiel JJ, Marshall VS, Jones JM (1998) Science 282:1145–1147
2. Gepstein L (2002) Circ Res 91:866–876
3. Takahashi K, Yamanaka S (2006) Cell 126:663–676
4. Richards M, Tan S, Fong CY, Biswas A, Chan WK, Bongso A (2003) Stem Cells 21:546–556
5. Meng G, Liu S, Rancourt DE (2012) Stem Cells Dev 21:2036–2048
6. Tavakoli T, Xu X, Derby E, Serebryakova Y, Reid Y, Rao MS, Mattson MP, Ma W (2009) BMC Cell Biol 10:44
7. Chang KH, Nelson AM, Fields PA, Hesson JL, Ulyanova T, Cao H, Nakamoto B, Ware CB, Papayannopoulou T (2008) Exp Cell Res 314:2930–2940
8. Allegrucci C, Young LE (2007) Hum Reprod Update 13:103–120
9. Xu C, Inokuma MS, Denham J, Golds K, Kundu P, Gold JD, Carpenter MK (2001) Nat Biotechnol 19:971–974
10. Hughes CS, Postovit LM, Lajoie GA (2010) Proteomics 10:1886–1890
11. Stewart MH, Bendall SC, Bhatia M (2008) J Mol Med 86:875–886
12. Xu C, Rosler E, Jiang J, Lebkowski JS, Gold JD, O'Sullivan C, Delavan-Boorsma K, Mok M, Bronstein A, Carpenter MK (2005) Stem Cells 23:315–323
13. Xu RH, Peck RM, Li DS, Feng X, Ludwig T, Thomson JA (2005) Nat Methods 2:185–190
14. Wang L, Li L, Menendez P, Cerdan C, Bhatia M (2005) Blood 105:4598–4603
15. Klagsbrun M, Baird A (1991) Cell 67:229–231
16. Dvorák P, Hampl A, Jirmanová L, Pacholíková J, Kusakabe M (1998) J Cell Sci 111:2945–2952

17. Yu J, Vodyanik MA, Smuga-Otto K, Antosiewicz-Bourget J, Frane JL, Tian S, Nie J, Jonsdottir GA, Ruotti V, Stewart R, Slukvin II, Thomson JA (2007) Science 318:1917–1920
18. Rowland TJ, Miller LM, Blaschke AJ, Doss EL, Bonham AJ, Hikita ST, Johnson LV, Clegg DO (2010) Stem Cells Dev 19:1231–1240
19. Jones MB, Chu C, Pendleton JC, Betenbaugh MJ, Shiloach J, Baljinnyam B, Rubin JS, Shamblott MJ (2010) Stem Cells Dev 19:1923–1935
20. Furue MK, Na J, Jackson JP, Okamoto T, Jones M, Baker D, Hata R, Moore HD, Sato JD, Andrews PW (2008) Proc Natl Acad Sci U S A 105:13409–13414
21. Stojkovic P, Lako M, Przyborski S, Stewart R, Armstrong L, Evans J, Zhang X, Stojkovic M (2005) Stem Cells 23:895–902
22. Dziadek M, Timpl R (1985) Dev Biol 111:372–382
23. Beattie GM, Lopez AD, Bucay N, Hinton A, Firpo MT, King CC, Hayek A (2005) Stem Cells 23:489–495
24. Li Y, Powell S, Brunette E, Lebkowski J, Mandalam R (2005) Biotechnol Bioeng 91:688–698
25. Amit M, Shariki C, Margulets V, Itskovitz-Eldor J (2004) Biol Reprod 70:837–845
26. Noaksson K, Zoric N, Zeng X, Rao MS, Hyllner J, Semb H, Kubista M, Sartipy P (2005) Stem Cells 23:1460–1467
27. Tsutsui H, Valamehr B, Hindoyan A, Qiao R, Ding X, Guo S, Witte ON, Liu X, Ho CM, Wu H (2011) Nat Commun 2:167
28. Preissner KT, Reuning U (2011) Semin Thromb Hemost 37:408–424
29. Hayman EG, Pierschbacher MD, Suzuki S, Ruoslahti E (1985) Exp Cell Res 160:245–258
30. Braam SR, Zeinstra L, Litjens S, ward-van Oostwaard D, van den Brink S, van Laake L, Lebrin F, Kats P, Hochstenbach R, Passier R, Sonnenberg A, Mummery CL (2008) Stem Cells 26:2257–2265
31. Chen G, Gulbranson DR, Hou Z, Bolin JM, Ruotti V, Probasco MD, Smuga-Otto K, Howden SE, Diol NR, Propson NE, Wagner R, Lee GO, Antosiewicz-Bourget J, Teng JM, Thomson JA (2011) Nat Methods 8:424–429
32. Swistowski A, Peng J, Han Y, Swistowska AM, Rao MS, Zeng X (2009) PLoS One 4:e6233
33. Miyazaki T, Futaki S, Hasegawa K, Kawasaki M, Sanzen N, Hayashi M, Kawase E, Sekiguchi K, Nakatsuji N, Suemori H (2008) Biochem Biophys Res Commun 375:27–32
34. Rodin S, Domogatskaya A, Ström S, Hansson EM, Chien KR, Inzunza J, Hovatta O, Tryggvason K (2010) Nat Biotechnol 28:611–615
35. Eastham AM, Spencer H, Soncin F, Ritson S, Merry CL, Stern PL, Ward CM (2007) Cancer Res 67:11254–11262
36. Soncin F, Ward CM (2011) Genes 2:229–259
37. Nagaoka M, Koshimizu U, Yuasa S, Hattori F, Chen H, Tanaka T, Okabe M, Fukuda K, Akaike T (2006) PLoS One 1:e15
38. Nagaoka M, Si-Tayeb K, Akaike T, Duncan SA (2010) BMC Dev Biol 10:60
39. Bongso A, Fong CY, Gauthaman K (2008) J Cell Biochem 105:1352–1360
40. Banu N, Rosenzweig M, Kim H, Bagley J, Pykett M (2001) Cytokine 13:349–358
41. Curran JM, Chen R, Hunt JA (2006) Biomaterials 27:4783–4793
42. Brafman DA, Chang CW, Fernandez A, Willert K, Varghese S, Chien S (2010) Biomaterials 31:9135–9144
43. Villa-Diaz LG, Nandivada H, Ding J, Nogueira-de-Souza NC, Krebsbach PH, O'Shea KS, Lahann J, Smith GD (2010) Nat Biotechnol 28:581–583
44. Melkoumian Z, Weber JL, Weber DM, Fadeev AG, Zhou Y, Dolley-Sonneville P, Yang J, Qiu L, Priest CA, Shogbon C, Martin AW, Nelson J, West P, Beltzer JP, Pal S, Brandenberger R (2010) Nat Biotechnol 28:606–610
45. Derda R, Musah S, Orner BP, Klim JR, Li L, Kiessling LL (2010) J Am Chem Soc 132:1289–1295
46. Klim JR, Li L, Wrighton PJ, Piekarczyk MS, Kiessling LL (2010) Nat Methods 7:989–994
47. Saha K, Mei Y, Reisterer CM, Pyzocha NK, Yang J, Muffat J, Davies MC, Alexander MR, Langer R, Anderson DG, Jaenisch R (2011) Proc Natl Acad Sci U S A 108:18714–18719
48. Meng Y, Eshghi S, Li YJ, Schmidt R, Schaffer DV, Healy KE (2010) FASEB J 24:1056–1065
49. Rosler ES, Fisk GJ, Ares X, Irving J, Miura T, Rao MS, Carpenter MK (2004) Dev Dyn 229:259–274
50. Peerani R, Rao BM, Bauwens C, Yin T, Wood GA, Nagy A, Kumacheva E, Zandstra PW (2007) EMBO J 26:4744–4755
51. Vallier L, Alexander M, Pedersen RA (2005) J Cell Sci 118:4495–4509
52. Xiao L, Yuan X, Sharkis SJ (2006) Stem Cells 24:1476–1486
53. Yao S, Chen S, Clark J, Hao E, Beattie GM, Hayek A, Ding S (2006) Proc Natl Acad Sci U S A 103:6907–6912
54. Levenstein ME, Ludwig TE, Xu RH, Llanas RA, VanDenHeuvel-Kramer K, Manning D, Thomson JA (2006) Stem Cells 24:568–574
55. Bendall SC, Stewart MH, Menendez P, George D, Vijayaragavan K, Werbowetski-Ogilvie T, Ramos-Mejia V, Rouleau A, Yang J, Bosse M, Lajoie G, Bhatia M (2007) Nature 448:1015–1021

Part II

Culture and Differentiation of Pluripotent Stem Cells

Chapter 8

Methods for Culturing Human Embryonic Stem Cells on Feeders

Jasmeet Kaur and Mary Lynn Tilkins

Abstract

Human embryonic stem cells can be cultured and maintained on fibroblast feeder cells of murine or human origin. Thorough protocols are provided for the growth and maintenance of human embryonic stem cells on either feeder cell type.

Key words Pluripotent stem cell, Embryonic fibroblast feeder cells, Mouse embryonic fibroblast, Human foreskin fibroblast

1 Introduction

Embryonic stem cells (ESC) are isolated from blastocyst stage embryos and are pluripotent and capable of differentiating into all three germ layers: ectoderm, endoderm, and mesoderm. The ESC field was launched in 1981, when two independent labs successfully derived mouse ESC (1, 2). Early protocols required feeder cells to enhance mouse ESC attachment and growth (3, 4), and the use of feeder cells is still the norm in current mouse ESC protocols. When derivation of human ESC was reported in 1998 (5) the researchers used well-established ESC techniques, including mouse embryonic fibroblast feeder cells, in their groundbreaking work. Human ESC can be successfully isolated, cultured, and maintained on fibroblast feeder cells of mouse or human origin (6–8). These feeder cells not only facilitate stem cell recovery from liquid nitrogen storage but also promote cell attachment and expansion by laying down extracellular matrix and by secreting growth factors that help to maintain stem cell pluripotency.

Uma Lakshmipathy and Mohan C. Vemuri (eds.), *Pluripotent Stem Cells: Methods and Protocols*, Methods in Molecular Biology, vol. 997, DOI 10.1007/978-1-62703-348-0_8, © Springer Science+Business Media New York 2013

The purpose of this chapter is to provide a basic protocol for the isolation and establishment of mouse embryonic fibroblast feeder (MEF) cells and/or the use of human fibroblast feeder (HFF) cells to successfully support the growth and maintenance of human ESC.

2 Materials

2.1 Reagents

1. CELLstart™ CTS™ defined, humanized substrate for cell culture.
2. D-PBS CTS™ with calcium chloride and magnesium chloride.
3. Dispase.
4. Dulbecco's Phosphate Buffered Saline (D-PBS) without calcium and magnesium; TrypLE™ Select CTS™, without Phenol Red.
5. ES Cell Qualified Fetal Bovine Serum (FBS).
6. FGF-basic Full Length CTS™ Recombinant Human (100 μg lyophilized powder); bFGF.
7. GlutaMAX™-I CTS™.
8. Human fibroblast feeders cells.
9. KnockOut™ CTS™ DMEM.
10. KnockOut™ Serum Replacement (KSR).
11. KnockOut™ SR XenoFree CTS.
12. KnockOut™ SR GF Cocktail CTS™.
13. KnockOut™ DMEM/F-12 CTS™.
14. 2-Mercaptoethanol.
15. MEM Non-Essential Amino Acids Solution (NEAA).
16. Mitomycin C.
17. Penicillin–streptomycin or gentamicin (optional).
18. 0.1% porcine gelatin solution.
19. STEMPRO® EZPassage™ Disposable Cell Passaging Tool.
20. Timed pregnant day 13.5 CD-1 mice or mouse embryonic fibroblast feeder cells.

2.2 Equipment

1. Biosafety cabinet.
2. Centrifuge (suitable for 15 mL and 50 mL centrifuge tubes).
3. Colony marker for microscope.
4. Freezer (−20°C).
5. Freezer (−80°C).

6. Gamma irradiator.
7. 37°C, humidified 5% CO_2 incubator.
8. Liquid nitrogen storage tank.
9. Mr. Frosty Nalgene Cryo 1°C Freezing Container.
10. Phase-contrast microscope.
11. Pipet-Aid.
12. Plastic cryovial rack.
13. 4°C refrigerator.
14. Water bath (37°C).

2.3 Supplies

1. 5 mL, 10 mL, and 25 mL serological pipets.
2. 15 mL and 50 mL centrifuge tubes.
3. 60 mm tissue culture-treated dishes.
4. 1.5 mL cryovials.
5. 60 mm and 100 mm non-tissue culture-treated dishes.

2.4 Media Preparation

1. Prepare 500 mL Feeder Medium by combining 440 mL KnockOut™ CTS™ DMEM, 50 mL ES Cell Qualified FBS, 5 mL NEAA, and 5 mL GlutaMAX™-I CTS™.
2. Prepare 10 μg/mL bFGF solution by adding 990 μL D-PBS and 10 μL 10% BSA to a 10 μg vial of FGF-basic Full Length.
3. Prepare a 2 mg/mL Dispase solution by adding 50 mL KnockOut™ DMEM/F-12 CTS solution to a 100 mg Dispase vial. Aliquot and store at 4°C for up to 2 weeks, or at –20°C for up to 6 months.
4. Prepare KSR hESC Medium (for culturing hESC on MEF) by combining 78 mL KnockOut™ CTS™ DMEM or D-MEM/F-12, 20 mL KSR, 1 mL NEAA, 1 mL GlutaMAX, and 40 μL bFGF. Add 55 mM 2-Mercaptoethanol just prior to use (see Note 10).
5. Prepare 100 mL MEF Freezing Medium by adding 20 mL DMSO to 80 mL Feeder Medium.
6. Prepare 500 mL Inactive HFF Medium by combining 440 mL KnockOut™ CTS™ DMEM, 50 mL KnockOut™ SR XenoFree CTS™, 5 mL NEAA, and 5 mL GlutaMAX™-I CTS™.
7. Prepare 100 mL hESC Freezing Medium by adding 20 mL DMSO to 80 mL KSR hESC Medium.
8. Prepare 100 mL KSR XenoFree Medium (for culturing hESC on HFF) by combining 83 mL KnockOut™ DMEM CTS™, 15 mL KnockOut™ SR XenoFree CTS™, 1 mL NEAA, 1 mL GlutaMAX™-I, 40 μL bFGF (see Note 10).

9. Prepare 100 mL hESC KSR XenoFree Freezing Medium by adding 20 mL DMSO to 80 mL KSR XenoFree Medium.
10. Prepare EB Medium by combining 83 mL of KnockOut™ DMEM, 15 mL KnockOut™ SR XenoFree, 1 mL GlutaMAX™, 1 mL NEAA.
11. Prepare 1:50 CELLstart solution by combining 49 mL D-PBS with calcium chloride and magnesium chloride with 1 mL CELLstart™ CTS™.

3 Methods

Workflow diagram for culturing hESC on feeders (Fig. 1) (see Note 1).

3.1 Mouse Embryonic Fibroblast (MEF) Methods

3.1.1 Isolation of MEF Cells

1. Sacrifice a day 13.5 CD-1 pregnant mouse by cervical dislocation.
2. Place the mouse on its back and spray the abdomen with 70% ethanol. Using sterile scissors, cut through the skin of the abdomen.
3. With sterile forceps, gently free the uterus from the abdomen and snip with sterile scissors to detach it.
4. Transfer the uterus to a Petri dish containing sterile D-PBS.
5. Carefully free the embryos from the embryonic sacs, and transfer them to a fresh Petri dish containing sterile D-PBS.
6. Under a dissecting microscope in a biosafety cabinet if possible, use two pairs of sterile forceps to remove the head and internal organs from the harvested embryos, leaving the limbs intact.
7. Transfer the dissected embryos to a sterile 50 mL tube. For every ten embryos, add 5 mL of warm 0.25% trypsin/EDTA. Dissociate embryos by triturating with a 10 mL serological pipet.
8. To further dissociate the embryonic tissue, transfer the tissue/trypsin solution to the barrel of an appropriately sized syringe with an attached 18 G 1½″ needle. Carefully replace the syringe plunger, invert the syringe, and expel any air from the syringe. With slow and steady pressure, push the tissue/trypsin solution through the needle, and collect in a 50 mL centrifuge tube.
9. Transfer the centrifuge tube containing cell suspension to a 37°C water bath, and incubate for 5 min. Remove tube from water bath, triturate the cell suspension a few times using a 10 mL pipet, and return tube to water bath. Repeat this incubation and trituration procedure 2 more times. The cell suspension should contain very few clumps of tissue.

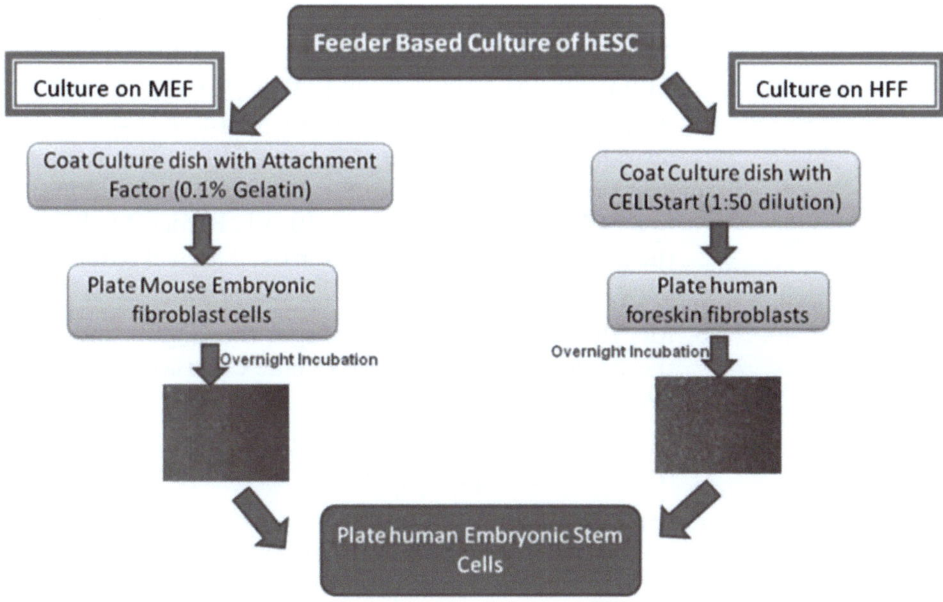

Fig. 1 Shows a workflow diagram of hESC cultured on MEF or HFF cells

10. Add an equal volume of Feeder Medium to inactivate the trypsin.
11. Label one T-225 flask for every 1.3 embryos dissociated (e.g., 26 embryos would require (20) T-225 flasks).
12. Mix the cell suspension thoroughly with a 10 mL pipet, transfer 1.3 embryos per T225 flask, and add Feeder Medium to make a final volume of 40 mL per flask. Place flasks in a 37°C incubator with a humidified atmosphere of 5% CO_2. Swirl flasks in a north-to-south and east-to-west pattern to dispense cells evenly throughout flasks.
13. Observe flasks the next day using a phase-contrast microscope. Cells should have attached and begun expanding. Remove and replace the medium with 50 mL fresh Feeder Medium/T-225 flask.
14. Cryopreserve MEF cells when flasks are nearly confluent (usually 3–4 days post-plating).

3.1.2 Cryopreservation of Primary MEF Cells

1. Processing no more than five flasks at a time, aspirate and discard Feeder Medium from T-225 flasks.
2. Rinse the cells once with D-PBS containing no calcium or magnesium, using 20 mL/flask. Discard D-PBS rinse.

3. Apply 3 mL warm 0.25% trypsin/EDTA solution per flask. Swirl flask to coat entire cell surface. Gently rap side of flask with palm. MEF sheet should slough off.
4. Add 5 mL Feeder Medium to each flask to inactivate trypsin. Swirl flask to coat surface. Triturate MEF in each flask and pool cells from the various flasks into one sterile vessel. Rinse four to five flasks with an additional 20 mL Feeder Medium, taking the same medium from flask to flask. This will reduce the number of cells left behind.
5. Mix the final MEF cell suspension thoroughly and determine the viable cell count by trypan blue exclusion method using a hemocytometer.
6. Viability should be ≥90%. Centrifuge the cell suspension at room temperature for 5 min at $200 \times g$.
7. Aspirate medium and resuspend the cells in Feeder Medium at 2× the desired final freezing concentration (e.g., 4×10^6/mL). Add an equal volume of MEF Freezing Medium (e.g., final concentration will be 2×10^6/mL) to yield a final concentration of 10% DMSO.
8. Dispense 1 mL MEF cell suspension per cryovial, mixing thoroughly between dispenses.
9. For best results, MEF vials should be cryopreserved using a controlled-rate freezing device (e.g., CryoMed® Freezer or Mr. Frosty Nalgene Cryo 1°C Freezing Container), following manufacturer's instructions.
10. The next day, transfer vials to permanent liquid nitrogen storage.

3.1.3 Thawing Active MEF Cells

1. Warm Feeder Medium before thawing the cells. Place 10 mL Feeder Medium in a 15 mL centrifuge tube.
2. Remove MEF vial from liquid nitrogen and rapidly thaw in a 37°C water bath by gently swirling until just a small frozen chunk remains in the vial.
3. Spray vial with 70% isopropanol to decontaminate it and transfer vial to the biosafety cabinet.
4. Aseptically transfer the entire contents of the vial into the centrifuge tube containing 10 mL warm Feeder Medium.
5. Rinse the vial with 1–2 mL Feeder Medium and add it to the same 15 mL centrifuge tube.
6. Gently pellet cells by centrifuging at $200 \times g$ at room temperature for 5 min.
7. Carefully aspirate the supernatant without disturbing the cell pellet and discard it.
8. Gently "flick" the tube at the base to fully dislodge the cell pellet from the tube bottom.

Table 1
Initial MEF seeding information (passage 0)

Flask size	MEF seeding density	Initial volume of pre-warmed MEF media (mL)
T-75	$2–2.5 \times 10^6$	15
T-160	5×10^6	30
T-225	7×10^6	50

9. Add 5 mL of warm Feeder Medium to the pellet and gently triturate. Using Table 1 as a guide, transfer the MEF cells to a vented tissue culture-treated flask.
10. Place flask in a 37°C incubator with a humidified atmosphere of 5% CO_2. Gently swirl flask north to south then east to west for homogeneous distribution of cells.
11. The next day, examine the cells under a phase-contrast microscope to confirm attachment and expansion.

3.1.4 Passaging MEF Cells

1. When the culture is nearly confluent, subculture at a 1:4 ratio.
2. Aspirate and discard the supernatant from the flask and wash twice with D-PBS that does not contain calcium or magnesium.
3. Add an appropriate volume of pre-warmed 0.25% Trypsin/EDTA per flask according to Table 2. Swirl flask to coat entire cell surface. Gently rap side of flask with palm. MEF sheet should slough off.
4. Add 5 mL Feeder Medium to flask to inactivate trypsin, and gently triturate to achieve a single-cell suspension.
5. Transfer MEF cells to a sterile centrifuge tube.
6. Centrifuge the cell suspension at room temperature for 5 min at $200 \times g$.
7. Discard the supernatant and gently flick the bottom of the tube to loosen the cell pellet.
8. Resuspend the cell pellet with warm Feeder Medium, and determine the viable cell count by trypan blue exclusion method using a hemocytometer.
9. Passage MEF at a 1:4 split ratio in new T flasks as recommended in Table 1 below.
10. Place flasks in a 37°C incubator with a humidified atmosphere of 5% CO_2. Gently swirl flask north to south then east to west for homogeneous distribution of cells. This is passage 1.

Table 2
Trypsin and TrypLE express volumes for dissociation

Flask size	Suggested volume (mL)
T-75	1.5
T-160	3
T-225	5

11. Observe the cells daily under a phase-contrast microscope to confirm expansion. If cells will not be passaged until day 4, add 50% fresh medium on day 3, without removing any supernatant from the flasks.
12. When cultures are nearly confluent, repeat the dissociation procedure by following steps 2–8, passaging flasks once again at 1:4. This is passage 2.
13. When passage 2 flasks are nearly confluent, MEF cells can be passaged once again, cryopreserved or treated with Mitomycin C to arrest cell division. Note: Active MEF cells should not be used beyond passage 6 for best hESC results.

3.1.5 Mitomycin C Preparation

Caution: Mitomycin C is highly toxic. Read the accompanying material safety data sheet and handle with care. Mitomycin C waste should be disposed of following institute policy for hazardous waste.

1. Using a syringe with attached 20 G 1½″ needle, add 2 mL Feeder Medium to a 2 mg Mitomycin C vial.
2. Let the vial sit at room temp with occasional swirling to ensure complete dissolution of the Mitomycin C.
3. With a fresh syringe and needle, carefully aspirate the reconstituted Mitomycin C solution and transfer it to a sterile plastic bottle. Add 198 mL Feeder Medium to yield a final concentration of 10 μg/mL. This is now MEF Inactivation Medium.
4. Aliquot and store the medium in the dark at 4°C for up to 6 weeks, or at −20°C for long-term storage.

3.1.6 Mitotic Inactivation of MEFs Using Mitomycin C

Note: Check each flask individually by observing the cells under a microscope to ensure cell growth and culture sterility. MEFs should be 90–95% confluent to begin Mitomycin C treatment.

1. Processing no more than five flasks at a time, carefully aspirate and discard MEF supernatant from T-225 flasks. Apply warm MEF Inactivation Medium at 22 mL per T-225 flask.

2. Return flasks to a 37°C incubator with a humidified atmosphere of 5% CO_2 for 2–2.5 h.
3. Aspirate the MEF Inactivation Medium from the flasks, setting aside for hazardous waste disposal.
4. Rinse MEF flasks three times with 20 mL D-PBS containing calcium and magnesium, pooling rinses with MEF Inactivation Medium waste.
5. Apply fresh Feeder Medium and incubate the cells in a 37°C incubator, with 5% CO_2 in humidified air, overnight. Alternatively, inactive MEF can be harvested and used immediately, or cryopreserved for later use.
6. To harvest inactive MEF, rinse each flask with 20 mL D-PBS that does not contain calcium nor magnesium. Add an appropriate volume of pre-warmed 0.25% Trypsin/EDTA per flask according to Table 2. Swirl flask to coat entire cell surface. Gently rap side of flask with palm. Cell sheet should slough off. Add 5 mL Feeder Medium to flask to inactivate trypsin, and gently triturate to achieve a single-cell suspension.
7. Transfer inactive MEF to a sterile centrifuge tube.
8. Centrifuge the cell suspension at room temperature for 5 min at $200 \times g$.
9. Discard the supernatant and gently flick the bottom of the tube to loosen the cell pellet.
10. Resuspend the cell pellet with warm Feeder Medium, and determine the viable cell count by trypan blue exclusion method using a hemocytometer.
11. Plate inactive MEF at 2.5×10^4 cells/cm^2 in the desired vessels, or alternatively, inactive MEF feeder cells can be cryopreserved for later use.
12. To cryopreserve inactive MEF following centrifugation, aspirate medium and resuspend the cells in Feeder Medium at 2× the desired final freezing concentration (e.g., 1×10^7/mL). Add an equal volume of MEF Freezing Medium (e.g., final concentration will be 5×10^6/mL) to yield a final concentration of 10% DMSO.
13. Dispense 1 mL inactive MEF cell suspension per cryovial, mixing thoroughly between dispenses.
14. For best results, inactive MEF vials should be cryopreserved using a controlled-rate freezing device (e.g., CryoMed® Freezer or Mr. Frosty Nalgene Cryo 1°C Freezing Container), following manufacturer's instructions.
15. The next day, transfer vials to permanent liquid nitrogen storage.

3.1.7 Coating Vessels for hESC Culture on Inactive MEF Cells

1. Label culture vessels (e.g., 60 mm dishes).
2. Cover the entire surface of the culture vessels with 0.1% gelatin (w/v) solution. Incubate the vessels for 30 min in a 37°C incubator, or for 2 h at room temperature.
3. Completely remove the gelatin solution from the culture vessels by aspiration. Note: It is not necessary to wash the culture surface before adding cells or medium.
4. Coated vessels may be used immediately or stored at room temperature for up to 24 h.

3.1.8 Thawing and Plating Inactive MEF Cells

1. Aspirate the gelatin solution from the coated culture vessels (see Subheading 3.1.7).
2. Add the appropriate volume of Feeder Medium into each culture vessel (refer to Table 1).
3. Plate inactive MEF feeder cells at 2.5×10^4 cells/cm^2.
4. Swirl vessel in a north-to-south and east-to-west pattern to dispense cell evenly throughout.
5. Incubate the cells in a 37°C incubator with a humidified atmosphere of 5% CO_2.
6. Use the inactive MEF culture vessel within 3–4 days after plating (see Note 8 and Fig. 2)

3.1.9 Thawing and Plating hESCs on Inactive MEF Cells

1. At 3–4 h before plating hESC, aspirate the Feeder Medium from the vessel. Gently apply KSR hESC Medium by trickling medium down the side of the vessel so as not to disturb the feeder layer; use 4 mL per 60 mm culture dish. Place dish in a 37°C incubator with a humidified atmosphere of 5% CO_2, until ready for use.
2. Remove a hESC cryovial from the liquid nitrogen storage tank and rapidly thaw in a 37°C water bath by gently swirling until just a small frozen chunk remains in the vial.
3. Spray vial with 70% isopropanol to decontaminate it, and transfer vial to the biosafety cabinet.
4. Aseptically transfer the entire contents of the vial into a 15 mL centrifuge tube using a 5 mL serological pipet (see Note 3).
5. Rinse the cryovial with 1–2 mL pre-warmed KSR hESC Medium. Transfer this rinse medium to the same 15 mL tube containing the cells.
6. Add 4 mL pre-warmed KSR hESC Medium dropwise to the cells. While adding the medium, gently move the tube back and forth to mix the hESCs. Note: Adding the medium slowly helps the cells to avoid osmotic shock.
7. Centrifuge the tube at $200 \times g$ for 2 min at room temperature.

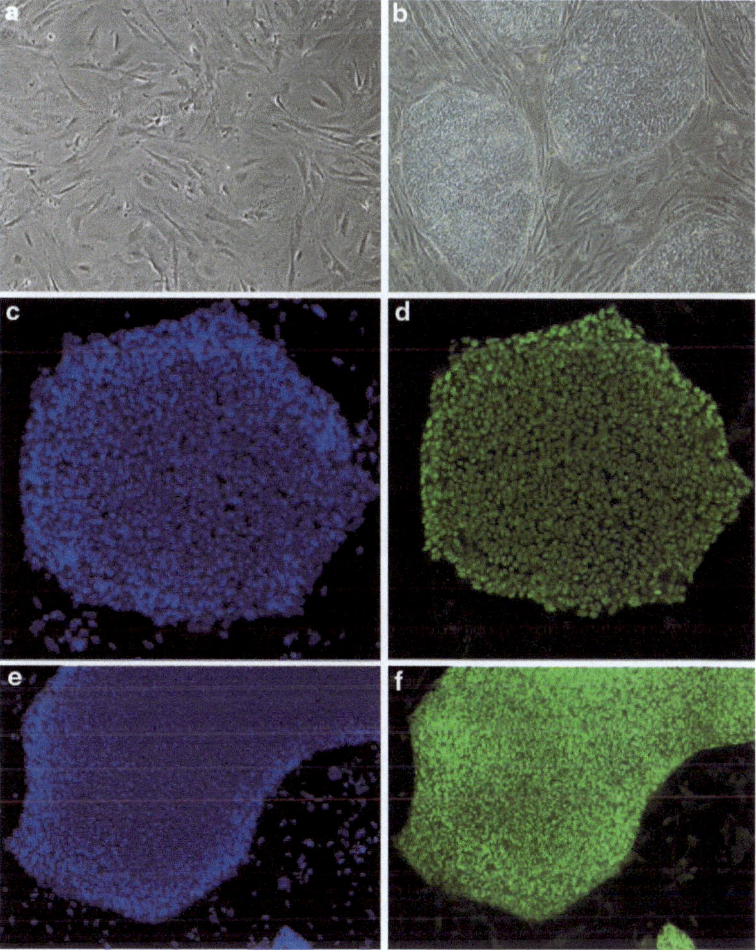

Fig. 2 Shows morphology image of MEF (**a**) and hESC cultured on MEF (**b**) and immunocytochemical staining of hESC for DAPI (**c**, **e**), Oct 4 (**d**), and Sox 2 (**f**)

8. Aspirate the supernatant.
9. Aspirate the conditioned hESC Medium from the 60 mm culture dish containing the inactive MEFs, and use it to gently resuspend the hESC pellet.
10. Slowly add the hESC suspension into the inactive MEF culture dish (see Note 7).
11. Place the vessel in a 37°C incubator with a humidified atmosphere of 5% CO_2.
12. Swirl vessel in a north to south and east-to-west pattern to evenly distribute the hESC.
13. Examine the cells under a microscope daily and change the medium daily. It may take up to a week for hESC colonies to become visible (see Note 2 and Fig. 2).

14. Passage the hESC before colonies become too crowded or overgrown. The split ratio depends on the total number of hESCs recovered, and can range from 1:1 to 1:3 at the first passage (see Note 5).

3.1.10 Passaging Human ESCs on Inactive MEF Cells

1. A day before passaging the hESC culture, prepare fresh inactive MEF culture vessels following the instructions for MEF feeder cultures in Subheading 3.1.7 and 3.1.8.
2. Before plating the hESC, aspirate the Feeder Medium and replace it with KSR hESC Medium. Transfer to a 37°C incubator for 3–4 h before plating the hESCs.
3. Remove the confluent hESC-MEF culture vessel to be passaged from the incubator. Observe the dish under a dissecting microscope and physically remove any differentiated colonies (see Note 6).
4. Aspirate the spent medium from the hESC-MEF culture vessels and add an appropriate volume of pre-warmed Dispase solution to the culture vessel (e.g., 2 mL/60 mm dish or 4 mL/100 mm dish).
5. Incubate the hESC-MEF culture vessels for 5–6 min in a 37°C incubator.
6. Remove dish from the incubator and observe under the microscope for rolled up colony edges. The colonies will not detach from the surface completely; see Fig. 3.
7. Carefully aspirate the Dispase solution from the hESC-MEF culture vessel and add an appropriate amount of KSR hESC Medium to each vessel (e.g., 5 mL to each 100 mm dish).
8. Use a cell scraper to gently detach the hESCs from the vessel.
9. Pool the hESCs into a 15 mL centrifuge tube and rinse the hESC-MEF culture vessel with an appropriate amount of KSR hESC Medium. Transfer this medium to the same 15 mL tube containing the hESCs.
10. Gently pipet the cells up and down a few times in the tube to further break up the cell colonies.
11. Centrifuge the cells at $200 \times g$ for 2 min at room temperature.
12. Gently aspirate the supernatant without disturbing the hESC pellet.
13. Gently flick the bottom of the tube to dislodge the cell pellet and gently resuspend the pellet in an appropriate amount of KSR hESC Medium to accommodate the number of MEF vessels that will receive hESC.
14. Invert the tube to mix the hESC suspension, and transfer hESC suspension to MEF vessels according to the desired split ratio.

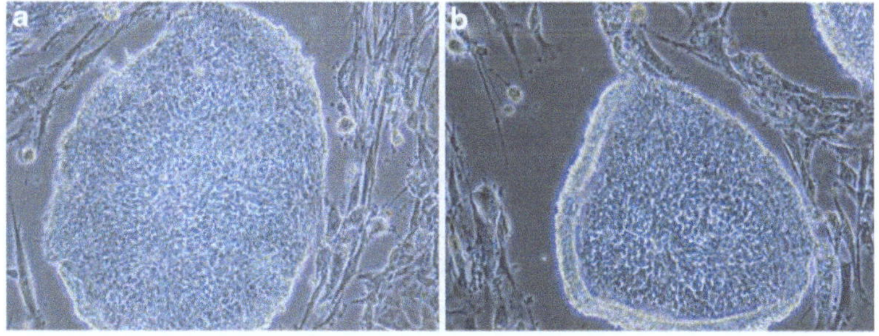

Fig. 3 Panels (**a**) and (**b**) show hESC-MEF colonies rolling up at the edges, during dissociation with Dispase

15. Place the vessels in a 37°C incubator with a humidified atmosphere of 5% CO_2. Swirl vessel s in a north-to-south and east-to-west pattern to evenly distribute hESC.
16. Observe the cells daily and replace the spent medium (see Note 2).
17. Passage cells before hESC colonies become too large or crowded (approximately every 4–7 days).

Cryopreservation of hESC on Inactive MEF Cells

1. Follow the protocol for passaging hESC with KSR hESC Medium mentioned in Subheading 3.1.10.
2. At step 13 in Subheading 3.1.10, gently resuspend the cell pellet with a 50/50 mix of KSR hESC Medium and hESC Freezing Medium (to yield 10% DMSO) using a 5 mL pipet and triturate.
3. Dispense 1 mL of the cell suspension per cryovial using a 5 mL pipet, mixing thoroughly between dispenses.
4. For best results, vials should be cryopreserved using a controlled-rate freezing device (e.g., CryoMed® Freezer or Mr. Frosty Nalgene Cryo 1°C Freezing Container), following manufacturer's instructions.
5. The next day, transfer vials to permanent liquid nitrogen storage. Expect some cell death at recovery, and freeze hESC at a higher density than would normally be passaged to facilitate recovery from liquid nitrogen storage. For example, if cells are routinely passaged at 1:5, a 1:3 or 1:4 split ratio is recommended for freezing hESC (see Note 7).

3.2 Human Foreskin Fibroblast (HFF) Methods

3.2.1 Recovery of HFF Cells from Liquid Nitrogen

1. Warm Feeder Medium before thawing the cells. Place 10 mL Feeder Medium in a 15 mL centrifuge tube.
2. Remove HFF vial (e.g., Cat. No. CRL-2429, American Type Culture Collection) from liquid nitrogen and rapidly thaw in a 37°C water bath by gently swirling until just a small frozen chunk remains in the vial.

3. Spray vial with 70% isopropanol to decontaminate it and transfer vial to the biosafety cabinet.
4. Aseptically transfer the entire contents of the vial into the 15 mL centrifuge tube containing 10 mL warm Feeder Medium using a 5 mL serological pipet.
5. Rinse the vial with 1–2 mL Feeder Medium and add it to the same 15 mL centrifuge tube.
6. Gently pellet cells by centrifuging at 200 × *g* for 3 min at room temperature.
7. Carefully aspirate the supernatant without disturbing the cell pellet and discard it.
8. Gently "flick" the tube at the base to fully dislodge the cell pellet from the tube bottom.
9. Add the required volume of Feeder Medium to the pellet and gently triturate. Transfer the HFF cells to appropriate-sized vessel(s). Incubate in a 37°C incubator with a humidified atmosphere of 5% CO_2.

3.2.2 Passaging HFF Cells

1. Observe the vessel(s) for confluence; subculture when confluent, but not tightly packed with cells (see Fig. 4a). Note: Be sure to scale up and bank cells at low passage.
2. Rinse cells twice with D-PBS and discard.
3. Add pre-warmed TrypLE Select to HFF cells and place the vessel in a 37°C incubator for 2–3 min. Remove the vessel from the incubator and tap the side with palm to dislodge the HFF cells.
4. Gently triturate the cells gently and transfer them to a sterile 15 mL centrifuge tube. Rinse flask twice with Feeder Medium and pool with cells in the tube.
5. Gently pellet cells by centrifuging at room temperature for 5 min at 200 × *g*.
6. Aseptically, aspirate the supernatant without disturbing the cell pellet and discard it.
7. Gently "flick" the bottom of the tube to fully dislodge the cell pellet from the tube bottom.
8. Add the required volume of pre-warmed Feeder Medium to the pellet and gently triturate.
9. Subculture active feeders between 1:4 and 1:8 in Feeder Medium for routine maintenance.

3.2.3 Irradiating HFF Cells

1. Remove the confluent vessel(s) from the incubator.
2. Aspirate the spent medium from the vessel(s).
3. Add pre-warmed TrypLE solution (see Table 2 for volume) and transfer to a 37°C incubator.

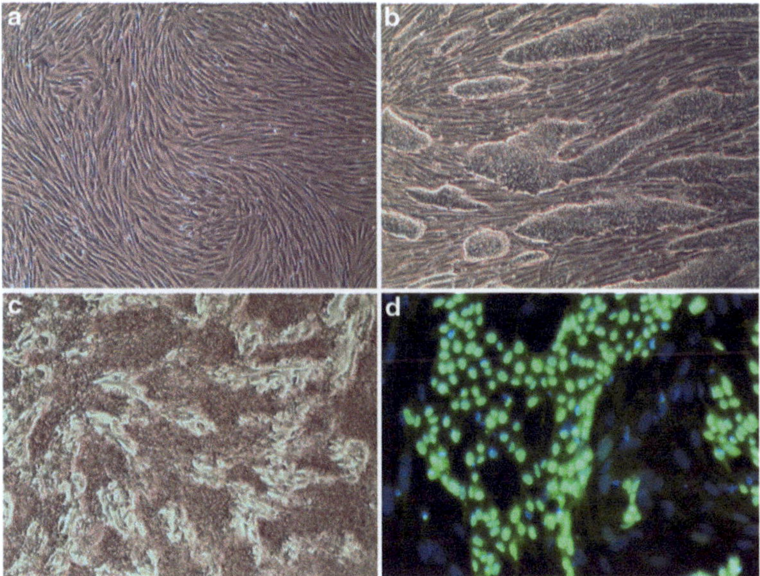

Fig. 4 Shows morphology of HFF (**a**) and hESC cultured on HFF (**b**). hESC colonies also stained positive for alkaline phosphatase (**c**), Oct 4 (**d**, *Green*), and DAPI (**d**, *Blue*)

4. After 1 min, remove the vessels and observe under the microscope.
5. When cells begin to detach, transfer dish to biosafety cabinet.
6. Gently aspirate the TrypLE solution.
7. Add pre-warmed Feeder Medium to the vessels and wash off the cells. Collect the cell suspension in a 50 mL centrifuge tube.
8. Centrifuge the tube at $200 \times g$ for 5 min at room temperature.
9. Gently aspirate the supernatant and resuspend the cell pellet in a small volume of pre-warmed Feeder Medium, obtaining a single-cell suspension.
10. Determine the viable cell count by trypan blue exclusion method using a hemocytometer.
11. To inactivate HFF, expose to gamma irradiation at a dose of 30 Gy.
12. Plate inactive HFF at 7×10^4 cells/cm^2 in the desired cell culture vessel(s) that have been pre-coated with CELLstart™ CTS (see below).

3.2.4 CELLstart CTS Coating of Culture Vessels to Receive Inactive HFF

1. Dilute CELLstart™ CTS™ at 1:50 in D-PBS CTS™ with calcium chloride and magnesium chloride. Pipet gently to mix. Do not vortex.
2. Add diluted CELLstart™ CTS™ to coat the entire surface of the culture vessel (e.g., 0.078 mL/cm^2). Swirl or gently tap the vessel to distribute the diluted CELLstart™ CTS™ evenly.

3. Place culture vessels containing diluted CELLstart™ CTS in a 37°C incubator with a humidified atmosphere of 5% CO_2 for 1–2 h.
4. After incubation, remove coated vessels from the incubator. For immediate use, place vessels at room temperature. For use the next day, carefully wrap vessels containing diluted CELLstart™ CTS™ with Parafilm® and store at 2–8°C.
5. Immediately before use, remove all CELLstart™ CTS diluent from the vessel. It is not necessary to rinse vessels following removal of CELLstart™ CTS™.

3.2.5 Thawing and Plating Inactive HFFs

1. Remove the cryovial containing inactivated HFFs from the liquid nitrogen storage tank.
2. Thaw vial of HFF cells in a 37°C water bath, until just a small frozen chunk remains in the vial.
3. Spray vial with 70% isopropanol to decontaminate it and transfer vial to the biosafety cabinet.
4. Aseptically transfer the entire contents of the vial into a 15 mL centrifuge tube containing 10 mL warm Inactive HFF Feeder Medium (containing KSR XenoFree) using a 5 mL serological pipet.
5. Rinse the vial with 1–2 mL Inactive HFF Medium and add it to the same 15 mL centrifuge tube.
6. Gently pellet cells by centrifuging for 5 min at $200\times g$ at room temperature.
7. Carefully aspirate the supernatant without disturbing the cell pellet and discard it.
8. Gently "flick" the tube at the base to fully dislodge the cell pellet from the tube bottom.
9. Resuspend the cell pellet with warm Inactive HFF Medium, and determine the viable cell count by trypan blue exclusion method using a hemocytometer.
10. Plate inactive HFF at 7×10^4 cells/cm^2 in the desired CELLstart-coated vessels. Place the vessel in a 37°C incubator with a humidified atmosphere of 5% CO_2.
11. Swirl vessel in a north-to-south and east-to-west pattern to evenly distribute feeder cells.
12. Use the inactive HFF culture vessels within 3–4 days after plating.

3.2.6 Recovery of Cryopreserved hESCs on Inactive HFF Cells

1. Rapidly thaw frozen vial of hESC in a 37°C water bath, until a small frozen piece remains in the vial.
2. Decontaminate vial with 70% isopropyl alcohol.
3. Aseptically transfer the entire contents of the vial into a 15 mL centrifuge tube.

4. Gently add 3 mL pre-warmed KSR XenoFree Medium dropwise to the centrifuge tube containing thawed hESC. Rinse the vial with 1–2 mL additional medium and add to the same centrifuge tube.
5. Gently pellet hESC by centrifuging at 200 × *g* at room temperature for 2 min.
6. Carefully aspirate the supernatant without disturbing the cell pellet and discard.
7. Gently “flick” the tube to fully dislodge the cell pellet from the tube bottom.
8. Carefully add the desired volume of pre-equilibrated KSR XenoFree Medium to the hESC pellet. Do *not* triturate cells.
9. Gently invert the centrifuge tube containing hESCs to mix cells. Using a pipet, transfer the cells to a HFF containing culture vessel (steps 13–14 in section 3.1.10) (see Note 7).
10. Place vessel in a 37°C incubator with a humidified atmosphere of 5% CO_2. Carefully swirl vessel in a north-to-south and east-to-west pattern to evenly distribute hESC.
11. Fluid change vessel 24 h post-thaw and daily thereafter, until approximately 70–80% confluent (see Notes 2 and 5).

3.2.7 Passaging hESC on Inactive HFF Cells

1. Observe hESC under the microscope and confirm that the cells are ready to be subcultured (when 70–80% confluent) (see Note 5).
2. Physically remove any differentiated hESC colonies prior to passaging the culture. A 20 G 1½″ needle attached to a syringe works well for removing differentiated hESCs (see Note 6).
3. Pre-warm the required volumes of TrypLE™ Select CTS™ to 37°C, and pre-equilibrate the required volume KSR XenoFree Medium to temperature and gases before use. Minimize dwell time.
4. Remove spent medium from culture vessel using a pipet, and discard.
5. Rinse hESCs twice with D-PBS CTS™ without calcium chloride, magnesium chloride (1×), liquid.
6. Gently add pre-warmed TrypLE™ Select CTS™ to the culture vessel (e.g., 1 mL/60 mm dish). Swirl vessel to coat the entire cell surface.
7. Place culture vessel in a 37°C incubator for 2–3 min.
8. Remove vessel from the incubator. Gently tap the sides of the dish to dislodge cells.
9. Transfer cells to a sterile 15 mL centrifuge tube.
10. Rinse dish twice with pre-warmed wash medium, gently “spraying off” any cells that have not detached, and pool with cells in tube. Do not triturate!

11. Pellet cells by centrifugation at 200 × *g* for 2 min at room temperature.
12. Carefully aspirate the supernatant without disturbing the cell pellet and discard it.
13. Gently "flick" the bottom of tube to fully dislodge the cell pellet from the tube bottom.
14. Gently resuspend the cells in pre-equilibrated complete medium using a 2 mL or 5 mL serological pipet. Do not triturate!
15. Transfer cells to a fresh CELLstart-coated, HFF-plated dish at the desired cell ratio or seeding density. A 1:2 split is recommended during adaptation, or 8×10^4 cells/cm^2. For routine maintenance, cells can be split at 1:4 to 1:8 or 4×10^4 cells/cm^2 using KSR XenoFree Medium. Adjust densities as needed to suit your particular hESC line.
16. Place vessel in a 37°C incubator, with a humidified atmosphere of 5% CO_2. Carefully swirl vessel in a north-to-south and east-to-west pattern to evenly distribute hESC.
17. Gently fluid change culture the next day to remove cell debris and to provide fresh nutrients, and daily thereafter.
18. Observe cells daily and passage by the above protocol whenever required (approximately every 3–5 days) (Table 3).

Note: Cultures can also be passaged manually by using Stempro EZ passaging tool to cut the colonies in small clumps, followed by collecting the colonies by using a cell scraper; resume with step 10.

3.2.8 Cryopreservation of hESC on Inactive HFF Cells

1. Follow the protocol for passaging hESC with KSR XenoFree Medium mentioned in Subheading 3.2.7.
2. At step 14 in Subheading 3.2.7, gently resuspend the cell pellet with a 50/50 mix of KSR XenoFree Medium and hESC KSR XenoFree Freezing Medium (to yield 10% DMSO) using a 5 mL pipet and triturate.
3. Dispense 1 mL of the cell suspension per cryovial using a 5 mL pipet, mixing thoroughly between dispenses (see Note 3).
4. For best results, vials should be cryopreserved using a controlled-rate freezing device (e.g., CryoMed® Freezer or Mr. Frosty Nalgene Cryo 1°C Freezing Container), following manufacturer's instructions.
5. The next day, transfer vials to permanent liquid nitrogen storage. Expect some cell death at recovery, and freeze hESC at a higher density than would normally be passaged to facilitate recovery from liquid nitrogen storage. For example, if cells are routinely passaged at 1:5, a 1:3 or 1:4 densities are recommended for freezing hESC (see Note 7).

Table 3
Establishing passage 1 HFF T-flasks

Initial vessel	Passage 1 vessels	Passage 2 vessels
T-75	(2) T-160 flasks	(8) T-160 flasks
T-160	(4) T-160 flasks or (3) T-225 flasks	(16) T-160 flasks or (12) T-225 flasks
T-225	(4) T-225 flasks	(16) T-225 flasks

3.2.9 EB Formation

1. Culture hESC in KSR hESC Medium or KSR XenoFree Medium as desired for a minimum of three passages.
2. When the culture is 70–80% confluent, dissociate hESC enzymatically as previously described, or manually using the Stempro EZ passaging tool and cell scraper (see Subheading 3.2.7).
3. Dilute the cell clumps at 1:4 in EB Medium and transfer to non-tissue culture-treated 60 mm dishes (4 mL/60 mm dish). Note: No pre-coating of dishes is necessary. Transfer 60 mm dishes to a 37°C incubator, with a humidified atmosphere of 5% CO_2.
4. On day 2, observe EBs (they will be floating/in suspension culture). Transfer the entire EB suspension from the non-tissue culture-treated dish to a 15 mL centrifuge tube (use a separate tube for every 60 mm dish). Let the EBs settle to the bottom of the tube.
5. When most of the EBs have settled out, aspirate and discard the supernatant, leaving the EBs behind.
6. Gently, resuspend EBs with 4 mL fresh EB Medium and return to the non-tissue culture-treated dishes. Transfer to a 37°C incubator, with a humidified atmosphere of 5% CO_2.
7. On day 4, repeat the fluid change procedure (steps 4 and 5 above), but this time, resuspend EBs in 10 mL EB Medium.
8. Transfer 5 mL EBs to each of two non-tissue culture-treated 100 mm dishes. Add 5 mL EB Medium to each 100 mm dish and return dishes to 37°C incubator.
9. Continue fluid-changing the 100 mm EB dishes every 2–3 days, as described above. Some of the EBs will attach to the dishes and expand. Continue fluid-changing the attached EBs.
10. Let the cells culture for 21 days (or desired time course).
11. Fix and stain EB outgrowths to examine by immunocytochemistry or lyse and harvest them (e.g., with TRIzol reagent) to examine gene expression by RT-PCR, microarray, etc. (Fig. 5)

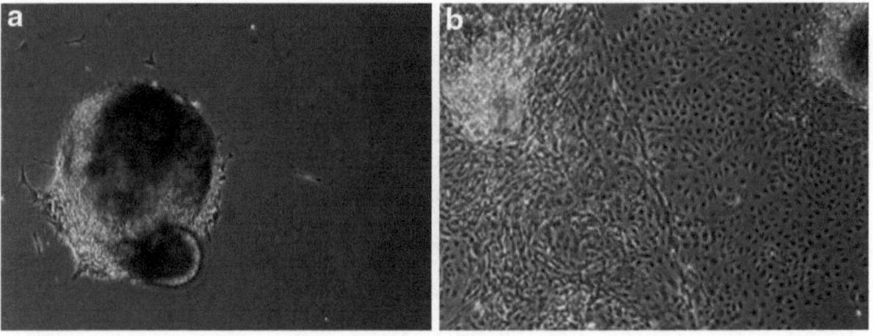

Fig. 5 EB formation from hESC. (**a**) Image of attached EB on culture dish. (**b**) Differentiated cell outgrowth from an attached hESC EB

4 Notes

1. All protocols described here must be performed under sterile conditions. For detailed information on sterile techniques, refer to reviews by Cote (9) and Phelan (10). All cultures should be routinely tested for mycoplasma infection.
2. hESC cultures must be fluid-changed daily for optimal growth without differentiation.
3. Use a 5 mL serological pipet when thawing cells from liquid nitrogen, and for passaging and plating hESC.
4. Always pre-warm hESC Medium before use.
5. Timing of passage is critical. For best results, hESC should be nearing confluence (70–80%) at the time of passage. If passaged too light or overgrown, hESC will differentiate.
6. Before passaging hESC, remove the differentiated colonies manually. The culture should be of high quality, 70–80% confluent, and contain no differentiated hESC.
7. Plate thawed hESC at a high seeding density; not all cells will be viable, attach, and expand.
8. Plate the MEF and HFF at recommended seeding density to maintain pluripotent hESC.
9. Prepare a frozen stock of cells in control medium prior to adaptation to a new culture condition.
10. Addition of antibiotics to hESC media is optional. Media for hESC culture can be stored at 4°C for about 2 weeks.

References

1. Evans MJ, Kaufman MH (1981) Establishment in culture of pluripotential cells from mouse embryos. Nature 292:154–156
2. Martin GR (1981) Isolation of a pluripotent cell line from early mouse embryos cultured in medium conditioned by teratocarcinoma stem cells. Proc Natl Acad Sci USA 78:7634–7638
3. Robertson EJ (ed) (1987) Teratocarcinomas and embryonic stem cells: a practical approach. IRL, Oxford
4. Thomas KR, Capecchi MR (1987) Site-directed mutagenesis by gene targeting in mouse embryo-derived stem cells. Cell 51:503–512
5. Thomson JA, Itskovitz-Eldor J, Shapiro SS, Waknitz MA, Swiergiel JJ, Marshall VS, Jones JM (1998) Embryonic stem cell lines derived from human blastocysts. Science 282:1145–1147
6. Richards M, Fong CY, Chan WK et al (2002) Human feeders support prolonged undifferentiated growth of human inner cell masses and embryonic stem cells. Nat Biotechnol 20:933–936
7. Lin S, Talbot P (2011) Methods for culturing mouse and human embryonic stem cells. Methods Mol Biol 690:31–56
8. Eiselleova L, Peterkova I, Neradil J, Slaninova I, Hampl A, Dvorak P (2008) Comparative study of mouse and human feeder cells for human embryonic stem cells. Int J Dev Biol 52(4):353–363. doi:10.1387/ijdb.082590le
9. Cote RJ (2001) Aseptic technique for cell culture. Curr Protoc Cell Biol Chapter 1: Unit 1.3
10. Phelan MC (2006) Techniques for mammalian cell tissue culture. Curr Protoc Hum Genet Appendix 3: Appendix 3G

Chapter 9

Methods for Culturing Human Embryonic Stem Cells in a Xeno-Free System

Jasmeet Kaur, Mary Lynn Tilkins, Richard Eckert, and Shayne Boucher

Abstract

Defined pluripotent stem cell culture media is a valuable tool for tracking and analyzing morphological, cell signaling and gene expression changes in human embryonic stem cells. Cultivation of hESC under xeno-free culture settings provides researchers with a consistent and reproducible environment to test experimental hypotheses and move towards translational and clinical research (Richards et al., Nat Biotechnol 20:933–936, 2002; Richards et al., Stem Cells 21:546–556, 2003). One of the primary concerns of the xenogeneic culture is the transfer of foreign pathogens or antigens that induce disease or immune response by the patient. These undesirable by-products may come from the use of murine-derived feeder cells, xenogeneic matrices, or from animal-derived components found in the cell culture medium or matrix used to isolate or expand the stem cells (Beattie et al., Stem Cells 23:489–495, 2005; Koivisto et al., Reprod Biomed Online 9:330–337, 2004). This chapter describes standardized protocols for obtaining consistent and reproducible results for culturing PSC under xeno-free, feeder-free, or feeder-based systems.

Key words Xeno-free culture, Pluripotent stem cell, Embryonic fibroblast feeder cells

1 Introduction

Human embryonic stem cells (hESC) were derived and maintained on inactivated murine embryonic fibroblast (MEF) feeder layers (1, 2). Use of MEFs and other components of animal origin for culturing hESC has been a concern owing to the risk of viral infection or pathogen transmission, thus making hESC unsafe for use in regenerative medicine. Defined conditions where all hESC culture components are "xeno-free," meaning human-derived or synthetic, are required in order to have hESC in a suitable state for therapeutic applications.

Several researchers have been able to successfully derive and propagate hESC lines on a variety of human fibroblast feeder (HFF) layers and also on a recombinant humanized matrix (CELLstart™) for feeder-free cultures. The hESC colonies on human feeders have angular or straight edges as compared to circular colonies of hESC observed on MEF. Cells in the feeder-free xeno-free condition

Uma Lakshmipathy and Mohan C. Vemuri (eds.), *Pluripotent Stem Cells: Methods and Protocols*, Methods in Molecular Biology, vol. 997, DOI 10.1007/978-1-62703-348-0_9,

Table 1
Initial HFF seeding information (passage 0)

Flask size	HFF seeding density	Initial volume of pre-warmed HFF medium (mL)
T-75	5.25×10^6	15
T-160	1.12×10^7	30
T-225	1.575×10^7	50

Table 2
Trypsin and TrypLE express volumes for dissociation

Flask size	Suggested volume (mL)
T-75	1.5
T-160	3
T-225	5

Table 3
Establishing passage 1 HFF T-flasks

Initial vessel	Passage 1 vessels	Passage 2 vessels
T-75	(2) T-160 flasks	(8) T-160 flasks
T-160	(4) T-160 flasks or (3) T-225 flasks	(16) T-160 flasks or (12) T-225 flasks
T-225	(4) T-225 flasks	(16) T-225 flasks

appear to be more flattened on the initial day of culture and colonies get more compact by days 4–5. The hESC derived and propagated in xeno-free conditions remain pluripotent over extended passages, express hESC morphology with high nuclear to cytoplasmic ratio, maintain stable karyotype, and are able to form embryoid bodies (EB) and differentiate into the three germ layers—ectoderm, endoderm, and mesoderm (3). Apart from the differences in gross colony appearance, all other characteristics of hESC appear to be similar. These protocols will serve to standardize methodologies for cultivating PSC (pluripotent stem cells) under xeno-free conditions (Tables 1, 2, 3).

2 Materials

For xeno-free culturing cells can be cultured on human feeders or a humanized matrix (CELLstart). The protocol will discuss the feeder-based culture followed by feeder-free culture. Unless otherwise specified, all materials are from Life Technologies.

2.1 Reagents

1. CELLstart™ CTS™ defined, humanized substrate for cell culture DPBS CTS™ with calcium chloride and magnesium chloride.
2. ES Cell Qualified FBS.
3. FGF-basic Full Length CTS™ Recombinant Human Protein (100 μg), lyophilized powder.
4. GlutaMAX™-I CTS™.
5. Gentamicin.
6. Human feeders.
7. KnockOut™ CTS™ DMEM (1×), liquid.
8. KnockOut™ SR XenoFree CTS.
9. KnockOut™ SR GF Cocktail CTS™.
10. KnockOut™ DMEM/F-12 CTS™.
11. 2-Mercaptoethanol (1,000×), liquid.
12. Nonessential amino acids (NEAA).
13. STEMPRO® EZPassage™ Disposable Cell Passaging Tool.
14. TrypLE™ Select CTS™ (1×), liquid, without Phenol Red.

2.2 Equipment

1. Biosafety cabinet.
2. Centrifuge (suitable for 15 mL and 50 mL centrifuge tubes).
3. Colony marker for microscope.
4. Freezer (−20°C).
5. Freezer (−80°C).
6. Gamma irradiator.
7. 37°C, humidified 5% CO_2 incubator.
8. Liquid nitrogen storage tank.
9. Mr. Frosty Nalgene Cryo 1°C Freezing Container.
10. Phase-contrast microscope.
11. Pipet aid.
12. Plastic cryovial rack.
13. 4°C refrigerator.
14. Water bath (37°C).

2.3 Supplies

1. 5 mL, 10 mL, and 25 mL serological pipets.
2. 15 mL and 50 mL centrifuge tubes.
3. 60 mm tissue culture-treated dishes.
4. 1.5 mL cryovials.
5. 60 mm and 100 mm non-tissue culture-treated dishes.

2.4 Media Preparation for Feeder and Feeder-Free Culture of Pluripotent Stem Cells (PSC)

1. Prepare 500 mL HFF Medium by adding the following: 440 mL KnockOut™ CTS™ DMEM, 50 mL ES Cell Qualified FBS, 5 mL of NEAA, and 5 mL GlutaMAX™-I CTS™.
2. Prepare 500 mL Inactive HFF Medium by combining 440 mL KnockOut™ CTS™ DMEM, 50 mL KnockOut™ SR XenoFree CTS™, 5 mL NEAA, and 5 mL GlutaMAX™-I CTS™.
3. Prepare 100 μg/mL bFGF solution by adding the following: 100 μg of Basic FGF, 990 μL of D-PBS, and 10 μL of 10% BSA.
4. Prepare 100 mL KnockOut SR XenoFree Feeder Medium by adding the following: 83 mL of KnockOut™ DMEM CTS™, 1 mL of GlutaMAX™-I, 15 mL of KnockOut™ SR XenoFree CTS™, 1 mL of NEAA, 20 μL of bFGF CTS™ (see Note 9).
5. Prepare 100 mL KnockOut SR XenoFree Feeder-free Medium by adding the following: 83 mL of KnockOut™ DMEM CTS™, 1 mL of GlutaMAX™-I, 15 mL of KnockOut™ SR XenoFree CTS™, 1 mL KnockOut SR GF Cocktail CTS, 20 μL of bFGF CTS™ (see Note 9).
6. Prepare EB Medium by adding the following: 83 mL of KnockOut™ DMEM, 15 mL of KnockOut™ SR XenoFree, 1 mL of GlutaMAX™, 1 mL of NEAA.
7. Prepare 1:50 CELLstart solution by adding the following: 49 mL of DPBS CTS™ with calcium chloride and magnesium chloride, 1 mL of CELLstart™ CTS™.
8. Prepare hESC KSR XenoFree Freezing Medium by adding the following: 73 mL of KnockOut™ DMEM CTS™, 1 mL of KnockOut™ SR GF Cocktail CTS™, 20 μl of bFGF CTS™, 1 mL of GlutaMAX™-I, and 25 mL of KnockOut™ SR XenoFree CTS™.

3 PSC Culture in Xeno-Free Condition

Workflow diagram of xeno-free culture of pluripotent stem cells (Fig. 1) (see Note 1).

3.1 Recovery of HFF Cells from Liquid Nitrogen

1. Warm HFF Medium before thawing the cells. Place 10 mL HFF Medium in a 15 mL centrifuge tube.
2. Remove HFF vial (e.g., Cat. No. CRL-2429, American Type Culture Collection) from liquid nitrogen and rapidly thaw in a 37°C water bath by gently swirling until just a small frozen chunk remains in the vial.
3. Spray vial with 70% isopropanol to decontaminate it and transfer vial to the biosafety cabinet.
4. Aseptically transfer the entire contents of the vial into the 15 mL centrifuge tube containing 10 mL warm HFF Medium using a 5 mL serological pipet.

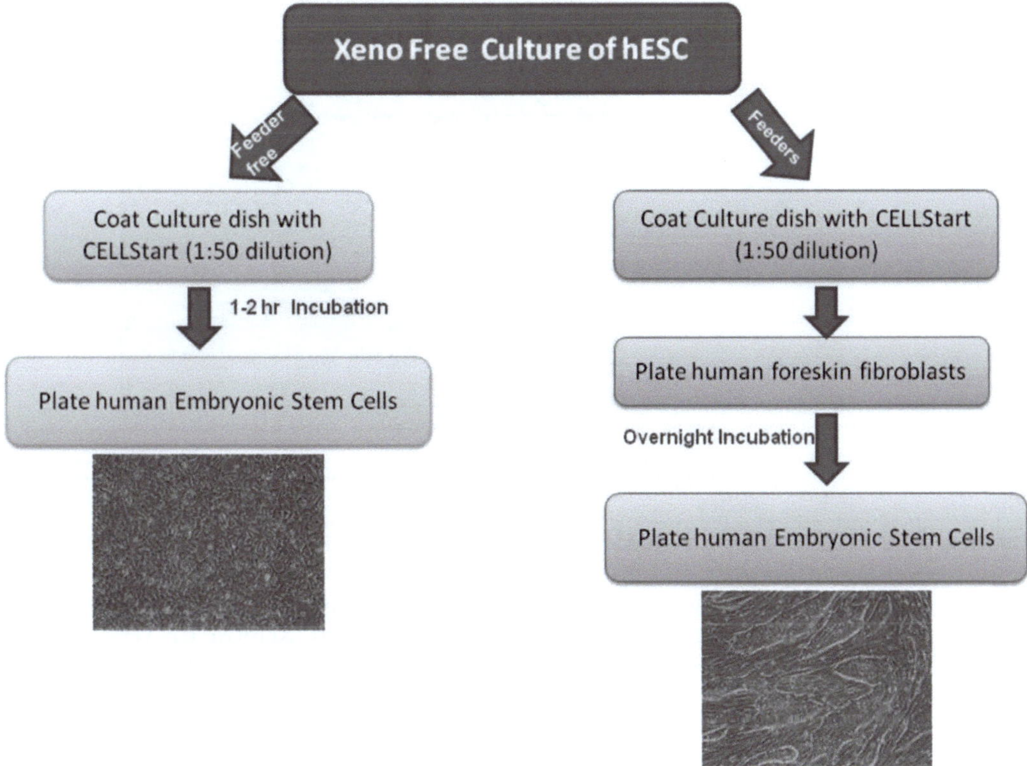

Fig. 1 The workflow diagram of culturing hESC on human feeders and in feeder-free culture with CELLstart™ CTS matrix

5. Rinse the vial with 1-2 mL HFF Medium and add it to the same 15 mL centrifuge tube.
6. Gently pellet cells by centrifuging at 200 × *g* for 3 min at room temperature.
7. Carefully aspirate the supernatant without disturbing the cell pellet and discard it.
8. Gently "flick" the tube at the base to fully dislodge the cell pellet from the tube bottom.
9. Add the required volume of HFF Medium to the pellet and gently triturate. Transfer the HFF cells to appropriate-sized vessel(s). Incubate in a 37°C incubator with a humidified atmosphere of 5% CO_2.

3.2 Passaging HFF

1. Observe the vessel(s) for confluence; subculture when confluent, but not tightly packed with cells (see Fig. 2). Note: Be sure to scale up and bank cells at low passage number.
2. Rinse cells twice with D-PBS and discard.

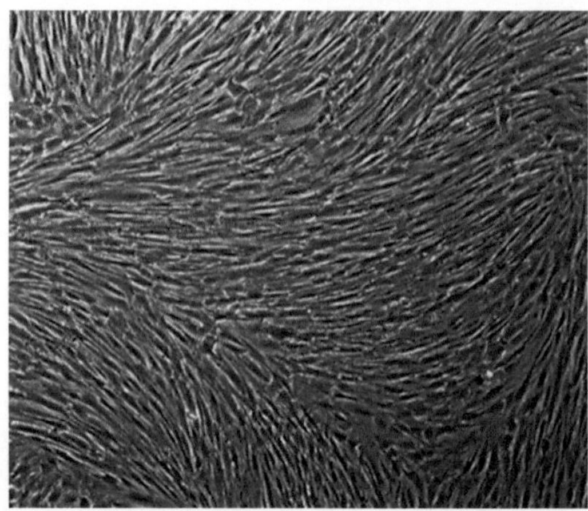

Fig. 2 Morphology image of human feeders cultured on CELLstart™ CTS-coated dishes

3. Add pre-warmed TrypLE Select to HFF cells and place the vessel in a 37°C incubator for 2–3 min. Remove the vessel from the incubator and tap the side with palm to dislodge the HFF cells.
4. Gently triturate the cells gently and transfer them to a sterile 15 mL centrifuge tube. Rinse flask twice with HFF Medium and pool with cells in the tube.
5. Gently pellet cells by centrifuging at room temperature for 5 min at 200×*g*.
6. Aseptically, aspirate the supernatant without disturbing the cell pellet and discard it.
7. Gently "flick" the bottom of the tube to fully dislodge the cell pellet from the tube bottom.
8. Add the required volume of pre-warmed HFF Medium to the pellet and gently triturate.
9. Subculture active feeders between 1:4 and 1:8 in HFF Medium for routine maintenance.

3.3 Irradiating and Cryopreserving Human Fibroblast Feeder Layers

1. Remove the confluent vessel(s) from the incubator.
2. Aspirate the spent medium from the vessel(s).
3. Add pre-warmed TrypLE solution (2 mL for a 60 mm dish) and transfer to a 37°C incubator.
4. After 1 min, remove the vessels and observe under the microscope.
5. When cells begin to detach, transfer dish to biosafety cabinet.
6. Gently aspirate the TrypLE solution.
7. Add pre-warmed HFF Medium to the vessels and wash off the cells. Collect the cell suspension in a 50 mL centrifuge tube.

8. Centrifuge the tube at 200 × *g* for 5 min at room temperature.
9. Gently aspirate the supernatant and resuspend the cell pellet in a small volume of pre-warmed HFF Medium, obtaining a single cell suspension.
10. Determine the viable cell count by trypan blue exclusion method using a hemocytometer.
11. To inactivate HFF, expose to gamma irradiation at a dose of 30 Gy.
12. Plate inactive HFF at $7 \times 10^4/\text{cm}^2$ in the desired cell culture vessel(s) that have been pre-coated with CELLstart™ CTS (see below) (see Note 7).

3.4 CELLstart CTS Coating of Culture Vessels for Feeder and Feeder-Free Culture of PSC

1. Dilute CELLstart™ CTS™ at 1:50 in D-PBS CTS™ with calcium chloride and magnesium chloride. Pipet gently to mix. Do not vortex.
2. Add diluted CELLstart™ CTS™ to coat the entire surface of the culture vessel (e.g., 0.078 mL/cm^2). Swirl or gently tap the vessel to distribute the diluted CELLstart™ CTS™ evenly.
3. Place culture vessels containing diluted CELLstart™ CTS in a 37°C incubator with a humidified atmosphere of 5% CO_2, for 1–2 h.
4. After incubation, remove coated vessels from the incubator. For immediate use, place vessels at room temperature. For use the next day, carefully wrap vessels containing diluted CELLstart™ CTS™ with Parafilm®, and store at 2–8°C.
5. Immediately before use, remove all CELLstart™ CTS diluent from the vessel. It is not necessary to rinse vessels following removal of CELLstart™ CTS™.
6. If growing cells in a feeder-free system proceed with plating hESCs on the CELLstart™ CTS™-coated vessel at the desired density.

3.5 Thawing and Plating Inactive HFFs

1. Remove the cryovial containing inactivated HFFs from the liquid nitrogen storage tank.
2. Thaw vial of HFF cells in a 37°C water bath, until just a small frozen chunk remains in the vial.
3. Spray vial with 70% isopropanol to decontaminate it and transfer vial to the biosafety cabinet.
4. Aseptically transfer the entire contents of the vial into a 15 mL centrifuge tube containing 10 mL warm Inactive HFF Medium (containing KSR XenoFree) using a 5 mL serological pipet.
5. Rinse the vial with 1–2 mL Inactive HFF Medium and add it to the same 15 mL centrifuge tube.

6. Gently pellet HFF by centrifuging for 5 min at 200 ×*g* at room temperature.
7. Carefully aspirate the supernatant without disturbing the cell pellet and discard it.
8. Gently "flick" the tube at the base to fully dislodge the cell pellet from the tube bottom.
9. Resuspend the cell pellet with warm Inactive HFF Medium, and determine the viable cell count by trypan blue exclusion method using a hemocytometer.
10. Plate inactive HFF at 7×10^4 cells/cm^2 in the desired CELLstart-coated vessels. Place the vessel in a 37°C incubator with a humidified atmosphere of 5% CO_2 (see Note 7).
11. Swirl vessel in a north to south, east to west pattern to evenly distribute feeder cells.
12. Use the inactive HFF culture vessels within 3–4 days after plating.

3.6 Recovery of Cryopreserved hESCs on Feeder and Feeder-Free System

1. Rapidly thaw frozen vial of cells in a 37°C water bath, until a small frozen piece remains in the vial.
2. Decontaminate vial with 70% isopropyl alcohol.
3. Aseptically transfer the entire contents of the vial into a 15 mL centrifuge tube (see Note 3).
4. Gently add 3 mL pre-warmed KnockOut SR XenoFree Feeder or Feeder-free Medium dropwise to the centrifuge tube containing thawed hESC (see Note 4). Rinse the vial with 1–2 mL additional medium and add to the same centrifuge tube (see Note 8).
5. Gently pellet cells by centrifuging at 200 ×*g* at room temperature for 2 min.
6. Carefully aspirate the supernatant without disturbing the cell pellet and discard.
7. Gently "flick" the tube to fully dislodge the cell pellet from the tube bottom.
8. Carefully add the desired volume of pre-equilibrated KnockOut SR XenoFree Feeder or Feeder-free Medium to the hESC pellet. Do *not* triturate cells.
9. Gently invert the centrifuge tube containing hESCs to mix cells. Using a pipet, transfer the cells to inactive HFF culture vessel for feeder-based system or CELLstart-coated only culture vessel for feeder-free system (see Subheadings 3.7.14–3.7.15) (see Note 6).
10. Place vessel in a 37°C incubator with a humidified atmosphere of 5% CO_2. Carefully swirl vessel in a north to south, east to west pattern to evenly distribute hESC.
11. Fluid change vessel 24 h post-thaw and daily thereafter, until approximately 70–80% confluent (see Fig. 3) (see Notes 2, 5).

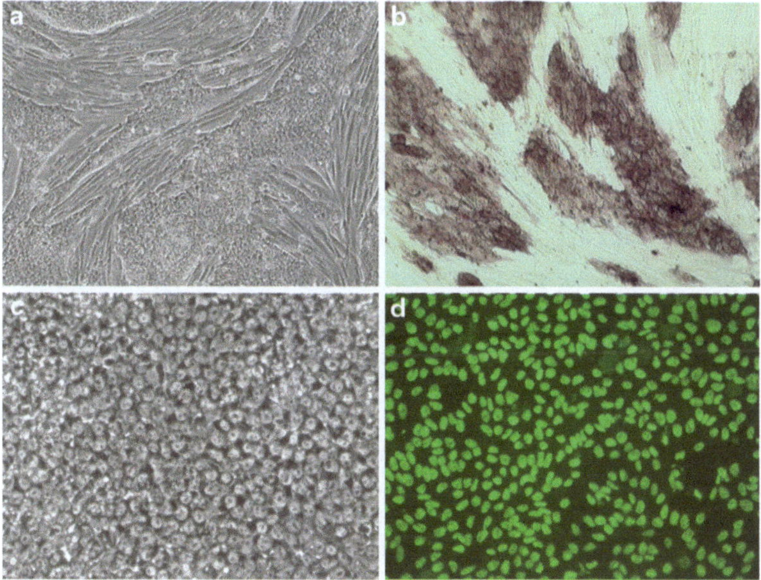

Fig. 3 Morphology image of hESC cultured on HFF in KSR XenoFree Medium (**a**) and stained with alkaline phosphatase (**b**). Morphology image of feeder-free hESC cultured in KSR XenoFree Medium on CELLstart matrix (**c**) and ICC (**d**) for cells expressing pluripotent marker Oct-4

3.7 Passaging hESC on Feeder and Feeder-Free Culture

1. Observe hESC under the microscope and confirm that the cells are ready to be subcultured (when 70–80% confluent).
2. Physically remove any differentiated hESC colonies prior to passaging the culture. A 20 G 1½″ needle attached to a syringe works well for removing differentiated hESCs.
3. Pre-warm the required volumes of TrypLE™ Select CTS™ to 37°C, and pre-equilibrate the required volume of appropriate KnockOut SR XenoFree Medium to temperature and gases before use. Minimize dwell time.
4. Remove spent medium from culture vessel using a pipet, and discard.
5. Rinse hESCs twice with DPBS CTS ™ without calcium chloride, magnesium chloride (1×), liquid.
6. Gently add pre-warmed TrypLE™ Select CTS™ to the culture vessel (e.g., 1 mL/60 mm dish). Swirl vessel to coat the entire cell surface.
7. Place culture vessel in a 37°C incubator for 2–3 min.
8. Remove vessel from the incubator. Gently tap the sides of the dish to dislodge cells.
9. Transfer cells to a sterile 15 mL centrifuge tube.
10. Rinse dish twice with pre-warmed wash medium, gently “spraying off” any cells that have not detached, and pool with cells in tube. Do not triturate!

11. Pellet hESCs by centrifugation at 200 × *g* for 2 min at room temperature.
12. Carefully aspirate the supernatant without disturbing the cell pellet and discard it.
13. Gently "flick" the bottom of tube to fully dislodge the cell pellet from the tube bottom.
14. Gently resuspend hESCs in pre-equilibrated complete medium using a 2 mL or 5 mL serological pipet. Do not triturate!
15. Transfer hESCs to a fresh inactive HFF-plated culture vessel or CELLstart-coated culture vessel at the desired cell ratio or seeding density. A 1:2 split is recommended during adaptation, or 8×10^4 cells/cm^2. For routine maintenance, cells can be split at 1:4–1:8, or 4×10^4 cells/cm^2 using appropriate KnockOut™ SR XenoFree CTS™ Medium. Adjust densities as needed to suit your particular hESC line (see Note 5).
16. Place vessel in a 37°C incubator, with a humidified atmosphere of 5% CO_2. Carefully swirl vessel in a north to south, east to west pattern to evenly distribute hESC.
17. Gently fluid-change culture the next day to remove cell debris and to provide fresh nutrients, and daily thereafter.
18. Observe hESCs daily and passage by the above protocol whenever required (approximately every 3–5 days).

Note: Cultures can also be passaged manually by using Stempro EZ passaging tool to cut the colonies in small clumps, followed by collecting the colonies by using a cell scraper; resume with step 10.

3.8 Cryopreservation of hESC Using KnockOut SR XenoFree

1. Follow the protocol for passaging hESC with appropriate KnockOut SR XenoFree Medium mentioned in Subheading 3.7.
2. At step 14 in Subheading 3.7, gently resuspend the cell pellet with a 50/50 mix of appropriate KnockOut SR XenoFree Medium and hESC KSR XenoFree Freezing Medium (to yield 10% DMSO) using a 5 mL pipet and triturate.
3. Dispense 1 mL of the cell suspension per cryovial using a 5 mL pipet, mixing thoroughly between dispenses (see Note 3).
4. For best results, vials should be cryopreserved using a controlled rate freezing device (e.g., CryoMed® Freezer or Mr. Frosty Nalgene Cryo 1°C Freezing Container), following manufacturer's instructions.
5. The next day, transfer vials to permanent liquid nitrogen storage. Expect some cell death at recovery, and freeze hESC at a higher density than would normally be passaged to facilitate recovery from liquid nitrogen storage. For example, if cells are routinely passaged at 1:5, a 1:3 or 1:4 densities are recommended for freezing hESC (see Note 6).

3.9 EB Formation

1. Culture hESC in KnockOut SR XenoFree Medium as desired, for a minimum of three passages.
2. When the culture is 70–80% confluent, dissociate hESC enzymatically as previously described, or manually using the Stempro EZ passaging tool and cell scraper (see Subheading 3.7).
3. Dilute the cell clumps at 1:4 in EB Medium and transfer to non-tissue culture-treated 60 mm dishes (4 mL/60 mm dish). Note: No pre-coating of dishes is necessary. Transfer 60 mm dishes to a 37°C incubator, with a humidified atmosphere of 5% CO_2.
4. On day 2, observe EBs (they will be floating/in suspension culture). Transfer the entire EB suspension from the non-tissue culture-treated dish to a 15 mL centrifuge tube (use a separate tube for every 60 mm dish). Let the EBs settle to the bottom of the tube.
5. When most of the EBs have settled out, aspirate and discard the supernatant, leaving the EBs behind.
6. Gently, resuspend EBs with 4 mL fresh EB Medium and return to the non-tissue culture treated dishes. Transfer to a 37°C incubator, with a humidified atmosphere of 5% CO_2.
7. On day 4, repeat the fluid change procedure (steps 4 and 5 above), but this time, resuspend EBs in 10 mL EB Medium.
8. Transfer 5 mL EBs to each of two non-tissue culture-treated 100 mm dishes. Add 5 mL EB Medium to each 100 mm dish and return dishes to 37°C incubators.
9. Continue fluid-changing the 100 mm EB dishes every 2–3 days, as described above. Some of the EBs will attach to the dishes and expand. Continue fluid-changing the attached EBs.
10. Let the cells culture for 21 days (or desired time course).
11. Fix and stain EB outgrowths to examine by immunocytochemistry or lyse and harvest them (e.g., with TRIzol reagent) to examine gene expression by RT-PCR, microarray, etc. (Fig. 4).

4 Notes

1. All protocols described here must be performed under sterile conditions. For detailed information on sterile techniques, refer to review by Cote (4) and Phelan (5). All cultures should be routinely tested for mycoplasma infection.
2. hESC cultures must be fluid changed daily for optimal growth without differentiation.
3. Use a 5 mL serological pipet when thawing cells from liquid nitrogen and for passaging and plating hESC.
4. Always pre-warm hESC medium before use.

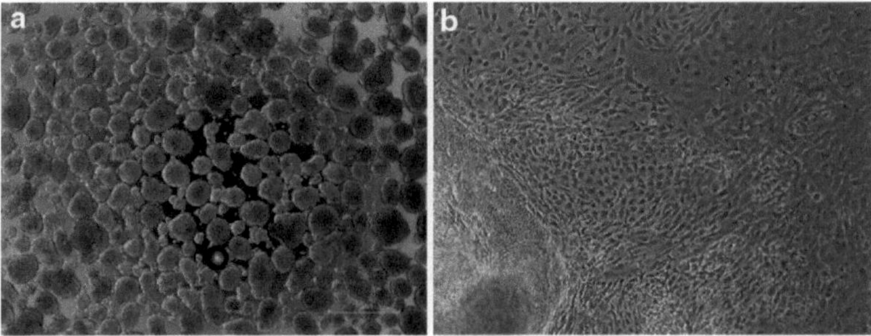

Fig. 4 EB formation image shows floating embryoid bodies at day 4 (**a**) and attached differentiating EB at day 21 (**b**)

5. Timing of passage is critical. For best results, hESC should be nearing confluence (70–80%) at the time of passage. If passaged too light or overgrown, hESC will differentiate.
6. Plate thawed hESC at a high seeding density; not all cells will be viable, attach, and expand.
7. Plate HFF at recommended seeding density to maintain pluripotent hESC.
8. Prepare a frozen stock of cells in control medium prior to adaptation to a new culture condition.
9. Addition of antibiotics to hESC media is optional. Media for hESC culture can be stored at 4°C for about 2 weeks.

Acknowledgements

This work was supported by Life Technologies. The authors thank Drs. Pauline Lieu and Soojung Shin for their feedback and guidance.

References

1. Amit M, Shariki C, Margulets V, Itskovitz-Eldor J (2004) Feeder layer- and serum-free culture of human embryonic stem cells. Biol Reprod 70:837–845
2. Thomson JA, Itskovitz-Eldor J, Shapiro SS, Waknitz MA, Swiergiel JJ, Marshall VS, Jones JM (1998) Embryonic stem cell lines derived from human blastocysts. Science 282:1145–1147
3. Rajala K, Hakala H, Panula S, Aivio S, Pihlajamaki H, Suuronen R et al (2007) Testing of nine different xeno-free culture media for human embryonic stem cell cultures. Hum Reprod 22:1231–1238
4. Cote RJ (2001) Aseptic technique for cell culture. Curr Protoc Cell Biol Chapter 1: Unit 1.3
5. Phelan MC (2007) Basic techniques in mammalian cell tissue culture. Curr Protoc Cell Biol 36:1.1.1–1.1.18

Chapter 10

Directed Differentiation of Human Pluripotent Stem Cells Along the Pancreatic Endocrine Lineage

Dennis Van Hoof and Muluye E. Liku

Abstract

Many research groups are engaged in using human pluripotent stem cells (hPSCs) to generate surrogate pancreatic β-cells for transplantation into diabetic patients. However, to our knowledge, there is no report on the successful generation of glucose-responsive insulin-producing β-cells from hPSCs in vitro.

Below, we outline a method that is based on published protocols as well as our own experience by which one can differentiate hPSCs along the pancreatic lineage to generate insulin-producing β-cell-like cells. The protocol, which spans five distinct stages, is an attempt to recapitulate the derivation of pancreatic β-cells in vitro as they form in the developing embryo. We included details on materials and techniques, suggest ways to customize it to your hPSC line of choice, added notes on how to monitor and analyze the cells during differentiation, and indicate what results can be expected.

Key words Human pluripotent stem cells, Definitive endoderm, Islets of Langerhans, Pancreatic β-cells, Insulin-producing cells, In vitro differentiation, Growth factors, Pancreas development

1 Introduction

The potential of human pluripotent stem cells (hPSCs), such as human embryonic stem cells (hESCs) and human induced pluripotent stem cells (hiPSCs), to differentiate into derivatives of the three primary germ layers holds great promise for regenerative medicine to replace lost or damaged tissues and organs. In the protocol described below, we provide guidelines as a starting point for the differentiation of hPSCs into insulin-producing β cell like cells that might ultimately be used in the clinics for transplantation into diabetic patients who lack a sufficient β-cell mass.

Of note is that large differences exist in the propensity of individual hPSC lines to efficiently differentiate into a particular lineage (1). Unfortunately, among the most difficult somatic cell types to generate in vitro are those belonging to the endocrine pancreas, including the insulin-producing β-cells of the islets of Langerhans. This might be due to the relatively late appearance of

Uma Lakshmipathy and Mohan C. Vemuri (eds.), *Pluripotent Stem Cells: Methods and Protocols*, Methods in Molecular Biology, vol. 997, DOI 10.1007/978-1-62703-348-0_10, © Springer Science+Business Media New York 2013

these cells during embryonic development and the numerous developmental stages that the pluripotent inner cell mass derivatives go through while developing along the pancreatic lineage. Interestingly, pancreatic β-cell-derived hiPSCs were found to have an increased ability to differentiate into β-cell-like cells in vitro as well as in vivo (2). A likely explanation is conservation of epigenetic memory of the β-cell state during reprogramming. In addition, reporter hPSC lines, expressing GFP when hallmark genes like *SOX17* (3) and *INSULIN* (4) become activated, have been developed, allowing us to monitor the progress and efficiency of the differentiation in process. Additional reporter lines expressing fluorescent proteins when other pancreatic genes, including *PDX1* and *NGN3*, become activated would be useful, but are currently not yet available. There is, however, an "empty" GFP reporter hESC line that is suitable for the generation of such reporters (5).

Most of the early protocols for generating pancreatic cells from pluripotent stem cells relied on spontaneous differentiation after an embryoid body formation step. However, this approach is difficult to manipulate and is neither specific nor efficient to generate homogeneous populations of cell types. Stepwise protocols, relying on the consecutive exposure of the differentiating cells to soluble factors, were initially developed for mouse embryonic stem cells (6), but these early version multistage protocols used conditions that do not accurately reflect the in vivo environment of these cells. Also, the insulin-containing cells generated at the end of the protocol were found to have taken up insulin from the medium rather than having produced this hormone de novo (7). Therefore, we advise to either avoid using insulin-containing medium supplements or to additionally check for expression of C-peptide, which is produced and secreted along with insulin in equimolar concentrations. Another complicating factor is that cells of the pancreatic endocrine lineage share many of the transcription factors associated with neuronal cells, making the analysis of intermediate cell types challenging. Collecting a heterogeneous pool of differentiating cells for global expression analysis is not sufficient to determine whether the differentiating cells follow the pancreatic developmental pathway; thus, we suggest using (fluorescence) microscopy in addition to qPCR to confirm the authenticity of the cells created.

A critical step to the generation of pancreatic cells is the proper formation of definitive endoderm. The first publication on efficient and directed differentiation of hESCs into "true" definitive endoderm was a major breakthrough (8). One of the challenges of characterizing these definitive endoderm cells is the lack of specific markers for definitive endoderm; many markers known to be expressed by these cells are also present in extraembryonic endoderm derivatives. Of the known endoderm markers, SOX7 is a transcription factor that is excluded from definitive endoderm, and can, thus, be used in negative screens.

Most of the many pancreatic differentiation protocols for hPSCs published to date are based on knowledge from pancreatic development as it occurs in vivo, providing guidelines for directed differentiation in vitro (for reviews, *see* refs. 9, 10). However, be aware that there is currently no universal protocol that works well for all different hPSC lines. To this end, we focus on the most widely used protocol developed by Novocell (currently Viacyte) (11) to which we included adaptations that we found to be beneficial for the efficiency of generating pancreatic cells (12). Also, note that, as of today, no group has reported the in vitro generation of truly functional glucose-responsive insulin-producing cells with all the properties and characteristics of mature β-cells as they reside in the islets of Langerhans in vivo. Therefore, we suggest you use the protocol outlined below as a reference and starting point, and encourage you to further explore the addition or exclusion of growth factors and other components in different combinations and concentrations as well as change the duration of individual stages as you judge best based on current literature and your own empirical findings.

2 Materials

2.1 Human Pluripotent Stem Cells

Any hPSC line can be used, although one might experience differences in the propensity and efficacy of different cell lines to generate cells along the pancreatic lineage. Enzymatic passaging of the human pluripotent stem cells is preferred (see Note 1).

2.1.1 Media and Components for Maintenance Prior to Differentiation

1. Gelatin: Dissolve 5.0 g Gelatin from Porcine Skin Type A (Sigma-Aldrich, Cat#G1890) in 500 ml PBS (Cellgro, Cat#21-040-CV) overnight at 37°C; filter the following day through a 0.22 μm Stericup (Millipore, Cat#SCGVU05RE). Store at 4°C for up to 3 weeks.
2. hPSC medium: 500 ml DMEM/F12 (Cellgro, Cat#10-090-CV) supplemented with 130 ml Knockout Serum Replacement (Invitrogen/Life Technologies, Cat#10-828-028), 6.5 ml Penicillin-Streptomycin (Invitrogen/Life Technologies, Cat#15140-122), 6.5 ml Nonessential Amino Acids (Invitrogen/Life Technologies, Cat#11140-050), 6.5 ml GlutaMAX (Invitrogen/Life Technologies, Cat#35050-061), 6.5 ml EmbryoMax ES Cell Qualified β-mercaptoethanol 100× (Millipore, Cat#ES-007-E). Filter through a 0.22 μm Stericup (Millipore, Cat#SCGVU05RE). Store at 4°C for up to 3 weeks.
3. Mouse embryonic fibroblast (MEF)-conditioned hPSC medium: Prepare MEF-conditioned hPSC medium according to standard procedures. For instance, plate mitotically inactive

MEF cells at densities appropriate for the maintenance of the human pluripotent stem cell line of choice using MEF medium. The following day, replace the MEF medium with the amount of hPSC medium that is normally used for maintaining the hPSC line, and incubate for 24 h at normal cell-growing conditions (i.e., humidified atmosphere, 37°C, 5% CO_2). Collect the MEF-conditioned hPSC medium after 24 h of incubation (see Note 2), pass through a 0.22 μm filter, and aliquot into single-use volumes. Store at −20°C. Add basic Fibroblast Growth Factor (FGF2; R&D Systems, Cat#233-FB/CF) shortly before use at a final concentration of 10 ng/ml (see Note 3).

2.1.2 Media and Components for Differentiation

1. Bovine Serum Albumin Fraction V (BSA-V) solution: Dissolve 1.0 g of BSA-V (Sigma-Aldrich, Cat#A9418) in 10 ml DMEM/F12 with GlutaMAX (Invitrogen/Life Technologies, Cat#10565) and Penicillin-Streptomycin (Invitrogen/Life Technologies, Cat#15140-122), and pass through a 0.22 μm filter. Store at 4°C for up to 3 weeks.
2. Stage 1a medium: DMEM/F12 with GlutaMAX (Invitrogen/Life Technologies, Cat#10565) and Penicillin-Streptomycin (Invitrogen/Life Technologies, Cat#15140-122), supplemented with final concentrations of 2.0 mg/ml BSA-V (see item 1 in Subheading 2.1.2 for details), 0.5× N2 (Invitrogen/Life Technologies, Cat#17502-048), 0.5× B27 (Invitrogen/Life Technologies, Cat#17504-044), 100 ng/ml activin A (R&D Systems, Cat#338-AC) (see Note 4), 50 ng/ml Wnt3a (R&D Systems, Cat#5036-WN/CF) (see Note 5), and 0.1 μM wortmannin (Sigma-Aldrich, Cat#W1628) (see Note 6).
3. Stage 1b medium: DMEM/F12 with GlutaMAX (Invitrogen/Life Technologies, Cat#10565) and Penicillin-Streptomycin (Invitrogen/Life Technologies, Cat#15140-122), supplemented with final concentrations of 2.0 mg/ml BSA-V (see item 1 in Subheading 2.1.2 for details), 0.5× N2 (Invitrogen/Life Technologies, Cat#17502-048), 0.5× B27 (Invitrogen/Life Technologies, Cat#17504-044), 100 ng/ml activin A (R&D Systems, Cat#338-AC) (see Note 4).
4. Stage 2 medium: RPMI 1640 with GlutaMAX (Invitrogen/Life Technologies, Cat#61870) and Penicillin-Streptomycin (Invitrogen/Life Technologies, Cat#15140-122), supplemented with final concentrations of 2.0% (vol/vol) Fetal Bovine Serum (FBS; Invitrogen/Life Technologies, Cat#10439), and 50 ng/ml Keratinocyte Growth Factor (FGF7; Sigma-Aldrich, Cat#K1757) (see Note 7).

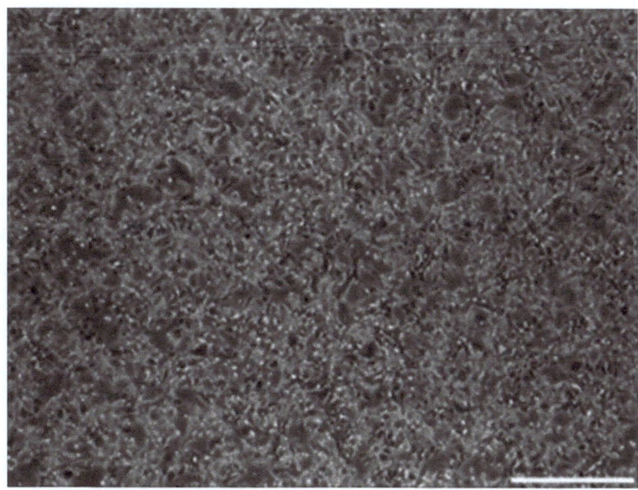

Fig. 1 CHO cells stably expressing activin A at the recommended density to start the conditioning. Scale bar, 500 μm

5. Stage 3 medium: DMEM/F12 with GlutaMAX (Invitrogen/Life Technologies, Cat#10565) and Penicillin-Streptomycin (Invitrogen/Life Technologies, Cat#15140-122), supplemented with final concentrations of 2.0 mg/ml BSA-V (see item 1 in Subheading 2.1.2 for details), 1.0× N2 (Invitrogen/Life Technologies, Cat#17502-048), 500 nM (–)-indolactam V (Sigma-Aldrich, Cat#I0661) (see Note 8), and 10 ng/ml FGF10 (R&D Systems, Cat#345-FG/CF) (see Note 9).
6. Stage 4 medium: DMEM with high glucose and GlutaMAX (Invitrogen/Life Technologies, Cat#10566) and Penicillin-Streptomycin (Invitrogen/Life Technologies, Cat#15140-122), supplemented with final concentrations of 2.0 mg/ml BSA-V (see item 1 in Subheading 2.1.2 for details), and 1.0× B27 w/o insulin (Invitrogen/Life Technologies, Cat#0050129SA) (see Note 10).
7. Stage 5 medium: CMRL medium-1066 (Invitrogen/Life Technologies, Cat#11530) with Penicillin-Streptomycin (Invitrogen/Life Technologies, Cat#15140-122), supplemented with final concentrations of 1.0× B27 w/o insulin (Invitrogen/Life Technologies, Cat#0050129SA) (see Note 10), and 50 ng/ml Hepatocyte Growth Factor (HGF; Peprotech, Cat#100-39) (see Note 11).

2.2 Activin A-Secreting CHO Cells

The use of activin A-secreting CHO cells is optional. As indicated in Note 4, recombinant activin A-containing medium used for Stages 1a and 1b can be substituted with medium conditioned by activin A-secreting CHO cells (Fig. 1) (12, 13). These activin A-secreting CHO cells were a kind gift from Drs. Koji Y. Arai and Toshio Nishiyama. Please contact either to obtain a cryovial of these cells.

2.2.1 Media and Components for Activin A-Secreting CHO Cells

1. CHO cell maintenance medium: CHO-S-SFM II (Invitrogen/Life Technologies, Cat#12052) with Penicillin-Streptomycin (Invitrogen/Life Technologies, Cat#15140-122) and final concentrations of 1.0% (vol/vol) FBS (Invitrogen/Life Technologies, Cat#10439) (see Note 12).
2. Activin A-secreting CHO cell-conditioning medium: DMEM/F12 with GlutaMAX (Invitrogen/Life Technologies, Cat#10565) and Penicillin-Streptomycin (Invitrogen/Life Technologies, Cat#15140-122) (see Note 13).

3 Methods

All cell incubations are to be performed in a humidified atmosphere at 37°C, 5% CO_2.

3.1 Preparation of Human Pluripotent Stem Cells for Differentiation

Differentiation should be conducted in the absence of MEFs as these might interfere with or inhibit efficient differentiation of the hPSCs. Plate the hPSCs in the desired format (e.g., 12-well plates or 6-well plates) under feeder-free conditions that have proven suitable for the hPSC line of choice (see Notes 14 and 15). The seeding density and the number of days that the cells are grown prior to the onset of differentiation should be determined empirically and may vary significantly from hPSC line to hPSC line (see Note 16). The day on which the differentiation is initiated can be referred to as the undifferentiated stage or Stage 0.

3.2 Stage 1a: Differentiation into Mesendoderm

Prepare a tube with Stage 1a medium; add all growth factors and components shortly before application to the cells, but do not yet add wortmannin (see Note 6). Replace the MEF-conditioned hPSC medium on the hPSCs with PBS. Add wortmannin to the Stage 1a medium and mix. Then aspirate the PBS from the hPSCs and replace it with the Stage 1a medium. Incubate overnight for 16–24 h (see Note 17).

3.3 Stage 1b: Differentiation into Definitive Endoderm

Prepare a tube with Stage 1b medium; add the growth factor and components shortly before application to the cells. Shake the plate or dish with differentiating hPSCs vigorously to suspend the dead cells and cell debris. Do not wash the cells with PBS or medium, but replace the Stage 1a medium directly with the Stage 1b medium. Incubate overnight and refresh the medium each following day for a period of 2–4 days with vigorous shaking of the plate or dish prior to each medium refreshment (see Note 18).

3.4 Stage 2: Differentiation into Primitive Gut Tube

Prepare a tube with Stage 2 medium; add the growth factor and components shortly before application to the cells. Do not wash the cells with PBS or medium, but replace the Stage 1b medium directly with the Stage 2 medium. Incubate overnight and refresh the medium each following day for a period of 2–4 days (see Note 19).

3.5 Stage 3: Differentiation into Posterior Foregut

Prepare a tube with Stage 3 medium; add all growth factors and components shortly before application to the cells. Do not wash the cells with PBS or medium, but replace the Stage 2 medium directly with the Stage 3 medium. Incubate overnight and refresh the medium each following day for a period of 2–4 days (see Note 20).

3.6 Stage 4: Differentiation into Pancreatic Endoderm

Prepare a tube with Stage 4 medium; add all components shortly before application to the cells. Do not wash the cells with PBS or medium, but replace the Stage 3 medium directly with the Stage 4 medium. Incubate overnight and refresh the medium each following day for a period of 2–5 days (see Note 21).

3.7 Stage 5: Differentiation into Pancreatic Endocrine Cells

Prepare a tube with Stage 5 medium; add the growth factor and all components shortly before application to the cells. Do not wash the cells with PBS or medium, but replace the Stage 4 medium directly with the Stage 5 medium. Incubate overnight and refresh the medium each following day for a period of 2–7 days or more (see Note 22).

4 Notes

1. If the cell line of choice is passaged manually (e.g., by the cut-and-paste method), it should first be adapted to enzymatic passaging using mild dissociation solutions, such as TrypLE Select (Invitrogen/Life Technologies, Cat#12563) or Accutase (Millipore, Cat#SF006).
2. The same MEF cells can be used for six additional conditioning cycles. After collecting the conditioned medium, incubate the MEF cells with the same volume of (unconditioned) hPSC medium for 24 h.
3. FGF2 should be reconstituted and stored according to the manufacturer's instruction in small 50 μg/ml aliquots that can be used up within 1 week. Once an aliquot is thawed, store at 4°C; do not refreeze.
4. Activin A should be reconstituted and stored according to the manufacturer's instruction in small 100 μg/ml aliquots that can be used up within 1 week. Once an aliquot is thawed, store at 4°C; do not refreeze. An economical and alternative source of activin A is the use of Chinese hamster ovary (CHO) cells that stably express and secrete recombinant activin A (12, 13). In practice, these CHO cells are cultured in DMEM/F12 to make the conditioned medium (see item 2 in Subheading 2.2.1 for details). Note that the concentration of activin A has not

been accurately determined. Thus, titration by diluting the conditioned medium might be necessary for optimal differentiation conditions. The estimated secretion of activin A is approximately 0.3–0.5 μg/2 × 10^6 cells in 48 h (13). We have not observed adverse effects on the differentiation or survival of the hPSCs when using estimated concentrations of ~400 ng/ml activin A.

5. Wnt3a should be reconstituted and stored according to the manufacturer's instruction in small 50 μg/ml aliquots that can be used up within 1 week. Once an aliquot is thawed, store at 4°C; do not refreeze.
6. Wortmannin should be reconstituted and stored according to the manufacturer's instruction in small 1 mM aliquots that can be used up within 1 week. Once an aliquot is thawed, store at 4°C; do not refreeze. Since wortmannin has a half-life of ~10 min in tissue culture, this component should be added to Stage 1a medium last and shortly before application.
7. FGF7 should be reconstituted and stored according to the manufacturer's instruction in small 50 μg/ml aliquots that can be used up within 1 week. Once an aliquot is thawed, store at 4°C; do not refreeze.
8. (−)-Indolactam V should be reconstituted and stored according to the manufacturer's instruction in small 500 μM aliquots that can be used up within 1 week. Once an aliquot is thawed, store at 4°C; do not refreeze.
9. FGF10 should be reconstituted and stored according to the manufacturer's instruction in small 10 μg/ml aliquots that can be used up within 1 week. Once an aliquot is thawed, store at 4°C; do not refreeze.
10. It is important to use insulin-free B27 at this stage to prevent uptake of insulin by the differentiating cells (7).
11. HGF should be reconstituted and stored according to the manufacturer's instruction in small 50 μg/ml aliquots that can be used up within 1 week. Once an aliquot is thawed, store at 4°C; do not refreeze.
12. The CHO cells might have to be adapted to 1% (vol/vol) through a stepwise process. With each passage, lower the concentration of FBS; e.g., start with 10% FBS, then with the next passage use 5%, then 2%, and with the last passage 1% FBS. The CHO cells can be expanded for cryopreservation at this stage.
13. Plate 30 × 10^6 CHO cells (Fig. 1) in a T75 flask using CHO cell maintenance medium. The following day, replace this medium with 25 ml of Activin A-secreting CHO cell-conditioning medium and incubate for 24 h. Collect the conditioned medium, pellet the floating cells in this conditioned medium,

and filter the supernatant through a 0.22 μm filter. Collection can generally be repeated two more times. The conditioned medium can be stored at 4°C for up to 3 weeks. Do not freeze the medium as this will render the activin A inactive. Add the other components for Stages 1a and 1b (see items 2 and 3 in Subheading 2.1.2 for details) prior to application.

14. We found that some hPSC lines can be maintained under feeder-free conditions in MEF-conditioned medium with a standard concentration of bFGF on Gelatin-coated plates that had been incubated overnight with MEF medium (12).

15. If the cells are to be grown on cover slips for staining purposes or other downstream experiments, we advise using Thermanox plastic cover slips (Nunc, Cat#174950) instead of glass cover slips for improved adherence at later stages during differentiation.

16. Generally, a seeding density of 0.5–2.0×10^5 cells/cm^2 and 1–3 days of feeder-free growth prior to initiation of differentiation can be considered as a starting point. Use lower or higher seeding densities to find the optimal cell density and recovery period after passaging for your hPSC line of choice. Fluorescent microscopic analysis of the cells at Stage 1b to asses SOX17 expression and concomitant downregulation of OCT4 is a good readout for the differentiation efficiency of the initial stage.

17. A dramatic reduction in cell numbers is likely to happen during Stage 1a. Fix cells at 16, 20, and 24 h after the onset of differentiation for analysis by fluorescence microscopy to find the optimal duration for Stage 1a. The efficiency of Stage 1a will depend on a balance between the highest achievable percentage of mesendoderm-like cells and the least cell death. Increasing or decreasing the concentration of wortmannin might result in higher yields of mesendoderm-like cells. We suggest comparing undifferentiated hPSCs (fixed on the day when the differentiation was initiated) with differentiating cells fixed at multiple time points within the first 24 h using OCT4, SOX17, and BRACHYURY T as markers; expression of BRACHYURY T indicates the formation of mesendoderm. Coexpression of OCT4 with BRACHYURY T can be expected, but SOX17 should not yet be visible. See Tables 1 and 2 for suggested antibodies and primers to assess the differentiation efficiency. See Fig. 2a, b for the changes in the expression profiles that can be expected to occur.

18. During Stage 1b, the cell numbers increase and definitive endoderm-like cells become apparent. Analyze the cells 2, 3, and 4 days after the initiation of Stage 1b using OCT4 and SOX17 as negative and positive markers, respectively, to find the optimal duration for Stage 1b. BRACHYURY T should be

Table 1
Antibodies that can be used for the assessment of differentiating cells by immunofluorescence microscopy and flow cytometry

Antigen	Host	Dilution	Vendor	Catalog #	Stage	Embryonic equivalent[a]
OCT4	Mouse	1:200	Santa Cruz	sc-5279	Undifferentiated	Inner cell mass of blastocyst
SOX2	Rabbit	1:200	Chemicon	ab5603	Undifferentiated	Inner cell mass of blastocyst
BRACHYURY T	Goat	1:200	R&D Systems	AF2085	Stage 1a	Mesendoderm
SOX17	Goat	1:1,000	R&D Systems	AF1924	Stage 1b–5	Definitive endoderm
HNF1B	rabbit	1:300	Santa Cruz	sc-22840	Stage 2–5	Primitive gut tube
PDX1	Goat	1:10,000	Abcam	ab47383	Stage 3–5	Posterior foregut
NGN3	Sheep	1:200	R&D Systems	AF3444	Stage 4	Pancreatic endoderm
NKX6.1	Mouse	1:50	Hybridoma	F55A10-c	Stage 4	Pancreatic endoderm
INSULIN	Guinea pig	1:500	Dako	A0564	Stage 5	Islet of Langerhans
GLUCAGON	Rabbit	1:500	Sigma-Aldrich	G2654	Stage 5	Islet of Langerhans

[a]Developmental stage at which expression can be detected for the first time

Table 2
Primers (Applied Biosystems) that can be used for the assessment of differentiation by qPCR (Taqman assay)

Transcript	Catalog #	Stage	Embryonic equivalent[a]
OCT4	Hs00999632_gl	Undifferentiated	Inner cell mass of blastocyst
SOX2	Hs01053049_sl	Undifferentiated	Inner cell mass of blastocyst
BRACHYURY T	Hs00610080_ml	Stage 1a	Mesendoderm
SOX7	Hs00846731_sl	Stage 1b	Extraembryonic endoderm
CXCR4	Hs00976734_ml	Stage 1b	Definitive endoderm
SOX17	Hs00751752_sl	Stage 1b–5	Definitive endoderm
HNF1B	Hs01001602_ml	Stage 2–5	Primitive gut tube
PDX1	Hs00426216_ml	Stage 3–5	Posterior foregut
SOX9	Hs00165814_ml	Stage 3	Posterior foregut
NKX6.1	Hs00232355_ml	Stage 4	Pancreatic endoderm
NGN3	Hs00360700_gl	Stage 4	Pancreatic endoderm
INSULIN	Hs00355773_ml	Stage 5	Islet of Langerhans
GLUCAGON	Hs01031536_ml	Stage 5	Islet of Langerhans

[a]Developmental stage at which expression can be detected for the first time

absent at this stage. See Tables 1 and 2 for suggested antibodies and primers to assess the differentiation efficiency. See Fig. 2b, c for the changes in the expression profiles that can be expected to occur.

19. Cell death should be minimal during Stage 2. Hence, vigorous shaking of the plate or dish prior to refreshment is not necessary. Analyze the cells 2, 3, and 4 days after the initiation of Stage 2 using SOX17 and HNF1B as positive markers to find the optimal duration for Stage 2. SOX17 levels might decline to some extent, but the transcription factor should remain present. Coexpression of SOX17 and HNF1B should be nearly 100%. See Tables 1 and 2 for suggested antibodies and primers to assess the differentiation efficiency.

20. Cell death should be minimal during Stage 3. Hence, vigorous shaking of the plate or dish prior to refreshment is not necessary. The cells might change morphology dramatically and become multilayered, depending on the type of cell line that is used. In most cases, the three-dimensional clusters that arise over the course of Stage 3 contain the posterior foregut-like cells. Analyze the cells 2, 3, and 4 days after the initiation of

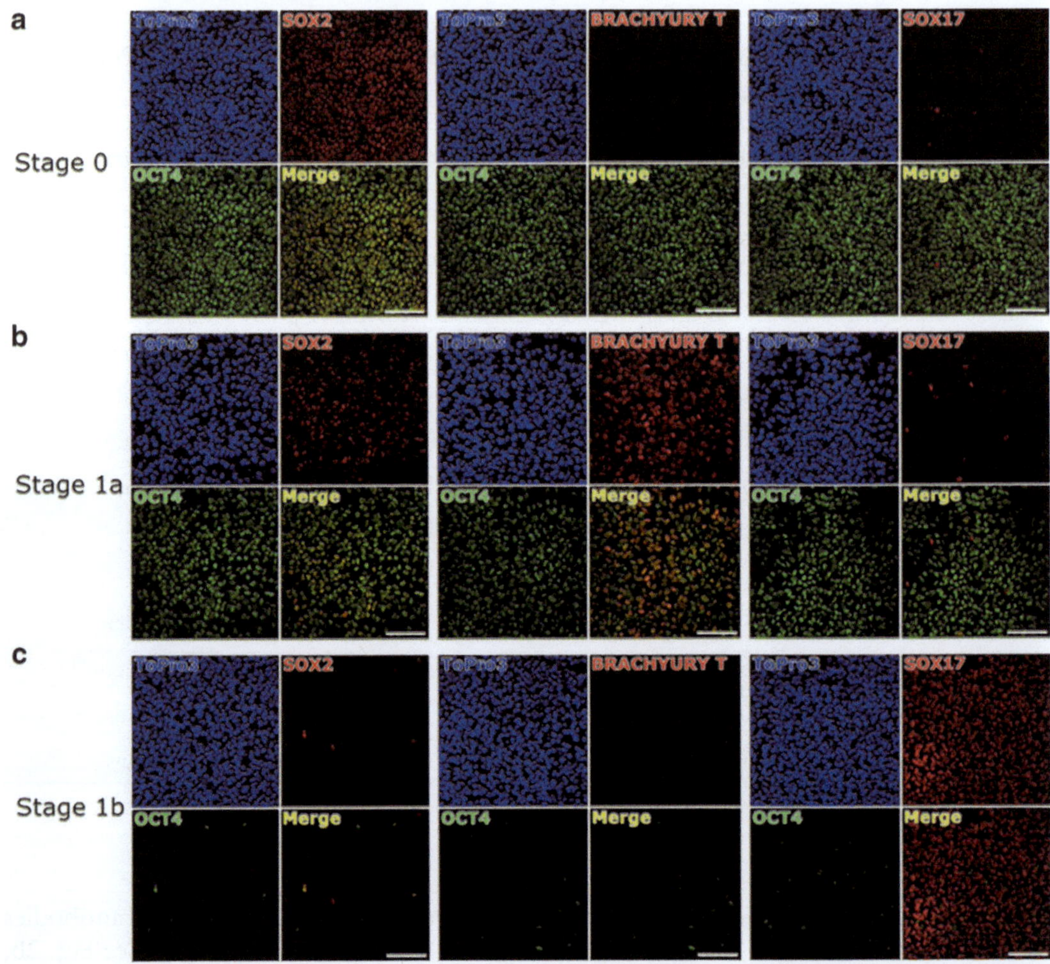

Fig. 2 Immunofluorescence stainings of hESCs differentiated into mesendoderm and definitive endoderm. (**a**) Undifferentiated hESCs (i.e., Stage 0), (**b**) hESCs differentiated into mesendoderm (i.e., Stage 1a), and (**c**) hESCs differentiated into definitive endoderm (i.e., Stage 1b). All samples were stained for ToPro3 (*blue*) to visualize nuclear DNA and OCT4 (*green*) in combination with either SOX2 (*red, left panel*), BRACHYURY T (*red, middle panel*), or SOX17 (*red, right panel*). A merged image of each sample was created by overlaying the OCT4 staining with either SOX2, BRACHYURY T, or SOX17. Scale bars, 100 μm

Stage 3 using SOX17, HNF1B, and PDX1 as positive markers to find the optimal duration for Stage 3. SOX17 levels might keep on declining, but the transcription factor should still be present. PDX1 is the hallmark transcription factor for this stage; its expression indicates that the cells have been specified for pancreatic fate. See Tables 1 and 2 for suggested antibodies and primers to assess the differentiation efficiency. Addition of varying concentrations of DAPT, all-trans-retinoic acid, and noggin has been reported to improve the differentiation efficiency of this stage (8, 11) (for reviews, see refs. 9, 10).

21. There are currently no potent growth factors or other signaling molecules known to further enhance the transition of posterior foregut-like cells into pancreatic endoderm-like cells, although some groups have reported slight improvement using a variety of molecules (for reviews, see refs. 9, 10). Cell death should be minimal during Stage 4. Hence, vigorous shaking of the plate or dish prior to refreshment is not necessary. The cells may further change morphology, forming spiderweb-like connective strands of cells, depending on the type of cell line that is used. In most cases, the "nodes" of these three-dimensional structures harbor the pancreatic endoderm-like cells. Analyze the cells 2, 3, 4, and 5 days after the initiation of Stage 4 using SOX17, HNF1B, PDX1, NGN3, and NKX6.1 as positive markers to find the optimal duration for Stage 4. NGN3 is the hallmark transcription factor for this stage. See Tables 1 and 2 for suggested antibodies and primers to assess the differentiation efficiency.
22. Cell death should be minimal during Stage 5. Hence, vigorous shaking of the plate or dish prior to refreshment is not necessary. The cells are not expected to change morphology dramatically at this stage. Analyze the cells every consecutive day for a week or longer, starting 2 days after the initiation of Stage 5 using SOX17, HNF1B, PDX1, INSULIN, and GLUCAGON as positive markers to find the optimal duration for Stage 5. INSULIN is the hallmark protein for this stage. All INSULIN-positive cells should coexpress PDX1, and may also coexpress GLUCAGON. Although these two endocrine hormones are not coexpressed by true, mature β-cells, they tend to be present at various ratios in pancreatic endocrine-like cells obtained through in vitro differentiation. In addition, release of INSULIN in response to exogenous glucose is expected to be minimal. See Tables 1 and 2 for suggested antibodies and primers to assess the differentiation efficiency.

Acknowledgements

This work was supported by the California Institute for Regenerative Medicine (CIRM Training Grant Number TG2-01153) and the Juvenile Diabetes Research Foundation (JDRF grant awards 3-2008-477, 35-2008-628, and 1-2010-152).

References

1. Osafune K, Caron L, Borowiak M et al (2008) Marked differences in differentiation propensity among human embryonic stem cell lines. Nat Biotechnol 26(3):313–315
2. Bar-Nur O, Russ HA, Efrat S, Benvenisty N (2011) Epigenetic memory and preferential lineage-specific differentiation in induced pluripotent stem cells derived from human pancreatic islet Beta cells. Cell Stem Cell 9(1):17–23
3. Wang P, Rodriguez RT, Wang J, Ghodasara A, Kim SK (2011) Targeting SOX17 in human embryonic stem cells creates unique strategies for isolating and analyzing developing endoderm. Cell Stem Cell 8(3):335–346
4. Nostro MC, Sarangi F, Ogawa S et al (2011) Stage-specific signaling through TGF{beta} family members and WNT regulates patterning and pancreatic specification of human pluripotent stem cells. Development 138(5):861–871
5. Macarthur CC, Xue H, Van Hoof D, et al (2012) Chromatin insulator elements block transgene silencing in engineered human embryonic stem cell lines at a defined chromosome 13 locus. Stem Cells Dev 21(2): 191–205
6. Lumelsky N, Blondel O, Laeng P, Velasco I, Ravin R, McKay R (2001) Differentiation of embryonic stem cells to insulin-secreting structures similar to pancreatic islets. Science 292(5520):1389–1394
7. Rajagopal J, Anderson WJ, Kume S, Martinez OI, Melton DA (2003) Insulin staining of ES cell progeny from insulin uptake. Science 299(5605):363
8. D'Amour KA, Agulnick AD, Eliazer S, Kelly OG, Kroon E, Baetge EE (2005) Efficient differentiation of human embryonic stem cells to definitive endoderm. Nat Biotechnol 23(12):1534–1541
9. Van Hoof D, D'Amour KA, German MS (2009) Derivation of insulin-producing cells from human embryonic stem cells. Stem Cell Res 3(2–3):73–87
10. Mfopou JK, Chen B, Mateizel I, Sermon K, Bouwens L (2010) Noggin, retinoids, and fibroblast growth factor regulate hepatic or pancreatic fate of human embryonic stem cells. Gastroenterology 138(7):2233–2245.e14
11. Kroon E, Martinson LA, Kadoya K et al (2008) Pancreatic endoderm derived from human embryonic stem cells generates glucose-responsive insulin-secreting cells in vivo. Nat Biotechnol 26(4):443–452
12. Van Hoof D, Mendelsohn AD, Seerke R, Desai TA, German MS (2011) Differentiation of human embryonic stem cells into pancreatic endoderm in patterned size-controlled clusters. Stem Cell Res 6(3):276–285
13. Arai KY, Tsuchida K, Li C et al (2006) Purification of recombinant activin A using the second follistatin domain of follistatin-related gene (FLRG). Protein Expr Purif 49(1):78–82

Chapter 11

Directed Differentiation of Pluripotent Stem Cells to Functional Hepatocytes

Philip Roelandt, Jolien Vanhove, and Catherine Verfaillie

Abstract

Differentiation of human stem cells to hepatocytes is crucial for industrial applications as well as to develop new therapeutic strategies for liver disease. The protocol described here, using sequentially growth factors known to play a role in liver embryonic development, efficiently differentiates human embryonic stem cells (hESC) as well as human-induced pluripotent stem cells (hiPSC) to hepatocytes by directing them through defined embryonic intermediates, namely, mesendoderm/definitive endoderm and hepatoblast and hepatocyte phenotype. After 28 days, the final differentiated progeny is a mixture of cells, comprising cells with characteristics of hepatoblasts and a smaller cell fraction with morphological and phenotypical features of mature hepatocytes. An extensive functional characterization of the stem cell progeny should be used to confirm that differentiated cells display functional characteristics of mature hepatocytes including albumin secretion, glycogen storage, and several detoxifying functions such as urea production, bilirubin conjugation, glutathione *S*-transferase activity, cytochrome activity and drug transporter activity.

Key words Pluripotent stem cells, Hepatocyte differentiation

1 Introduction

Primary hepatocytes are currently used for a number of applications: drug development, drug metabolism and drug toxicity studies (ADMET studies), study of hepatitis virus infection, and creation of antiviral drugs and investigation of new therapies for liver diseases such as development of artificial liver devices and cell transplantation. However, hepatocytes isolated from human liver can only be maintained in culture for a limited time, as they rapidly lose their phenotypic and functional characteristics. Stem- or progenitor-derived functional hepatocytes have been suggested as a good alternative for primary hepatocytes and the optimization of hepatic differentiation protocols is considered to be crucial for the further expansion of hepatocyte-like cells for applications in ADMET and therapeutic indications.

Uma Lakshmipathy and Mohan C. Vemuri (eds.), *Pluripotent Stem Cells: Methods and Protocols*, Methods in Molecular Biology, vol. 997, DOI 10.1007/978-1-62703-348-0_11, © Springer Science+Business Media New York 2013

The 28 days differentiation protocol described here is modeled on liver embryonic development. To mimic Nodal/Cripto signaling and canonical β-catenin activity involved in primitive streak (PS) and definitive endoderm (DE) formation, cells were first treated with Activin-A and Wnt3a, respectively. Growth factors produced by the cardiac and septum transversum mesenchyme to induce formation of hepatic endoderm from anterior endoderm were mimicked in vitro by the stimulation of cells with bone morphogenic protein-4 (BMP4) for 4 days. Fibroblast growth factor-1 (FGF1) was used to mimic the growth signal that induces liver bud growth (4 days), and finally hepatocyte growth factor (HGF) was used to enhance maturation of hepatoblasts to hepatocytes (16 days). The main differences between this protocol and the initially published protocol (1) are the absence of fetal bovine serum, FGF2, FGF4, FGF8b, and follistatin, shorter exposure to Activin-A and Wnt3a, and longer exposure to HGF (2).

2 Materials

2.1 Solutions

2.1.1 BSA-PBS Solution, 0.2%

Dissolve 100 mg of bovine serum albumin (BSA, Sigma A8806) in a final volume of 50 ml of PBS and filter using Millipore Express plus 0.22 μm filter to produce a 0.2% BSA-PBS solution.

2.1.2 BSA-HCl Solution, 0.2%

Dilute 200 μl HCl (stock concentration 2 M) in 100 ml distilled water to produce a 4 mM HCl solution. Dissolve 200 mg BSA in a final volume of 100 ml 4 mM HCl solution. Filter using Millipore Express plus 0.22 μm filter.

2.1.3 Activin-A Solution

Dissolve 5 μg Activin-A (R&D Systems 338-AC) in 500 μl 0.2% BSA-PBS to produce a 10 ng/μl solution.

2.1.4 Wnt3a Solution

Dissolve 5 μg Wnt3a (R&D Systems 1324-WN) in 500 μl 0.2% BSA-PBS to produce a 10 ng/ml solution.

2.1.5 BMP4 Solution

Dissolve 10 μg BMP (R&D Systems 314-BP) in 1,000 μl 0.2% BSA-HCl to produce a 10 ng/μl solution.

2.1.6 FGF1 Solution

Dissolve 25 μg FGF1 (R&D Systems 232-FA) in 2,500 μl 0.2% BSA-PBS to produce a 10 ng/μl solution.

2.1.7 HGF Solution

Dissolve 25 μg HGF (R&D Systems 294-HGN) in 2,500 μl 0.2% BSA-PBS to produce a 10 ng/μl solution.

2.1.8 Dexamethasone Solution

Dissolve 1.52 g dexamethasone (Sigma D2915) in 1,000 ml distilled H_2O or sterile PBS to produce a 0.25 mM solution. Filter using Millipore Express plus 0.22 μm filter.

2.1.9 L-Glutamine Solution with β-Mercaptoethanol

Dissolve 0.146 g L-Glutamine (Sigma G8540) in 5 ml PBS and add 7 μl β-mercaptoethanol (Sigma M6250) to produce 200 mM L-Glutamine with 20 mM β-mercaptoethanol.

2.1.10 L-Ascorbic Acid Solution

Dissolve 1.45 g L-ascorbic acid (Sigma A8960) in 500 ml PBS to produce a 10 mM solution. Stir with magnetic stir bar for 10–20 min in dark and filter using Millipore Express plus 0.22 μm filter.

2.1.11 MCDB Solution

Dissolve one vial MCDB (Sigma M-6770) in 1,000 ml MilliQ water in a flat bottom erlenmeyer flask. Stir with magnetic stir bar until completely dissolved (yellow suspension). Measure pH and adjust with NaOH or HCl dropwise until pH 7.2. Filter the solution using Millipore Express plus 0.22 μm filter.

2.2 Media Composition

2.2.1 MEF Expansion Medium (500 ml)

1. 417 ml of DMEM high glucose (Invitrogen 41965039).
2. 75 ml of fetal bovine serum (Hyclone CH30160.03?).
3. 4 mM of L-glutamine (Invitrogen 25030024).
4. 0.5 U of Penicillin/Streptomycin (Invitrogen 15140122).
5. 2× of minimum essential media (MEM) non-essential amino acids (NEAA; Invitrogen 11140035).
6. 0.1 mM of β-mercaptoethanol (Invitrogen 31350010).

2.2.2 hESC/iPSC Expansion Medium (500 ml)

1. 400 ml of DMEM/F-12 with HEPES (Invitrogen 31330).
2. 100 ml of Knockout Serum Replacement (Invitrogen 10828?).
3. 2.5 ml of L-glutamine solution with β-mercaptoethanol.
4. 1× of MEM non-essential amino acids (NEAA; Invitrogen 11140035).
5. 2 μg of FGF2 (R&D Systems 233-FB).
6. 0.5 U of Penicillin/Streptomycin (Invitrogen 15140122).

2.2.3 Hepatic Differentiation Medium (500 ml)

1. 285 ml of DMEM low glucose (Invitrogen 31885023).
2. 200 ml of MCDB-201-water (Sigma M-6770).
3. 0.25× of Linoleic acid—Bovine serum albumin (LA-BSA) (Sigma L-9530).
4. 0.25× of Insulin–transferrin–selenium (ITS) (Sigma I-3146).
5. 50 U of Penicillin/Streptomycin (Invitrogen 15140122).
6. 100 nM of L-ascorbic acid solution.
7. 1 μM of dexamethasone solution.
8. 50 μM of β-mercaptoethanol (Invitrogen 31350010).

2.3 Cell Culture Plates

2.3.1 MEF-Coated Plated

Mouse embryonic fibroblasts (MEFs) (Global Stem Inc, Rockville, USA) were maintained in MEF expansion medium and immortalized with Mitomycin C (KYOWA Mitomycin 2 mg). Mitomycin-treated

MEFs were plated at a density of 30,000 cells/cm^2 on 0.1% gelatine (Ultrapure water 0.1% gelatine, Chemicon ES-006-B)-coated 6-well plates (Sigma, CLS3506).

2.3.2 Matrigel-Coated Plates

Thaw matrigel (VWR, BDAA354230) slowly on ice. Dissolve 400 µl matrigel in 25 ml ice-cold DMEM-F12 without HEPES (Invitrogen 11320074) (final concentration 1,6%) and add 0.5 ml to every well of a 12-well plate (Sigma CLS3512). Incubate for at least 30 min at 37°C or 2–3 h at room temperature (see Note 1).

3 Methods

1. hESC/iPSC are cultured as described (3) in hESC/hiPSC expansion medium in a 21% O_2—5% CO_2—37°C incubator and passaged 1:3 using collagenase IV (Invitrogen 17104-019) every 5–7 days in iMEF-coated 6-well plates. hESC/hiPSC were transferred to a feeder-free differentiation matrigel-coated 12-well plates in mTesR1-medium (StemCell Technologies Inc 5875) (4).
2. Once 70–80% confluent, remove mTeSR1-medium, rinse wells with PBS (see Notes 2 and 3).
3. Add 1 ml differentiation medium containing 10 µl Activin-A solution (final concentration 100 ng/ml) and 5 µl Wnt3a solution (final concentration 50 ng/ml) per well. Keep in 21% O_2—5% CO_2—37°C incubator until day 2 (see Fig. 1).
4. On day 2, remove all medium, rinse with PBS and add 1 ml differentiation medium containing 10 µl Activin-A solution (see Note 2). Keep in 21% O_2—5% CO_2—37°C incubator until day 4.
5. On day 4, remove all medium, rinse with PBS and add 1 ml differentiation medium containing 5 µl BMP4 solution (final concentration 50 ng/ml) (see Note 2). Keep in 21% O_2—5% CO_2—37°C incubator until day 6 (see Fig. 1).
6. On day 6, remove 750 µl medium and add 1 ml differentiation medium containing 5 µl BMP4 solution. Keep in 21% O_2—5% CO_2—37°C incubator until day 8.
7. On day 8, remove all medium, rinse with PBS and add 1 ml differentiation medium containing 2 µl FGF1 solution (final concentration 20 ng/ml). Keep in 21% O_2—5% CO_2—37°C incubator until day 10 (see Fig. 1).
8. On day 10, remove 750 µl medium and add 1 ml differentiation medium containing 2 µl FGF1 solution. Keep in 21% O_2—5% CO_2—37°C incubator until day 12.
9. On day 12, remove all medium, rinse with PBS and add 1 ml differentiation medium containing 2 µl HGF solution (final concentration 20 ng/ml). Keep in 21% O_2—5% CO_2—37°C incubator until day 14 (see Fig. 1).

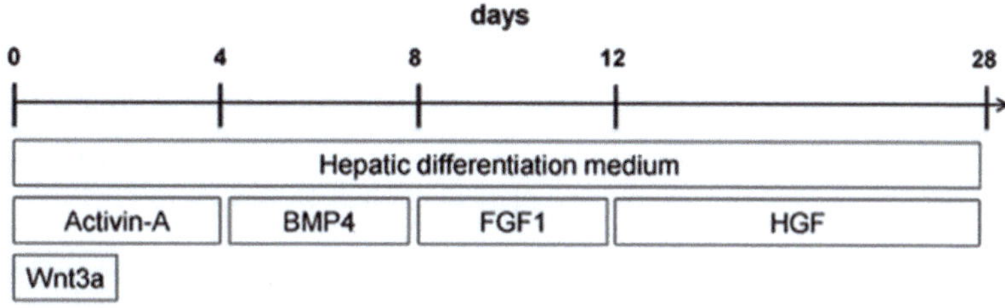

Fig. 1 Schematic overview of the hepatocyte differentiation protocol

10. On day 14, remove 750 μl medium and add 1 ml differentiation medium containing 2 μl HGF solution. Keep in 21% O_2—5% CO_2—37°C incubator until day 16.
11. On day 16, remove 750 μl medium and add 1 ml differentiation medium containing 2 μl HGF solution. Keep in 21% O_2—5% CO_2—37°C incubator until day 18.
12. On day 18, remove 750 μl medium and add 1 ml differentiation medium containing 2 μl HGF solution. Keep in 21% O_2—5% CO_2—37°C incubator until day 20.
13. On day 20, remove 750 μl medium and add 1 ml differentiation medium containing 2 μl HGF solution. Keep in 21% O_2—5% CO_2—37°C incubator until day 22.
14. On day 22, remove 750 μl medium and add 1 ml differentiation medium containing 2 μl HGF solution. Keep in 21% O_2—5% CO_2—37°C incubator until day 24.
15. On day 24, remove 750 μl medium and add 1 ml differentiation medium containing 2 μl HGF solution. Keep in 21% O_2—5% CO_2—37°C incubator until day 26.
16. On day 26, remove 750 μl medium and add 1 ml differentiation medium containing 2 μl HGF solution. Keep in 21% O_2—5% CO_2—37°C incubator until day 28 (end of protocol) (see Note 4) (see Fig. 1).

4 Notes

1. Store matrigel (with phenol red, growth factor reduced, VWR BDAA354230)-coated plates at 4°C for at least 7 days. It is advisable to freeze down the matrigel at –80°C in separate vials of smaller volume (for example, 400 μl) to avoid unnecessary freeze-thaw cycles. Matrigel polymerizes at room temperature (20°C), so it is necessary to thaw the bottle of matrigel on ice and use frozen pipette tips and cryotubes. If thawed too quickly, matrigel will partially polymerize and give rise to clumps.

2. Make sure to wash away the floating cells carefully because they may have a negative impact on the first days of differentiation, especially during the Activin-A and Wnt3a step.
3. It is advisable to place growth factors at 4°C and use within 10–14 days after thawing.
4. A single test cannot effectively evaluate the hepatocyte characteristics of hESC/iPSC progeny. It is important to assess the gene expression profile as well as protein synthesis and functional analysis of the cells. This can be done in a number of ways:
   (a) *RT-qPCR*: α_1-antitrypsine (AAT), α-foetoprotein (AFP), albumin (ALB), arginase-1 (ARG1), asialoglycoprotein-1 (ASGR1), carbamoyl phosphate synthase-1 (CPS1), connexin-32 (CX32), cytochrome P450 family members (CYP2C9, CYP3A4, CYP3A7, CYP7A1), factor V (F5), factor VII (F7), factor IX (F9), glucose-6-phosphatase (G6P), γ-glutamyl carboxylase (GGCX), hepatocyte nuclear factor 1α/1β/4α/6 (HNF1α/1β/4α/6), multidrug resistance-associated protein-2 (MRP2), Na^+/taurocholate cotransporting polypeptide (NTCP), organic anion transporter member 1B1 (OATP1B1), phosphoenolpyruvate carboxukinase-1 (PEPCK1), protein C (PROC), transthyretin (TTR), vitamin K epoxide reductase complex subunit-1 (VKORC1)
   (b) *Immunocytochemistry*: AAT, AFP, ALB, ASGR1, CYP3A4, HNF4α, MRP2, PEPCK
   (c) *Electron microscopy*: detection of multiple mitochondria, glycogen droplets, tight junctions, desmosomes, bile canaliculi
   (d) Functional analysis:
      - Albumin secretion (ELISA)
      - Urea production (colorimetric assay)
      - Bile acids production (colorimetric assay)
      - Drug transporter activity (fluorescence assay)
      - Glycogen storage (spectrophotometrical method of Seifter et al. (5))
      - Glutathione *S*-transferase activity (spectrophotometrical method of Habig et al. (6))
      - Cytochrome P450 subtype CYP2C9 and CYP3A4 (Luciferin luminescence assay)
      - Bilirubin conjugation (high pressure liquid chromatography (HPLC) according to the method of Muraca et al. (7))

References

1. Roelandt P, Pauwelyn KA, Sancho-Bru P et al (2010) Human embryonic and rat adult stem cells with primitive endoderm-like phenotype can be fated to definitive endoderm, and finally functional hepatocyte-like cells. PLoS One 5:e12101
2. Roelandt P, Obeid S, Paeshuyse J et al (2012) Human pluripotent stem cells derived hepatocytes support complete replication of hepatitis C virus. J Hepatol 57(2):246–251
3. Abeyta MJ, Clark AT, Rodriguez RT, Bodnar MS, Pera RA, Firpo MT (2004) Unique gene expression signatures of independently-derived human embryonic stem cell lines. Hum Mol Genet 13:601–608
4. Ludwig TE, Bergendahl V, Levenstein ME, Yu J, Probasco MD, Thomson JA (2006) Feeder-independent culture of human embryonic stem cells. Nat Methods 3:637–646
5. Seifter S, Dayton S, Novic B, Muntwyler E (1950) The estimation of glycogen with the anthrone reagent. Arch Biochem 25:191–200
6. Habig WH, Pabst MJ, Jakoby WB (1974) Glutathione S-transferases. The first enzymatic step in mercapturic acid formation. J Biol Chem 249:7130–7139
7. Muraca M, Blanckaert N (1983) Liquid-chromatographic assay and identification of mono- and diester conjugates of bilirubin in normal serum. Clin Chem 29:1767–1771

Chapter 12

Highly Efficient Directed Differentiation of Human Induced Pluripotent Stem Cells into Cardiomyocytes

Paul W. Burridge and Elias T. Zambidis

Abstract

Human-induced pluripotent stem cell (hiPSC)-derived cardiomyocytes are a novel source of cells for patient-specific cardiotoxicity drug testing, drug discovery, disease modeling, and regenerative medicine. We describe a versatile and cost-effective protocol for in vitro cardiac differentiation that is effective for a wide variety of hiPSC and human embryonic stem cell (hESC) lines. This highly optimized protocol produces contracting human embryoid bodies (hEB) with a near total efficiency of $94.7 \pm 2.4\%$ in less than 9 days, and minimizes the variability in cardiac differentiation commonly observed between various hiPSC and hESC lines. The contracting hEB derived using these methods contain high percentages of pure functional cardiomyocytes, highly reproducible electrophysiological profiles, and pharmacologic responsiveness to known cardioactive drugs.

Key words Human embryonic stem cell, Induced pluripotent stem cell, Heart, Cardiac, Cardiomyocyte, Differentiation, Forced aggregation

1 Introduction

The cardiac differentiation of human pluripotent stem cells provides a source of cardiomyocytes with great promise in drug screening, the study of the otherwise inaccessible biology of human cardiac development, and for cell replacement strategies. The advent of human-induced pluripotent stem cells (hiPSC), derived via somatic cell reprogramming, allows the production of pluripotent stem cells from patients with a known history—either carrying a specific genetic disease or from healthy patients. The cardiac differentiation of these cell lines thus allows for cardiac disease modeling along with patient-specific cardiotoxicity testing and novel drug discovery. In the future the use of non-diseased or disease-corrected hiPSC may provide a promising source of cells for regenerative medicine.

Once suitable pluripotent cell sources are available, the next step is to further refine cardiac differentiation. Three major techniques

Uma Lakshmipathy and Mohan C. Vemuri (eds.), *Pluripotent Stem Cells: Methods and Protocols*, Methods in Molecular Biology, vol. 997, DOI 10.1007/978-1-62703-348-0_12, © Springer Science+Business Media New York 2013

are commonly used for the differentiation of hESC or hiPSC to the cardiac lineage. They include differentiation involving cultivation of hPSC with a stromal cell layer possessing lineage inductive properties (1, 2), differentiation of cells in a monolayer (3, 4), and mechanical or enzymatic manipulation to form spherical clusters of cells termed human embryoid bodies (hEB) (5–7). Each of these techniques includes variables that influence the reproducibility of differentiation. These variables include the quality and culture method of the pluripotent stem cell line, contamination with murine embryonic fibroblast (MEF) feeders, the size of the pluripotent cell clumps used for hEB formation, and variation in the secreted factors produced from stromal cell coculture.

We describe here a highly efficient method for the generation of cardiac progenitors and cardiomyocytes from hESC and hiPSC (Fig. 1).

This protocol is further optimized from our earlier reports (8, 9), and is composed of four major steps: (1) culture of hESC/hiPSC in a rigorous and controlled manner in conditioned medium, (2) reproducible formation of hEB from a known number of cells using forced aggregation, (3) exposure to FBS for cardiac induction, and (4) adhesion and formation of contracting cardiomyocytes.

One of the most important problems with reproducibly differentiating hESC/hiPSC is that each cell line has been derived using different methods and often cultured using different conditions. Common culture conditions involve the growth of hESC/hiPSC as colonies, either on a MEF feeder layer or with feeder-free approaches on Matrigel/Geltrex or synthetic matrices. Colonies inherently have cells that are at different densities, which, due to contact inhibition, will usually be in a different part of their growth phase. To eliminate the variability in growth rate we developed a monolayer growth method (10). Using this method, cells are passaged as single cells every 3 days, and seeded at a known density, allowing tight control of growth rate prior to differentiation. Additionally, differentiation experiments are set up from cells that have been passaged on the previous day (11). We have shown that this technique substantially enhances differentiation by ensuring that only live and exponentially growing cells are incorporated into hEB. The differentiation of hESC/hiPSC as a hEB allows the formation of multidimensional cell-cell interactions more closely mimicking in vivo conditions whilst also producing cardiomyocytes electrically coupled in three dimensions. Simple methods for hEB formation, such as scraping colonies and suspending them in 20% FBS (5), produce hEB of a wide variety of sizes which results in poor yields of beating hEB. Here we use a forced aggregation method, first demonstrated for hESC differentiation (12), to form a known number of single cells into an embryoid body. We first described this for cardiac differentiation in 2007 (8) using a chemically defined medium (CDM) for the 96-well V-bottom plate

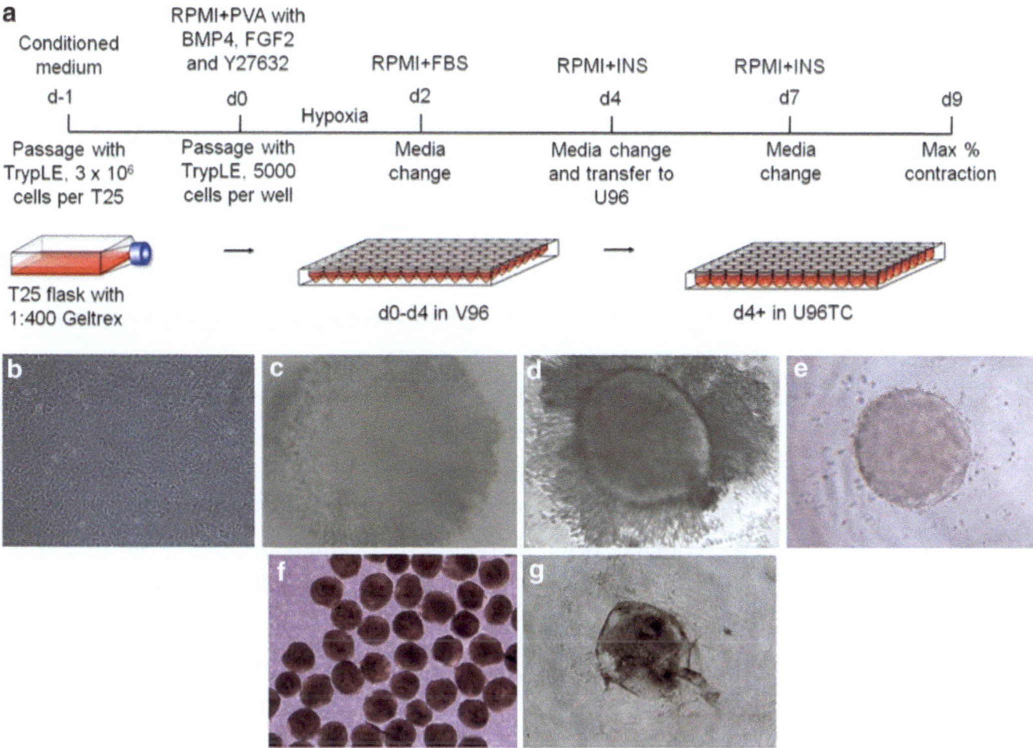

Fig. 1 Schematic of the forced aggregation directed cardiac differentiation method. (**a**) *Top row*: media used and timing of media changes; *middle*: physical tasks performed at each time point; *bottom*: cell format showing move from T25 flasks to V96 plates and U96 tissue culture-treated plates. (**b**) Growth of human pluripotent cells as a monolayer showing 100% confluence. (**c**) 5,000 cells per well of a V96 plate after 1 day. (**d**) On day 2 prior to first media change showing clear hEB formation. (**e**) Day 4 after media change and transfer to U96 plates showing removal of dead cells. (**f**) hEB on day 4 after transfer to bulk culture demonstrating uniformity of hEB size and shape. (**g**) Contracting hEB after 9 days of culture showing minimal surrounding fibroblasts

aggregation step followed by culture in 20% FBS. In this iteration of our differentiation methodology for hEB formation and mesoderm induction, we use a simplified version of the CDM (13) medium, based on RPMI 1640, along with the growth factors BMP4 and FGF2. The mechanism of action of BMP4 and FGF2 synergy has recently been investigated, which highlighted a role for the FGF2-induced maintenance of *NANOG* (14). We have recently demonstrated that during the initial d0–d2 stage, low level of Y27632 (ROCK inhibitor) (15), high levels of polyvinyl alcohol (PVA), and exposure to physiological oxygen (5%) levels combine to enhance cardiac differentiation. The protocol described here contains fetal bovine serum (FBS) during the d2–d4 stage, which provided the greatest reproducibility of cardiomyocyte generation, whilst the inclusion of L-ascorbic acid-2-phosphate further increased the final number of cardiomyocytes. After the d2–d4 exposure to

FBS, hEB are maintained in a simple CDM. By day 9 of differentiation, 94.7 ± 2.4% of hEB contracted, and each beating hEB routinely contained 60–90% pure cardiomyocytes as assessed by cardiac troponin I expression (TNNI3). hEB produced using this method will continue to contract for over 3 months.

2 Materials

2.1 Cells

1. E13.5 DR4 × CF1 mouse embryonic fibroblasts, passage 2, frozen at 3×10^6 cells per vial.
2. hESC (such as H9 (WA09), H7 (WA07), HES2 (ES02), or HES3 (ES03), supplied by WiCell Research Institute) (16, 17).
3. Viral construct-integrated hiPSC (such as iPS(IMR90)-1 or iPS(IMR90)-4, supplied by WiCell Research Institute) (18).
4. Nonviral, nonintegrated hiPSC (such as Gibco Episomal hiPSC line, derived from neonatal $CD34^+$ cord blood, Invitrogen) (9).

2.2 Reagents

1. Dulbecco's phosphate-buffered saline (D-PBS), pH 7.4, without $CaCl_2$ and $MgCl_2$.
2. TrypLE Express, store at RT, there is no need to warm to 37°C before use.
3. Geltrex Reduced Growth Factor Basement Membrane Matrix, store at −20°C, thaw at 4°C overnight, and subsequently store at 4°C.
4. 0.1% gelatin (1 g/L in distilled H_2O and autoclaved), store at 4°C.
5. DMEM, high glucose.
6. DMEM/F-12 (1×), liquid, 1:1 (with 2.5 mM glutamine).
7. RPMI 1640 (with 2.5 mM glutamine).
8. Fetal bovine serum.
9. KnockOut Serum Replacement.
10. Nonessential amino acids.
11. 2-Mercaptoethanol.
12. Chemically defined lipid concentrate.
13. L-ascorbic acid 2- phosphate (6.4 mg/mL in UltraPure water, store at 4°C for up to 3 weeks) (see Note 1).
14. UltraPure DNase/RNase-Free Distilled Water.
15. Human recombinant insulin, liquid.
16. 1-thioglycerol dilute to 150 mM in UltraPure water (13 μL in 987 μL), make fresh or store at −20°C and use once.
17. Polyvinyl alcohol (see Note 2).

18. Human serum albumin.
19. Recombinant human bone morphogenic protein 4 (BMP4), reconstitute to 100 ng/μL in 4 mM HCl with 0.1% HSA (1 mg/mL). Make aliquots and store at -20°C.
20. Recombinant human fibroblast growth factor 2 (FGF2), reconstitute to 100 ng/μL in D-PBS with 0.1% HSA (1 mg/mL). Make aliquots and store at -20°C.
21. Y27632 (ROCK inhibitor), 10 mM stocks in DMSO, store at -20°C.
22. Mitomycin C.
23. Hemocytometer.

2.3 Equipment

1. 6-well plates.
2. T25 and T175 flasks.
3. 15 and 50 mL conical tubes.
4. 2 mL serological pipettes.
5. 5, 10, 25 and 50 mL serological pipettes.
6. 150 mL PES media filters.
7. 250 and 500 mL and 1 L PES media filters.
8. 96-well V-bottom plates.
9. 96-well U-bottom tissue culture treated plates.
10. Reagent reservoirs.
11. Multichannel Pipette.
12. 8-channel aspirator.
13. P200 pipette tips without filters for aspirator.
14. Automatic cell counter.
15. Tissue culture incubator at 37°C, 5% CO_2, 85% relative humidity Hypoxic incubator or hypoxia chamber capable of 5% O_2, 5% CO_2.
16. Centrifuge.
17. Inverted tissue culture microscope with heated stage.
18. Cesium source irradiator capable of 5,000 cGy.

2.4 Reagents Setup

2.4.1 MEF Medium

DMEM, 10% (v/v) FBS, 1% (v/v) NEAA, 55 μM 2-mercaptoethanol. Filter sterilize and store at 4°C for up to 2 weeks (Table 1).

2.4.2 hESC Medium

DMEM/F 12, 15% (v/v) KSR, 1% (v/v) NEAA, 100 μM 2-mercaptoethanol. Filter sterilize, add 4 ng/mL FGF2, and store at 4°C for up to 2 weeks (Table 2).

Table 1
Mouse embryonic fibroblast medium formulation

MEF medium	100 mL	200 mL	562 mL
DMEM high glucose (with 4 mM L-glutamine)	89 mL	178 mL	500 mL
10% FBS	10 mL	20 mL	56.2 mL
1% NEAA (100×)	1 mL	2 mL	5.62 mL
55 μM 2-mercaptoethanol (55 mM)	100 μL	200 μL	562 μL
	Filter sterilize		

Table 2
hESC medium formulation

hESC medium	200 mL	595.6mL	1,074mL
DMEM/F-12 (with 2.5 mM L-glutamine)	167.6mL	500 mL	900 mL
15% KnockOut SR	30 mL	89.3mL	161 mL
1% NEAA (100×)	2 mL	5.97mL	10.61mL
100 μM 2-mercaptoethanol (55 mM)	364 μL	1.08mL	1.954mL
	Filter sterilize		
4 ng/mL FGF2 (100 ng/mL)	8 μL	23.8 μL	43 μL

2.4.3 Conditioned Medium (CM)

The above hESC medium conditioned (0.5 mL/cm²) on irradiated MEF seeded at 6×10^4 cells/cm² for 22–26 h. Filter sterilize and add an additional 4 ng/mL FGF2 (see below for full protocol). Store at 4°C for up 1 week or make aliquots and store at −20°C for up to 6 months.

2.4.4 RPMI + PVA Medium

RPMI 1640, 4 mg/mL PVA, 1% (v/v) chemically defined lipid concentrate, 10 μg/mL human recombinant insulin, 450 μM 1-thioglycerol. Filter sterilize and make fresh before each use. Add growth factors immediately before use (Table 3).

2.4.5 RPMI + FBS Medium

RPMI 1640, 20% (v/v) FBS, 221 μM L-ascorbic acid 2- phosphate, 450 μM 1-thioglycerol. Filter sterilize and store at 4°C for up to 2 weeks (Table 4).

2.4.6 RPMI + INS Medium

RPMI 1640, 1% (v/v) chemically defined lipid concentrate, 10 μg/mL human recombinant insulin, 221 μM L-ascorbic acid 2- phosphate, 450 μM 1-thioglycerol. Filter sterilize and store at 4°C for up to 2 weeks (Table 5).

Table 3
RPMI + PVA medium formulation

RPMI+PVA medium	50 mL	100 mL	200 mL
RPMI 1640 (with 2 mM L-glutamine)	47.5 mL	95 mL	190 mL
4 mg/mL PVA (100 mg/mL)	2 mL	4 mL	8 mL
1× chemically defined lipid concentrate	500 μL	1 mL	2 mL
450 μM 1-thioglycerol (150 mM)	150 μL	300 μL	600 μL
10 μg/mL recombinant human insulin	50 μL	100 μL	200 μL
Filter sterilize			
25 ng/mL recombinant human BMP4 (100 ng/μL)	12.5 μL	25 μL	50 μL
5 ng/mL recombinant human FGF2 (100 ng/μL)	2.5 μL	5 μL	10 μL
1 μM Y27632 (10 mM)	5 μL	10 μL	20 μL

Table 4
RPMI + FBS medium formulation

RPMI+FBS	100	200 mL	635 mL
RPMI 1640 (with 2 mM L-glutamine)	77 mL	158 mL	500 mL
20% FBS	20 mL	40 mL	127 mL
64 μg/mL L-ascorbic acid 2- phosphate (64 mg/mL)	1 mL	2 mL	6.4 mL
450 μM 1-thioglycerol (150 mM)	300 μL	600 μL	1.9 mL
Filter sterilize			

Table 5
RPMI + INS medium formulation

RPMI+INS medium	100	200 mL	512 mL
RPMI 1640 (with 2 mM L-glutamine)	98 mL	196 mL	500 mL
1× chemically defined lipid concentrate	1 mL	2 mL	5.1 mL
10 μg/mL recombinant human insulin (10 mg/mL)	100 μL	200 μL	510 μL
64 μg/mL L-ascorbic acid 2- phosphate (64 mg/mL)	1 mL	2 mL	5.1 mL
450 μM 1-thioglycerol (150 mM)	300 μL	600 μL	1.5 mL
Filter sterilize			

3 Methods

3.1 Compatible Methods for Pluripotent hESC/hiPSC Growth

All cultures should be maintained at 37°C in a humidified incubator with 5% CO_2 and atmospheric O_2 unless otherwise stated. This method is not optimized for hESC/hiPSC cultured on MEF, as colonies, or in defined media such as mTeSR1 (StemCell Technologies) or Nutristem (Stemgent). If cells have been cultured using these methods, they must be adapted to the monolayer conditions outlined below for at least 2 weeks prior to differentiation.

3.1.1 Production of MEF Conditioned Medium in a Reproducible Manner

Thaw a vial of p2 MEF frozen at 3×10^6 cells per vial in a T175 flask in 35 mL of MEF medium and grow to high confluence (~3–5 days) changing medium on day 1 and then every 2 days. The day before MEF will reach confluence, treat two to three T175 flasks with 35 mL of 0.1% gelatin solution and incubate at 37°C overnight (see Note 3). Aspirate medium from MEF and wash cells with D-PBS to remove any dead cells, add 3 mL of room temperature TrypLE, and incubate at 37°C for 1 min. Tap flask gently to release cells, wash bottom of flask with 20 mL of MEF medium, and transfer cell suspension to a 50 mL conical tube. Irradiate cells with 5,000 cGy (see Note 4). Centrifuge at $200 \times g$ for 5 min at RT. Aspirate supernatant and resuspend cells in 20 mL MEF medium, and count the number of cells using an automated cell counter or hemocytometer. Aspirate gelatin solution from treated T175 flasks and plate at 6×10^4 cells/cm^2 (10.5×10^6 cells in a T175) and top up to 35 mL with MEF medium. Allow MEF to attach for 24 h, rinse with D-PBS, and replace with 0.5 mL/cm^2 (87.5 mL in a T175) of hESC media (see Note 5). Condition hESC media for 22–26 h. Remove conditioned medium, filter sterilize, and add an additional 4 ng/mL FGF2 (see Note 6). Replace media with fresh hESC media every day for 7 days. CM can be stored at 4°C for up to 1 week or at −20°C for at least 6 months (see Note 7).

3.1.2 Making Geltrex-Coated Flasks

Allow a bottle of growth factor-reduced Geltrex (see Note 8) to thaw at 4°C overnight (see Note 9). Add 125 μL of Geltrex to 50 mL of cold (4°C) DMEM/F-12 (a 1:400 dilution) and mix thoroughly, plate 5 mL into each T25 flask (see Note 10), allow to polymerize at 37°C for at least 1 h, and store flasks at 4°C for up to 2 weeks. Before use allow flasks to warm to RT and aspirate media; washing with D-PBS is not required. To transfer hESC/hiPSC grown on MEF to monolayer culture, treat one confluent well of hPSC from a 6-well plate with TrypLE, wash with DMEM/F-12, centrifuge, resuspend in conditioned medium, and transfer cells to Geltrex-coated flask. It will take approximately two to three passages for the MEF to be removed and for the cells to be suitable for forced aggregation differentiation.

Table 6
Mouse embryonic fibroblast medium formulation

Plating density	Time to confluence
1×10^6	4–5 days
1.25×10^6	3 days
1.5×10^6	2–3 days
2×10^6	2 days
2.5×10^6	1 day

3.1.3 Passaging hPSC on Geltrex for Reproducible Cell Growth

On the first day that cells are confluent, aspirate medium, wash cells with D-PBS, add 1 mL of room temperature TrypLE for 1 min at 37°C, and gently tap the side of the flask once to free cells from the surface. Add 10 mL per flask DMEM/F-12 and using a 10 mL pipette jet the medium against the surface, transfer to a 15 mL conical tube and centrifuge at $200 \times g$ for 5 min at RT, aspirate supernatant, and resuspend pellet in 5 mL of conditioned medium. Count cells using an automated cell counter or hemocytometer. Plate 1.25×10^6 cells into a Geltrex-coated flask and top up to 5 mL of conditioned medium (see Note 11). 1.25×10^6 cells/T25 will result in confluence in 3 days and growth to $4–5 \times 10^6$ cells (see Table 6 and Note 12). Change media every day with fresh conditioned medium.

3.2 Forced Aggregation, Directed Cardiac Differentiation Procedure

1. *Day 1*—The day before intended forced aggregation split confluent hESC/hiPSC to $2.5–3.5 \times 10^6$ cells per T25 flask (see Note 13).
2. *Day 0*—Aspirate medium, wash cells with D-PBS, and add 1 mL of room temperature TrypLE into each flask. Incubate at 37°C for 1 min, and tap the side of the flask once to free cells from the surface. Add 10 mL DMEM/F-12 per flask and using a 10 mL pipette jet the medium against the surface and transfer to a 15 mL conical tube.
3. Centrifuge at $200 \times g$ for 5 min at RT, resuspend pellet in 1 mL RPMI + PVA (including growth factors) per flask, pipette up and down to confirm cells are single cell, and count cells using an automated cell counter.
4. Add 5×10^5 cells (5,000 cells per well) per 10 mL of RPMI + PVA, mix well, and transfer to a sterile reagent reservoir.
5. Aliquot 100 μL of the cell suspension into each well of a 96-well V-bottom plate using a 12-channel pipette (see Note 14).

6. Place the plates at 5% O_2, 5% CO_2, and 37°C for 2 days; hEB should be visible after 24 h, usually surrounded by a small amount of dead cells.
7. *Day 2*—On day 2 of differentiation (48 h after plating) aspirate media using a Corning 8-channel aspirator held at 45° placing the pipette tip into the opposite corner of the well (approximately 15 μL of media will remain in the bottom of the V shape). Replace media with 100 μL of RPMI + FBS and place plates in a regular ambient O_2, 5% CO_2 incubator.
8. *Day 4*—On day 4 of differentiation aspirate media as above and replace with 100 μL of RPMI + INS (see Note 15). Set the 12-channel pipette to ~130 μL; in one motion suck up ~60 μL of media, then return it to the well to resuspend the hEB, then take up all the media/hEB, and transfer to a U96 tissue culture-treated plate.
9. *Day 7*—hEB will begin to contract on d7 or d8. All hEB that are going to contract will be beating by d9 and should be counted on this day for comparative analysis. Plates must be kept at 37°C to maintain contraction.
10. From day 7 onwards media should be changed every 2–3 days.

4 Notes

1. L-ascorbic acid-2-phosphate is used rather than L-ascorbic acid as stock solutions maintain 80% activity for up to 4 weeks at 4°C (information from Sigma-Aldrich).
2. We have noted some variability in the solubility of PVA. Some batches may be made into solution by simply adding 2 g of PVA to a cold 500 mL bottle of RPMI 1640 and placing at 4°C for 72 h, shaking to mix periodically. The RPMI/PVA mix may then be kept at 4°C until use. Other batches of PVA will not go into solution at 4°C even after 1 month at 4°C. If this is the case with your batch an alternative method is as follows: Make a 10% w/v solution of PVA in cold, sterile distilled water in a glass bottle by slowly adding the PVA (5 g per 50 mL) to the water, trying to prevent the formation of clumps. Mix thoroughly and heat to 85°C for 30 min with agitation; a hybridization oven is ideal for this. Store at 4°C (Information from Sigma-Aldrich).
3. Coating T175 flasks with gelatin is essential for the MEF to stay attached for 7 days whilst conditioning the hESC media.
4. If you do not have access to an irradiator, confluent flasks can be treated with 10 μg/mL mitomycin C (in 0.2 mL/cm^2 DMEM) for 2.5 h at 37°C, then washed three times with D-PBS, passaged and counted, and then seeded.

5. hESC media to be placed on MEF to make CM must contain FGF2 for MEF to secrete/remove desired factors.
6. It is not necessary to supplement the media with additional 2-mercaptoethanol
7. Freezing CM already supplemented with FGF2 does not affect the capacity for hESC/hiPSC growth or maintenance of pluripotency.
8. We use Geltrex instead of Matrigel (BD Biosciences) as it is supplied in narrower range of concentrations (12–18 mg/mL) reducing potential variation. No difference in performance was noted for either matrix.
9. Geltrex can be subsequently maintained at 4°C for at least 2 months; aliquoting and storage at −20°C is not necessary.
10. We use double the recommended volume of media to dilute the Geltrex to prevent drying out and allow even coverage. Geltrex comes at a concentration of 12–18 mg/mL, using a 1:400 dilution and double volume results in 6–9 μg/cm^2, and 0.5 mg per 6-well plate (8 μg/cm^2) of Matrigel is commonly used.
11. We did not find that the addition of Y27632 (ROCK inhibitor) either 1 h before and/or for 24 h after passaging enhanced cell survival when using conditioned medium and plating cells at these recommended densities.
12. These growth rates are based on consistent growth. Cells do not respond well to changes in seeding density or being overgrown (greater than 8×10^6 cells per T25 flask). Cells will begin to grow faster after approximately 12 passages and seeding density will have to be reduced to prevent over-confluent growth that negatively affects subsequent differentiation. This method of growing hESC as a monolayer is unconventional but has been demonstrated to be successful for the growth of hESC for over 40 passages without karyotypic abnormality. We commonly grow cells for up to 12 passages in this manner to minimize concerns of karyotypic abnormality.
13. Splitting cells the day before forced aggregation assures that all cells are actively growing. Splitting to $2.5–3.5 \times 10^6$ cells per flask assures that cells are not in the lag phase seen when splitting cells to 1.25×10^6.
14. Although both uncoated U- and V-bottom plates can be used for this procedure we have found that it is easier to change the media on V-bottom plates whilst minimizing the chance of aspirating the hEB. Plates must be new as used plates will induced the formation of multiple hEB
15. Both RPMI + INS and RPMI + FBS media work the d4 onwards step. RPMI + FBS may produce more robust results for some cell lines but increases the proportion of fibroblasts growing around the hEB.

Acknowledgements

This study was supported by the Maryland Stem Cell Research Fund (E.T.Z.), and grants from the NHLBI Progenitor Biology Consortium (National Institutes of Health U01HL099775 and U01HL100397 (E.T.Z)). P.W.B. was supported by a postdoctoral fellowship grant from the Maryland Stem Cell Research Fund. We are grateful to Michal Millrod for careful editing of this manuscript.

References

1. Mummery C, Ward-van Oostwaard D, Doevendans P, Spijker R, van den Brink S, Hassink R, van der Heyden M, Opthof T, Pera M, de la Riviere AB, Passier R, Tertoolen L (2003) Differentiation of human embryonic stem cells to cardiomyocytes: role of coculture with visceral endoderm-like cells. Circulation 107:2733–2740
2. Passier R, Oostwaard DW, Snapper J, Kloots J, Hassink RJ, Kuijk E, Roelen B, de la Riviere AB, Mummery C (2005) Increased cardiomyocyte differentiation from human embryonic stem cells in serum-free cultures. Stem Cells 23:772–780
3. Laflamme MA, Chen KY, Naumova AV, Muskheli V, Fugate JA, Dupras SK, Reinecke H, Xu C, Hassanipour M, Police S, O'Sullivan C, Collins L, Chen Y, Minami E, Gill EA, Ueno S, Yuan C, Gold J, Murry CE (2007) Cardiomyocytes derived from human embryonic stem cells in pro-survival factors enhance function of infarcted rat hearts. Nat Biotechnol 25:1015–1024
4. Zhu WZ, Van Biber B, Laflamme MA (2011) Methods for the derivation and use of cardiomyocytes from human pluripotent stem cells. Methods Mol Biol 767:419–431
5. Kehat I, Kenyagin-Karsenti D, Snir M, Segev H, Amit M, Gepstein A, Livne E, Binah O, Itskovitz-Eldor J, Gepstein L (2001) Human embryonic stem cells can differentiate into myocytes with structural and functional properties of cardiomyocytes. J Clin Invest 108:407–414
6. Yang L, Soonpaa MH, Adler ED, Roepke TK, Kattman SJ, Kennedy M, Henckaerts E, Bonham K, Abbott GW, Linden RM, Field LJ, Keller GM (2008) Human cardiovascular progenitor cells develop from a KDR+ embryonic-stem-cell-derived population. Nature 453:524–528
7. Kattman SJ, Witty AD, Gagliardi M, Dubois NC, Niapour M, Hotta A, Ellis J, Keller G (2011) Stage-specific optimization of activin/nodal and BMP signaling promotes cardiac differentiation of mouse and human pluripotent stem cell lines. Cell Stem Cell 8:228–240
8. Burridge PW, Anderson D, Priddle H, Barbadillo Munoz MD, Chamberlain S, Allegrucci C, Young LE, Denning C (2007) Improved human embryonic stem cell embryoid body homogeneity and cardiomyocyte differentiation from a novel V-96 plate aggregation system highlights interline variability. Stem Cells 25:929–938
9. Burridge PW, Thompson S, Millrod MA, Weinberg S, Yuan X, Peters A, Mahairaki V, Koliatsos VE, Tung L, Zambidis ET (2011) A universal system for highly efficient cardiac differentiation of human induced pluripotent stem cells that eliminates interline variability. PLoS One 6:e18293
10. Denning C, Allegrucci C, Priddle H, Barbadillo-Munoz MD, Anderson D, Self T, Smith NM, Parkin CT, Young LE (2006) Common culture conditions for maintenance and cardiomyocyte differentiation of the human embryonic stem cell lines, BG01 and HUES-7. Int J Dev Biol 50:27–37
11. Ng ES, Davis R, Stanley EG, Elefanty AG (2008) A protocol describing the use of a recombinant protein-based, animal product-free medium (APEL) for human embryonic stem cell differentiation as spin embryoid bodies. Nat Protoc 3:768–776
12. Ng ES, Davis RP, Azzola L, Stanley EG, Elefanty AG (2005) Forced aggregation of defined numbers of human embryonic stem cells into embryoid bodies fosters robust, reproducible hematopoietic differentiation. Blood 106:1601–1603
13. Wiles MV, Johansson BM (1999) Embryonic stem cell development in a chemically defined medium. Exp Cell Res 247:241–248
14. Yu P, Pan G, Yu J, Thomson JA (2011) FGF2 sustains NANOG and switches the outcome of

BMP4-induced human embryonic stem cell differentiation. Cell Stem Cell 8:326–334

15. Watanabe K, Ueno M, Kamiya D, Nishiyama A, Matsumura M, Wataya T, Takahashi JB, Nishikawa S, Muguruma K, Sasai Y (2007) A ROCK inhibitor permits survival of dissociated human embryonic stem cells. Nat Biotechnol 25:681–686
16. Thomson JA, Itskovitz-Eldor J, Shapiro SS, Waknitz MA, Swiergiel JJ, Marshall VS, Jones JM (1998) Embryonic stem cell lines derived from human blastocysts. Science 282:1145–1147
17. Reubinoff BE, Pera MF, Fong CY, Trounson A, Bongso A (2000) Embryonic stem cell lines from human blastocysts: somatic differentiation in vitro. Nat Biotechnol 18:399–404
18. Yu J, Hu K, Smuga-Otto K, Tian S, Stewart R, Slukvin II, Thomson JA (2009) Human induced pluripotent stem cells free of vector and transgene sequences. Science 324:797–801

Chapter 13

Generation of Transgene-Free iPSC Lines from Human Normal and Neoplastic Blood Cells Using Episomal Vectors

Kejin Hu and Igor Slukvin

Abstract

Human induced pluripotent stem cells (iPSCs) have become an important tool for modeling human diseases and are considered a potential source of therapeutic cells. Original methods for iPSC generation use fibroblasts as a cell source for reprogramming and retroviral vectors as a delivery method of the reprogramming factors. However, fibroblasts require extended time for expansion and viral delivery of transgenes results in the integration of vector sequences into the genome which is a source of potential insertion mutagenesis, residual expressions, and reactivation of transgenes during differentiation. Here, we provide a detailed protocol for the efficient generation of transgene-free iPSC lines from human bone marrow and cord blood cells with a single transfection of non-integrating episomal plasmids. This method uses mononuclear bone marrow and cord blood cells, and makes it possible to generate transgene-free iPSCs 1–3 weeks faster than previous methods of reprogramming with fibroblasts. Additionally, we show that this approach can be used for efficient reprogramming of chronic myeloid leukemia cells.

Key words Epstein–Barr virus, Episomal plasmids, Reprogramming, Induced pluripotent stem cells, Human bone marrow, Cord blood, Chronic myeloid leukemia

1 Introduction

The iPSC technology offers a novel opportunity for basic research, disease modeling, drug screening, and cell therapies. Conventional reprogramming methods rely on gamma retroviral or lentiviral vectors for the delivery of the reprogramming factors (1–3). Virus-based techniques cause insertion mutagenesis, residual expression, and reactivation of transgenes and are of low efficiency. Many approaches have been investigated to avoid or eliminate integration of transgenes in the reprogrammed cells including non-integrating adenoviral vectors (4), Sendai RNA viral vectors (5, 6), repeated transient transfection (7), protein transduction (8, 9), RNA transfection (10), Cre-LoxP excision system (11), PiggyBac transposon system (12, 13), and small molecules (14–20). However, these alternative approaches still have their limitations such as low efficiency, multiple rounds of transfection, additional expertise for

Uma Lakshmipathy and Mohan C. Vemuri (eds.), *Pluripotent Stem Cells: Methods and Protocols*, Methods in Molecular Biology, vol. 997, DOI 10.1007/978-1-62703-348-0_13,

RNA and protein preparations, instability of RNA and protein samples, extra steps for excision of pre-integrated sequences and subsequent screening, and incomplete removal of the exogenous sequences.

EBV-based plasmids exist in mammalian cells as an extrachromosomal entity. EBV-plasmids require only two viral elements for maintenance in the cells: (1) a short cis sequence which is the latent origin of plasmid replication (*oriP*), and (2) a transelement of EBNA1 (Epstein-Barr Nuclear Antigen 1) (21). Because up to 5% of the cells lose EBV plasmids during each cell division (22, 23), transgene-free cells can be obtained simply by passaging and subcloning (24, 25). For these reasons, many laboratories, including our own, favor the EBV-based episomal vector system for the generation of iPSCs free of foreign sequences (24–31). For the establishment of patient-specific iPSC line, starting cells are critical due to the varied accessibility and reprogrammability of cells. Although fibroblasts have been traditionally used as a source of cells for reprogramming, the efficiency of fibroblast reprogramming, especially adult fibroblasts, is relatively low. Moreover, fibroblasts require 4–6 weeks for isolation and expansion (32, 33). In contrast, mature blood cells and their progenitors are the most accessible sources of cells in our body.

We demonstrate that transgene-free human iPSCs can be obtained from human bone marrow, human cord blood, or purified human $CD34^+$ cells using the non-integrating episomal system (25). This protocol can reprogram blood cells previously frozen in liquid nitrogen for over 6 years. While many non-integrating protocols require multiple rounds of transfections, our protocol uses only one transfection and does not require purification of particular blood subset although the purification of $CD34^+$ progenitors can help reprogram cells with higher efficiency. Unlike other protocols for blood reprogramming that predominantly reprogram T cell populations (5, 34–36), the population reprogrammed with our method is neither T nor B cells and thus the reprogrammed genome is free of recombined genomic DNA that results from gene rearrangements following maturation of T or B cells. Additionally, we demonstrate that the episomal vector-based approach can be used to generate iPSCs from neoplastic bone marrow cells from patients with chronic myeloid leukemia (CML) to model leukemia development in vitro (25). The iPSC-based model provides numerous advantages for the study of neoplastic blood diseases. It can be used to examine leukemia stem cell potentials at various stages of differentiation for which it may be difficult to obtain samples from patients, for example, at the hemangioblast stage. It also provides a unique opportunity to explore the role of epigenetic changes in the activation of the oncogene-induced aberrant regulatory circuits and to identify cell subsets with distinct tumor-initiating potential and drug sensitivity.

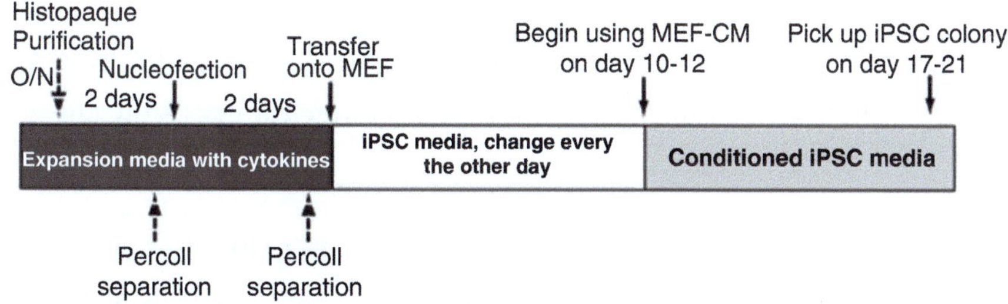

Fig. 1 Summary of protocols for reprogramming blood cells using episomal plasmids. *Arrows* indicate time points at which the indicated steps are carried out. *Broken arrows* designate the optional steps. O/N means overnight recovery by culturing of frozen whole bone marrow or whole cord blood overnight in hematopoietic expansion medium and fractioning for mononuclear cells with Histopaque gradient centrifugation. CM is conditioned media

In this chapter, we describe a detailed protocol for the reprogramming of archived normal and neoplastic blood cells with episomal constructs. Figure 1 outlines the major steps of the reprogramming protocol. The procedure starts with a short expansion of the mononuclear cells in the serum-free hematopoietic expansion media followed by nucleofection, an additional expansion for 2 days in the hematopoietic media and a transfer of cells onto mouse embryonic fibroblast (MEF) feeders. Reprogramming proceeds on MEF feeders in the standard iPSC medium for 10 days. Additional culture in MEF-conditioned media for another 10 days is required to grow typical iPSC colonies which can be handpicked for expansion and the final establishment of iPSC lines. Although we initially developed this method for reprogramming mononuclear cells from bone marrow and cord blood (25), we found that the same protocol works well for reprogramming of $CD34^+$ cells isolated from these sources.

2 Materials

All reagents should be cell-culture grade. Aseptic practice should be observed for all steps.

2.1 Reagents (See Note 1)

1. Recombinant human IL-3 (PeproTech).
2. Recombinant human IL-6 (PeproTech).
3. Recombinant human SCF (PeproTech).
4. Recombinant human Flt3L (PeproTech).
5. Recombinant human FGF-basic (PeproTech).
6. Zebra fish FGF-basic (gift from James Thomson, made in house).
7. StemSpan SFEM (Serum-Free Medium for Expansion of Hematopoietic cells) (Stemcell Technologies).
8. Histopaque-1077® (Sigma-Aldrich).
9. Percoll® (Sigma-Aldrich).

10. EX-CYTE® growth enhancement media supplement (Celliance).
11. Human cord blood mononuclear cells (AllCells, CA, USA).
12. Human bone marrow mononuclear cells (AllCells, CA, USA).
13. Human bone marrow cells from patient with CML (AllCells, CA, USA).
14. Human cord blood $CD34^+$ cells (AllCells, CA, USA).
15. Human bone marrow $CD34^+$ cells (AllCells, CA, USA).
16. DNase I (Promega).
17. L-Glutamine (Gibco).
18. γ-Irradiated MEF (WiCell).
19. Knockout serum replacement for ESC/iPSC (Gibco).
20. β-mercaptoethanol (Sigma).
21. D-MEM/F-12 (Gibco).
22. MEM nonessential amino acids solution (100×, 10 mM, HyClone®).
23. Penicillin-Streptomycin solution (100×, Cellgro®).
24. Hyclone® Fetal Bovine Serum (defined) (Thermo Scientific).
25. Collagenase type IV (Gibco).
26. Dimethyl sulfoxide (DMSO) (Sigma).
27. Gelatin (Sigma).
28. PBS (without calcium, without magnesium) (Hyclone®).
29. Sodium bicarbonate (Fisher Scientific).
30. Thiazovivin (Stemgent®).

2.2 Transfection Kit (See Note 2)

1. Amaxa® Human $CD34^+$ Cell Nucleofector® Kit (Lonza).

2.3 Plasmids

1. pEP4-EO2S-ET2K (Addgene).
2. pEP4-EO2S-EN2K (Addgene).
3. pCEP4-M2L (Addgene).

2.4 Key Equipment (See Note 2)

1. Nucleofector® II (Amaxa Biosystem).

2.5 iPSC Medium (See Note 3)

The iPSC medium is, in essence, ESC growth medium with a higher concentration of FGF2. It consists of 20% Knockout serum replacement (KOSR), 80% D-MEM/F-12, 1 mM L-glutamine, 0.1 mM β-mercaptoethanol, 1× nonessential amino acid (NEAA) (0.1 mM). We use 100 ng/ml of recombinant zebra fish FGF-basic, or 10 ng/ml of recombinant human FGF2. It can be stored for up to 2 weeks at 4°C.

2.6 Hematopoietic Expansion Medium

The hematopoietic expansion medium consists of StemSpan SFEM (serum-free expansion media) supplemented with 0.2% Ex-Cyte (Celliance), recombinant human IL-3 (10 ng/ml), recombinant human IL-6 (100 ng/ml), recombinant human SCF (100 ng/ml), and FMS-related tyrosine kinase-3 ligand (Flt-3 L, 100 ng/ml), 100 IU of penicillin, and 100 μg/ml of streptomycin. Upon arrival, StemSpan SFEM is aliquoted into 50-ml tubes, and stored at −20°C or −80°C. Cytokine, growth factors, lipid supplement, and antibiotics are added immediately before use.

2.7 MEF-Conditioned Medium (MEF-CM) (See Note 4)

For the preparation of MEF-conditioned medium, we use the modified method of Xu et al. (37). Add 15 ml of iPSC medium composed of 10% KOSR, 90% D-MEM/F-12, 0.5 mM L-glutamine, 0.5× NEAA, without FGF2 to a 10-cm tissue dish preseeded with MEF at a density of 2×10^4/cm^2. After 24 h of conditioning, medium is collected. The same dish of MEF can be used for up to 10 times. The collected conditioned media can be stored at −20°C for over 1 month. To make 500 ml of a working MEF-conditioned medium for reprogramming, combine the following: 450 ml of conditioned medium, 50 ml of fresh KOSR, 2.5 ml of 100 mM L-glutamine, 2.5 ml of 100× NEAA, 2 μl of β-mercaptoethanol, and appropriate amount of FGF to make a final concentration of 100 ng/ml for zebra fish FGF, or 10 ng/ml for recombinant human FGF2. The MEF-CM is sterilized by filtering through a 0.22-μm filter.

2.8 2× iPSC Freezing Medium

2× iPSC freezing medium is composed of 60% Hyclone® FBS, 20% DMSO, and 20% basal iPSC media. Prepare freshly when needed and keep chilled on ice.

2.9 0.1% Gelatin

Weigh 1 g of gelatin and put it into a 1-L Pyrex bottle; add 1 L of pure water. Autoclave to sterilize the solution. The gelatin can be stored at 4°C for over a year.

2.10 Preparation of Collagenase IV Solution

The working concentration of collagenase IV is 1 mg/ml in plain D-MEM/F-12 medium. Weigh 45 mg of collagenase IV; put it into a 50-ml tube; add 45 ml of D-MEM/F-12 medium into the 50-ml tube containing the collagenase IV; close the tube with a cap; mix well by shaking; remove the cap; sterilize the collagenase IV solution by filtration with a Steriflip® 50-ml filter (0.22 μm, Millipore). The solution can be stored for up to 14 days at 4°C.

2.11 MEF Preparation for Culture of Human Reprogrammed Cells (See Note 5)

Coat the 6 well plate or 10-cm dish with 0.1% gelatin (2 ml per well, and 10 ml per dish) overnight at 37°C. Next day, remove the plate/dish from the incubator and aspirate the gelatin solution from the plate/dish. Seed 2.5 ml of irradiated MEF at 0.75×10^5/ml into each well of a 6-well plate (around 2×10^4/cm^2), or 15 ml of the irradiated MEF (0.75×10^5 cells/ml) into one 10-cm tissue culture dish. Culture the MEF overnight at 37°C, 5% CO_2 before use.

2.12 Preparation of Percoll® Solution

5× Percoll solution: Mix 45 ml of Percoll® (Sigma, p1644) (sterile) with 5 ml of 10× PBS (sterilized by filtration), which results in 90% Percoll® in 1×PBS. Store the 5× Percoll stock solution at 4°C.

1× Percoll solution: Mix 10 ml of 5× Percoll® solution with 40 ml of sterile 1×PBS solution. The final Percoll® solution is 18% in 1× PBS. Smaller volume can be prepared similarly. The 1× Percoll solution can be prepared before use.

3 Methods

3.1 Preparation of Cells for Nucleofection (See Note 6)

1. Take one vial of blood mononuclear cells from liquid nitrogen tank, and thaw the cells quickly in a 37°C water bath; transfer the cells into the 15-ml tube containing 10 ml of plain SFEM medium (see Note 7).

Centrifuge the cells at 400 × *g* for 8–10 min.

2. Aspirate the supernatant, add 10 ml of fresh plain SFEM to the cell pellet, and repeat the washing.
3. Resuspend cells in 1 ml of hematopoietic expansion medium and count the cells with trypan blue. Add appropriate volume of hematopoietic expansion medium to the cell suspension to make the final concentration 1–2 millions cells per ml (see Note 8).
4. Put the cells in one well of a 6-well plate if the cell suspension volume is 4 ml or less; culture the cells in a T-25 tissue flask if cell suspension volume is 5–10 ml.
5. Culture at 37°C and 5% CO_2 for 2 days.

3.2 Nucleofection of Blood Cells and Reprogramming

1. On day 2 (48 h of culture in hematopoietic expansion media), transfer the cell suspension into a 15 ml tube (see Note 9).
2. Underlay cell suspension with 1.0–1.5 ml of Percoll solution.
3. Centrifuge at 300 × *g* for 20 min at room temperature.
4. During the centrifugation, take out the nucleofection kit from 4°C, and DNA from −80°C. Put 82 μl of nucleofection buffer into a sterile 1.5-ml tube and add 18 μl of the supplement (both are supplied as a components of Amaxa® Human $CD34^+$ Cell Nucleofector® Kit) into the buffer; mix well. During this period of time, the buffer can be brought to room temperature (see Notes 10 and 11).
5. After centrifugation, aspirate the supernatant and interface containing dead cells and debris without disturbing cell pellet; resuspend the cell pellet with 10 ml of SFEM, and centrifuge at 400 × *g* for 8 min (see Note 12).
6. During the washing, turn on the Nucleofector II. Set up the program to U-008 (U-08 for Nucleofector® I). Prepare one Amaxa cuvette and one Amaxa transfer pipette in the biosafety hood.

7. Add the following DNA plasmids into the buffer mixture prepared in step 4: 9 μg of pEP4-EO2S-ET2K, 9 μg of pEP4-EO2S-EN2K, and 6 μg of pCEP4-M2L. Mix well by gentle pipetting, being careful to avoid introducing bubbles. (See Note 13).
8. Carefully aspirate the entire media from cells in step 5 without disturbing cells (see Note 14).
9. Add buffer containing DNA (step 7) to the 15-ml tube containing the cells to be transfected (step 8); mix well by gentle pipetting, being careful to avoid introducing bubbles. Carefully transfer the cells and DNA in buffer into the transfection cuvette supplied with Amaxa® Nucleofector® Kit. Do not introduce any bubbles. Gently but quickly tap the cuvette five to ten times immediately to remove any bubbles.
10. Nucleofect the blood cells using program U-008 (see Note 15)
11. Add 500 μl of hematopoietic expansion medium into the cuvette; mix by one gentle pipetting motion, and transfer cells into dish or wells containing hematopoietic expansion medium (see Note 16).
12. Culture for 2 days in hematopoietic expansion medium at 37°C, 5% CO_2.
13. On day 2 after transfection, remove dead cells using Percoll® separation as described in steps 1–3 and 5. Transfer the cells onto a 10-cm dish preseeded with MEFs.
14. Change medium every other day for the first 10 days (see Note 17).
15. On day 10, start to use MEF-conditioned medium and change media every day.
16. iPSC colonies should appear between days 17 and 21. Under a microscope in the biosafety hood, scrape off and pick up the entire single iPSC colony (see Fig. 2) with a P200 tip. Transfer colony into a sterile 1.5-ml tube and separate cells by pipetting five times. Put one individual iPSC colony into one well of a 6-well plate. The first culture from this picked colony is considered a passage 1 iPSCs (P_1) (see Note 19).
17. Grow the P1 cells in iPSC medium at 37°C, 5% CO_2. Change the medium every day.
18. When new colonies reach the normal size, repeat step 16 if any incompletely reprogrammed colonies exist, and grow the P_2 iPSCs at 37°C, 5% CO_2 (see Note 20).
19. Between days 5 and 10 after culture, depending on the density and colony size, passage the new iPSC line using standard protocol for passaging human ESC/iPSCs on MEFs (now P_3).
20. Between days 5 and 7, freeze the newly established lines (see freezing protocol below). Passage the remaining wells

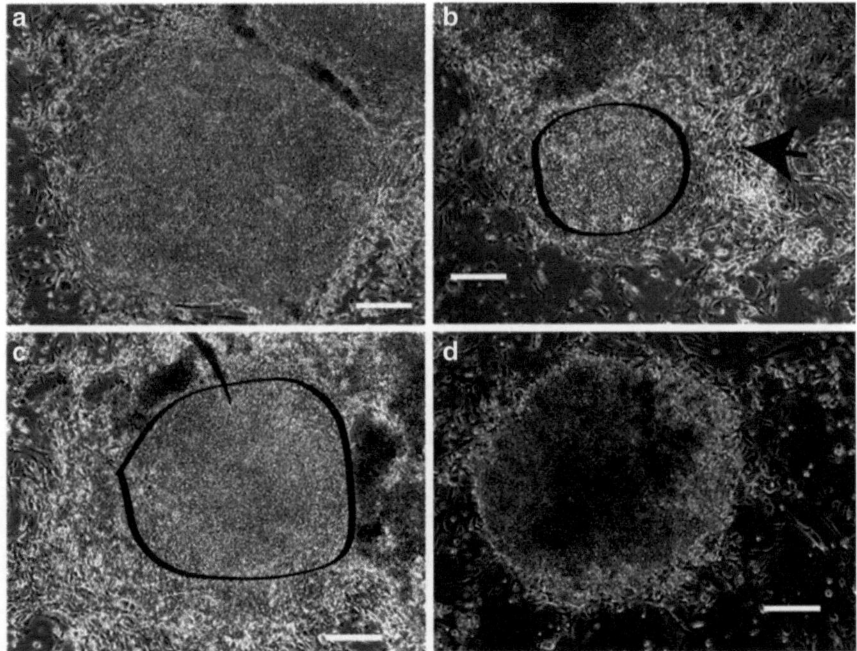

Fig. 2 Morphology of reprogrammed cell colonies. Panel (**a**) shows a high-quality colony. The colony is bright and composed of small, tightly packed cells. Only a few incompletely reprogrammed spindle cells are present at the periphery of the colony. Panels (**b**) and (**c**) are other examples of well-reprogrammed colonies. Completely reprogrammed cells within colonies are circled. However, more incompletely reprogrammed spindle-like cells can be seen at the periphery of colony (*arrow*). Panel (**d**) shows a colony at a less advanced stage of reprogramming. However, this colony can be picked up and used to establish iPSC lines after transferring to new MEFs to facilitate complete transition to pluripotency. Bars, 300 μm. Images of transfected human bone marrow mononuclear cells were taken on day 17 of culture on MEFs (day 19 after transfection)

for characterization and further freezing at higher passages. Established iPSC lines should be analyzed for expression of pluripotency markers by flow cytometry, immunofluorescence, RT-PCR, or gene profiling. Pluripotent potential of these cells should be evaluated using functional tests including teratoma generation, and in vitro differentiation into different lineages. The established lines should also be karyotyped to ensure genomic integrity. Characterization of iPSCs obtained from patient with chronic myeloid leukemia is shown in Fig. 3.

3.3 Freezing iPSCs

1. Warm up collagenase IV in a 37°C water bath for 10 min.
2. Take out the iPSC plate from the incubator.
3. Use 1.5 ml of collagenase IV solution for one well of iPSCs of a 6-well plate.
4. Incubate at 37°C for 5 min.
5. Aspirate the collagenase IV solution.

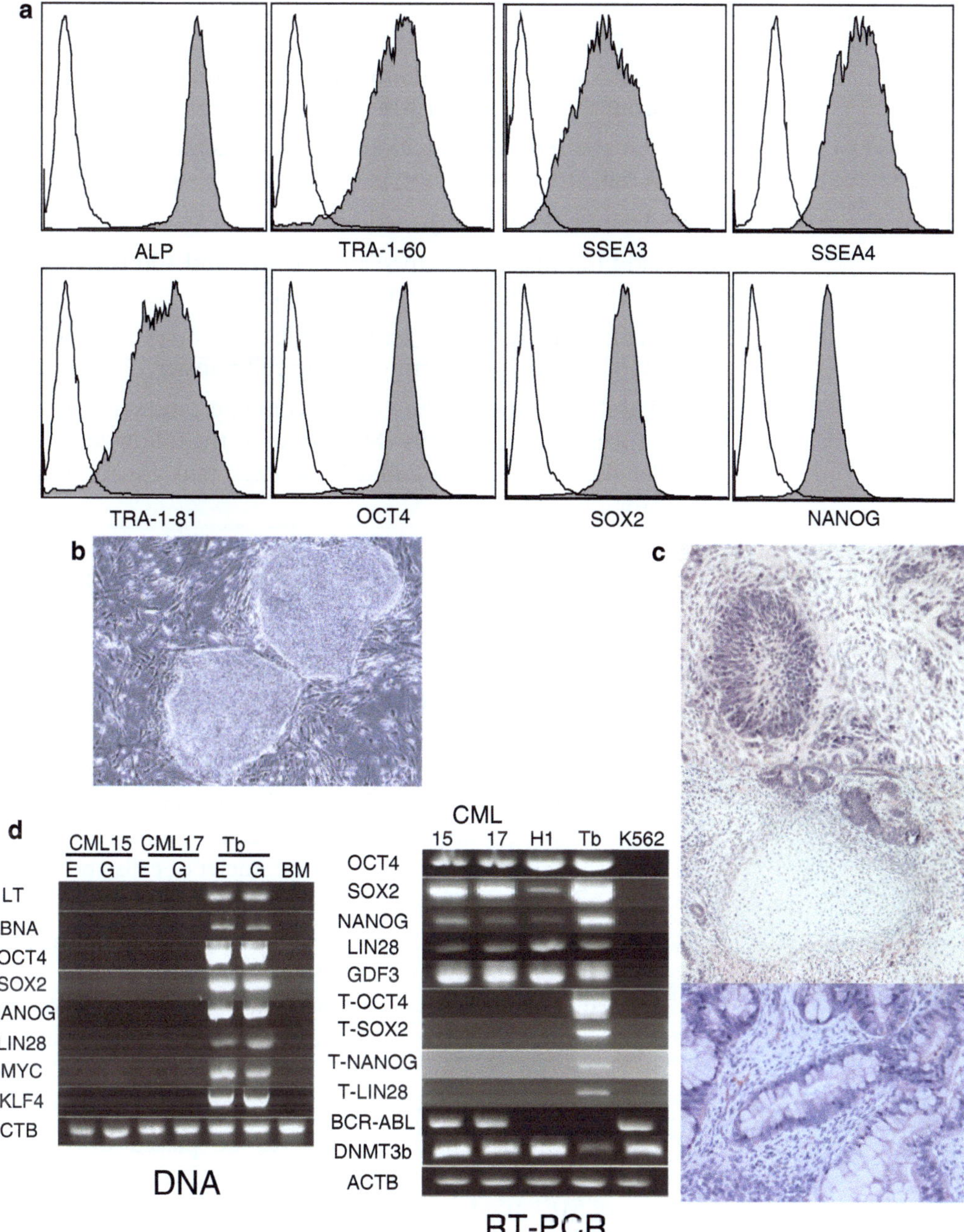

Fig. 3 Demonstration of pluripotency of iPSC lines generated from bone marrow of a patient with chronic myeloid leukemia (CML) in chronic phase. (**a**) Flow cytometric analysis of hESC-specific marker expression in CMLiPSC15 line (surface staining is shown for ALP, TRA-1-60, SSEA3, SSEA4, and TRA-1-81; intracellular staining is shown for OCT4, SOX2, and NANOG). (**b**) Typical morphology of CMLiPSC15 colony growing on MEFs. (**c**) H&E staining of teratoma from CMLiPSC15 line. Neural rosettes (*upper panel*), cartilage (*middle panel*), and intestine-like structure (*lower panel*) are shown. (**d**) PCR analysis of genomic (G) and episomal (E) DNA demonstrates that CMLiPSCs are transgene-free. Vector (EBNA)- and transgene (T)-specific primers were used as indicated. RT-PCR analysis of expression of transgenes (T) and endogenous pluripotency genes and BCR-ABL in CMLiPSC lines 15 and 17. H1 is human embryonic stem cell line H1. BM is bone marrow mononuclear cells. K562 is myeloid leukemia line established from a patient with chronic myeloid leukemia in blast crisis. Tb is positive controls, BM cells transfected with the same reprogramming plasmids

6. Add 3 ml of iPSC medium and gently wash off the iPSC colonies from the MEF feeder with a 5-ml pipette.
7. Put the iPSCs into a 15-ml tube and centrifuge at $200 \times g$ for 5 min.
8. Aspirate the washing medium.
9. Add 0.5 ml of iPSC media (ice cold), and add 0.5 ml of iPSC freezing media (ice cold) to cell pellet.
10. Resuspend cells and transfer cells into a cryotube (1.8 ml). Avoid excessive pipetting during cell transfer.
11. Put the cryotube in the freezing container, and start the freezing process immediately in −80°C freezer; the following day, transfer the iPSCs into a liquid N_2 storage tank for the long-term storage.

4 Notes

1. All stock solutions of cytokines are prepared at 1,000×, aliquoted and kept at −80°C.
2. Blood progenitors are difficult to transfect using the traditional nonviral transfection methods due to their quiescent nature and suspension growth (38). Nucleofection is critical for successful reprogramming of blood progenitor cells. Nucleofection directly delivers the DNA into the nuclei of cells and results in early and strong expression of transgenes.
3. Recombinant zebra fish FGF-basic can be used instead of human FGF2. However it is less stable at 37°C and must be used at concentration of 100 ng/ml. For commercial recombinant human FGF2, we recommend a concentration of 8–12 ng/ml.
4. Standard protocol for preparation of MEF-conditioned media requires high density of MEF to be plated in T-75 flask ($5.4 \times 10^4/\text{cm}^2$). We found that the routine $2 \times 10^4/\text{cm}^2$ density of MEF used for ESCs culturing works well for preparation of MEF-conditioned media and simplifies the procedure.
5. The complete MEF preparation protocol can be found in WiCell protocols: SOP-CC-003B; SOP-CC-006D; SOP-CC-031D; SOP-CC-021A; SOP-CC-009A; SOP-CC-013A (https://www.wicell.org).
6. We grow bone marrow and cord blood mononuclear cells for 2 days before transfection to amplify hematopoietic progenitors. We also recommend a 2-day expansion before transfection for $CD34^+$ cells isolated using magnetic beads. The electroporator creates an electromagnetic field in the cell solution which could affect viability of the purified $CD34^+$ cells

when magnetic beads attached to them. However, after 2 days of culture, the beads detach and degrade so that the cells can be safely electroporated.

7. If unprocessed bone marrow or cord blood is used, a standard Histopaque gradient centrifugation step should be carried out to obtain the mononuclear cell fraction. The current protocol is based on mononuclear cells.
8. For bone marrow mononuclear cells and $CD34^+$ cells, $1–2 \times 10^6$ is typically sufficient for reprogramming. However, for cord blood mononuclear cells, $5–10 \times 10^6$ cells are required for reprogramming.
9. Mature blood cells die in hematopoietic expansion medium and result in cell clumps. If cell clumping occurs, treat the cells with DNase I at a concentration of 200 units/ml for 30 min before Percoll® purification. DNase I treatment removes the DNA released by the dead cells and alleviate the clumping.
10. The shelf life of buffer after mixture with supplement is 3 months. We usually mix these two parts of the reaction just before transfection to ensure high transfection efficiency.
11. The optimal volume of the reaction buffer for the Nucleofector device is 100 μl. Significant changes to the reaction volume will result in transfection failure.
12. If purified $CD34^+$ cells are used for reprogramming, Percoll® separation is not required due to the low number of dead cells.
13. Concentration of plasmid DNA in stock solution should be >1.5 μg/μl to avoid significant dilution of nucleofection buffer following DNA addition. Excessive dilution of nucleofection buffer decreases transfection efficiency.
14. Washing medium must be completely removed before adding nucleofection buffer. This should be done very carefully to avoid aspiration and loss of cells. Tilt the tube so that the bottom is elevated to let liquid to flow down toward the mouth of the tube. The liquid can be collected by keeping the tip of the running vacuum pipette immediately downstream of the flow. This method allows for effective liquid removal and avoids disruption of the cell pellet.
15. Do not keep cells in the transfection buffer longer than 20 min. Extended contact time with buffer reduces the viability of cells and the gene transfer efficiency. After adding buffer, proceed with electroporation as quick as possible.
16. After nucleofection procedure, cells become friable. It is critical to use a pipette with larger bore size supplied with the nucleofection kit for gentle handling of cells.
17. Blood cells grow in suspension and transfected cells gradually become adherent. However, during the first few days following

reprogramming, many cells still grow in suspension. To avoid loss of cells following medium change, aspirated medium can be centrifuged to collect floating cells and reseeded back onto the same MEF dish.

18. MEF cultures should be prepared fresh for plating transfected cells. Because reprogramming requires extended cultures, the use of 3-day or older MEF plates will compromise the experiment due to the deterioration of the MEFs.
19. Sometimes no typical colonies appear following the first round of transfected cells on MEFs (P_0). Passaging P_0 cells onto new MEFs may produce high-quality iPSCs colonies. The culture ratio may vary from 1:3 to 1:6 depending on the density of the colonies in the P_0 dishes/plates. The reprogramming efficiency can be significantly increased by adding thiazovivin (1 μM) to reprogramming cultures. Thiazovivin can be added as early as the transfected cells are transferred onto MEF feeders or after the first passage.
20. The second picking will eliminate the contaminated incompletely reprogrammed cells from the first culture. Multiple colonies can be cultured in the second round of culture.

Acknowledgements

We thank Professor James Thomson at Morgridge Institute for Research, Madison, WI, for providing reprogramming plasmids and FGF2, and Patricia Liu for editorial assistance. This work was supported by funds from the National Institute of Health (P01 GM081629 and P51RR000167).

References

1. Yu J, Vodyanik MA, Smuga-Otto K, Antosiewicz-Bourget J, Frane JL, Tian S, Nie J, Jonsdottir GA, Ruotti V, Stewart R, Slukvin II, Thomson JA (2007) Induced pluripotent stem cell lines derived from human somatic cells. Science 318:1917–1920
2. Takahashi K, Tanabe K, Ohnuki M, Narita M, Ichisaka T, Tomoda K, Yamanaka S (2007) Induction of pluripotent stem cells from adult human fibroblasts by defined factors. Cell 131:861–872
3. Takahashi K, Yamanaka S (2006) Induction of pluripotent stem cells from mouse embryonic and adult fibroblast cultures by defined factors. Cell 126:663–676
4. Stadtfeld M, Nagaya M, Utikal J, Weir G, Hochedlinger K (2008) Induced pluripotent stem cells generated without viral integration. Science 322:945–949
5. Seki T, Yuasa S, Oda M, Egashira T, Yae K, Kusumoto D, Nakata H, Tohyama S, Hashimoto H, Kodaira M, Okada Y, Seimiya H, Fusaki N, Hasegawa M, Fukuda K (2010) Generation of induced pluripotent stem cells from human terminally differentiated circulating T cells. Cell Stem Cell 7:11–14
6. Ban H, Nishishita N, Fusaki N, Tabata T, Saeki K, Shikamura M, Takada N, Inoue M, Hasegawa M, Kawamata S, Nishikawa S (2011) Efficient generation of transgene-free human induced pluripotent stem cells (iPSCs) by temperature-sensitive Sendai virus vectors. Proc Natl Acad Sci USA 108: 14234–14239
7. Okita K, Nakagawa M, Hyenjong H, Ichisaka T, Yamanaka S (2008) Generation of mouse induced pluripotent stem cells without viral vectors. Science 322:949–953

8. Zhou H, Wu S, Joo JY, Zhu S, Han DW, Lin T, Trauger S, Bien G, Yao S, Zhu Y, Siuzdak G, Scholer HR, Duan L, Ding S (2009) Generation of induced pluripotent stem cells using recombinant proteins. Cell Stem Cell 4:381–384
9. Kim D, Kim CH, Moon JI, Chung YG, Chang MY, Han BS, Ko S, Yang E, Cha KY, Lanza R, Kim KS (2009) Generation of human induced pluripotent stem cells by direct delivery of reprogramming proteins. Cell Stem Cell 4:472–476
10. Warren L, Manos PD, Ahfeldt T, Loh YH, Li H, Lau F, Ebina W, Mandal PK, Smith ZD, Meissner A, Daley GQ, Brack AS, Collins JJ, Cowan C, Schlaeger TM, Rossi DJ (2010) Highly efficient reprogramming to pluripotency and directed differentiation of human cells with synthetic modified mRNA. Cell Stem Cell 7:618–630
11. Kaji K, Norrby K, Paca A, Mileikovsky M, Mohseni P, Woltjen K (2009) Virus-free induction of pluripotency and subsequent excision of reprogramming factors. Nature 458:771–775
12. Woltjen K, Michael IP, Mohseni P, Desai R, Mileikovsky M, Hamalainen R, Cowling R, Wang W, Liu P, Gertsenstein M, Kaji K, Sung HK, Nagy A (2009) piggyBac transposition reprograms fibroblasts to induced pluripotent stem cells. Nature 458:766–770
13. Yusa K, Rad R, Takeda J, Bradley A (2009) Generation of transgene-free induced pluripotent mouse stem cells by the piggyBac transposon. Nat Methods 6:363–369
14. Liang G, Taranova O, Xia K, Zhang Y (2010) Butyrate promotes induced pluripotent stem cell generation. J Biol Chem 285:25516–25521
15. Mali P, Chou BK, Yen J, Ye Z, Zou J, Dowey S, Brodsky RA, Ohm JE, Yu W, Baylin SB, Yusa K, Bradley A, Meyers DJ, Mukherjee C, Cole PA, Cheng L (2010) Butyrate greatly enhances derivation of human induced pluripotent stem cells by promoting epigenetic remodeling and the expression of pluripotency-associated genes. Stem Cells 28:713–720
16. Ichida JK, Blanchard J, Lam K, Son EY, Chung JE, Egli D, Loh KM, Carter AC, Di Giorgio FP, Koszka K, Huangfu D, Akutsu H, Liu DR, Rubin LL, Eggan K (2009) A small-molecule inhibitor of tgf-Beta signaling replaces sox2 in reprogramming by inducing nanog. Cell Stem Cell 5:491–503
17. Lin T, Ambasudhan R, Yuan X, Li W, Hilcove S, Abujarour R, Lin X, Hahm HS, Hao E, Hayek A, Ding S (2009) A chemical platform for improved induction of human iPSCs. Nat Methods 6:805–808
18. Shi Y, Do JT, Desponts C, Hahm HS, Scholer HR, Ding S (2008) A combined chemical and genetic approach for the generation of induced pluripotent stem cells. Cell Stem Cell 2:525–528
19. Huangfu D, Maehr R, Guo W, Eijkelenboom A, Snitow M, Chen AE, Melton DA (2008) Induction of pluripotent stem cells by defined factors is greatly improved by small-molecule compounds. Nat Biotechnol 26:795–797
20. Lyssiotis CA, Foreman RK, Staerk J, Garcia M, Mathur D, Markoulaki S, Hanna J, Lairson LL, Charette BD, Bouchez LC, Bollong M, Kunick C, Brinker A, Cho CY, Schultz PG, Jaenisch R (2009) Reprogramming of murine fibroblasts to induced pluripotent stem cells with chemical complementation of Klf4. Proc Natl Acad Sci USA 106:8912–8917
21. Lindner SE, Sugden B (2007) The plasmid replicon of Epstein-Barr virus: mechanistic insights into efficient, licensed, extrachromosomal replication in human cells. Plasmid 58:1–12
22. Kirchmaier AL, Sugden B (1995) Plasmid maintenance of derivatives of oriP of Epstein-Barr virus. J Virol 69:1280–1283
23. Sugden B, Warren N (1988) Plasmid origin of replication of Epstein-Barr virus, oriP, does not limit replication in cis. Mol Biol Med 5:85–94
24. Yu J, Hu K, Smuga-Otto K, Tian S, Stewart R, Slukvin II, Thomson JA (2009) Human induced pluripotent stem cells free of vector and transgene sequences. Science 324:797–801
25. Hu K, Yu J, Suknuntha K, Tian S, Montgomery K, Choi KD, Stewart R, Thomson JA, Slukvin II (2011) Efficient generation of transgene-free induced pluripotent stem cells from normal and neoplastic bone marrow and cord blood mononuclear cells. Blood 117:e109–e119
26. Chou BK, Mali P, Huang X, Ye Z, Dowey SN, Resar LM, Zou C, Zhang YA, Tong J, Cheng L (2011) Efficient human iPS cell derivation by a non-integrating plasmid from blood cells with unique epigenetic and gene expression signatures. Cell Res 21:518–529
27. Chen G, Gulbranson DR, Hou Z, Bolin JM, Ruotti V, Probasco MD, Smuga-Otto K, Howden SE, Diol NR, Propson NE, Wagner R, Lee GO, Antosiewicz-Bourget J, Teng JM, Thomson JA (2011) Chemically defined conditions for human iPSC derivation and culture. Nat Methods 8:424–429
28. Marchetto MC, Yeo GW, Kainohana O, Marsala M, Gage FH, Muotri AR (2009) Transcriptional signature and memory retention of human-induced pluripotent stem cells. PLoS One 4:e7076

29. Okita K, Matsumura Y, Sato Y, Okada A, Morizane A, Okamoto S, Hong H, Nakagawa M, Tanabe K, Tezuka K, Shibata T, Kunisada T, Takahashi M, Takahashi J, Saji H, Yamanaka S (2011) A more efficient method to generate integration-free human iPS cells. Nat Methods 8:409–412
30. Yu J, Chau KF, Vodyanik MA, Jiang J, Jiang Y (2011) Efficient feeder-free episomal reprogramming with small molecules. PLoS One 6:e17557
31. Rajesh D, Dickerson SJ, Yu J, Brown ME, Thomson JA, Seay NJ (2011) Human lymphoblastoid B cell lines reprogrammed to EBV-free induced pluripotent stem cells. Blood 118:1797–1800
32. Raya A, Rodriguez-Piza I, Navarro S, Richaud-Patin Y, Guenechea G, Sanchez-Danes A, Consiglio A, Bueren J, Izpisua Belmonte JC (2010) A protocol describing the genetic correction of somatic human cells and subsequent generation of iPS cells. Nat Protoc 5:647–660
33. Park IH, Lerou PH, Zhao R, Huo H, Daley GQ (2008) Generation of human-induced pluripotent stem cells. Nat Protoc 3:1180–1186
34. Staerk J, Dawlaty MM, Gao Q, Maetzel D, Hanna J, Sommer CA, Mostoslavsky G, Jaenisch R (2010) Reprogramming of human peripheral blood cells to induced pluripotent stem cells. Cell Stem Cell 7:20–24
35. Loh YH, Hartung O, Li H, Guo C, Sahalie JM, Manos PD, Urbach A, Heffner GC, Grskovic M, Vigneault F, Lensch MW, Park IH, Agarwal S, Church GM, Collins JJ, Irion S, Daley GQ (2010) Reprogramming of T cells from human peripheral blood. Cell Stem Cell 7:15–19
36. Brown ME, Rondon E, Rajesh D, Mack A, Lewis R, Feng X, Zitur LJ, Learish RD, Nuwaysir EF (2010) Derivation of induced pluripotent stem cells from human peripheral blood T lymphocytes. PLoS One 5:e11373
37. Xu C, Inokuma MS, Denham J, Golds K, Kundu P, Gold JD, Carpenter MK (2001) Feeder-free growth of undifferentiated human embryonic stem cells. Nat Biotechnol 19:971–974
38. Aggarwal R, Pompili VJ, Das H (2010) Genetic modification of ex-vivo expanded stem cells for clinical application. Front Biosci 15:854–871

Part III

Pluripotent Stem Cell Characterization

Chapter 14

Cellular Characterization of Human Pluripotent Stem Cells

Rene H. Quintanilla Jr.

Abstract

Human pluripotent stem cells (PSCs), in particular induced PSCs, are very difficult to derive, grow, and bank. They require extensive amounts of resources and time to render them useful for basic and applied research. As the derivation methods, culture systems and tissues of origin differ, so does the quality of the PSCs themselves. Consequently, there are generally accepted molecular and cellular markers that serve as benchmarks of pluripotency. PSCs undergo rigorous qualification before they can be truly considered a stem cell or completely reprogrammed into a stem-like cell as in the case of iPSCs. Morphology is a good indicator of PSCs but the further qualification of cellular markers of pluripotency and differential potential is necessary. The standard methods enclosed in this chapter delineate the techniques necessary to qualify PSCs at the cellular level.

Key words hESC, iPSC, Immunofluorescence, Cellular characterization, Human pluripotent stem cells

1 Introduction

Human pluripotent stem cells (hPSCs) are a specific category of cells that share the ability to regenerate indefinitely and differentiate into every cell type in the human body. This includes embryonic stem cells (hESCs) which are typically isolated from the inner cell mass of blastocysts or other early stage embryos (1), embryonic germ (EG) cells, and embryonal carcinoma (EC) cells (2). This category of cells was further expanded with the introduction of the reprogramming of human somatic cells into induced pluripotent stem cells (iPSC) (3). These PSCs share general characteristic which can be used to qualify their pluripotency and differentiation potential.

PSCs share common molecular and cellular signatures, particularly the up-regulation of pluripotent genes and increased protein expression of pluripotent markers, specifically SSEA4, TRA-1-60, TRA-1-81, Oct-4, Nanog, and Alkaline Phosphatase (4). The measurement of these pluripotent markers is key in ensuring that the derived cell lines are truly pluripotent and remain so over the extended use of these lines. In particularly, as scientific innovations

Uma Lakshmipathy and Mohan C. Vemuri (eds.), *Pluripotent Stem Cells: Methods and Protocols*, Methods in Molecular Biology, vol. 997, DOI 10.1007/978-1-62703-348-0_14, © Springer Science+Business Media New York 2013

have progressed to include various forms of reprogramming methods, culture systems, and different somatic cell types used to create iPSCs, there exists is a strong need to keep qualifying these cells. In this chapter we detail the typical methods used to characterize hPSCs at the cellular level.

The methods described in this chapter utilize the expression of normal pluripotent markers through surface protein recognition via immunofluorescence and flow cytometry. PSCs are also characterized via intracellular pluripotent protein determination through immunocytochemistry (5, 6). In addition, we will cover a traditional method of determining alkaline phosphatase (AP) activity (6) and introduce a novel nondestructive, transient method of visualizing (AP) activity that may be used during the derivation of iPSC (7).

PSCs must also be characterized for their potential to differentiate into every cell type. This is also a further measure of true pluripotency which is typically measured via teratoma formation in vivo or through standard embryoid body (EB) formation and undirected differentiation in vitro (8). The content in the immunocytochemistry method in this chapter will cover the same techniques necessary to determine the differentiation potential of the PSCs by qualifying the resulting cell types of the in vitro differentiation to represent the three typical embryonic germ layers: endoderm, mesoderm, and ectoderm (9).

2 Materials

2.1 Immunofluorescence Components

1. Basal Media: DMEM/F-12 with GlutaMAX (Life Technologies, #10565-018) (see Note 1).
2. Pluripotent Surface Marker Antibodies: mouse monoclonal primary antibodies raised against Human SSEA4 (Life Technologies, #414000), TRA-1-60 (Life Technologies, #411000), and TRA-1-81 (Life Technologies, #411100) (see Note 2).
3. Fluorophore-conjugated secondary antibodies: Alexa Fluor 488 goat-anti-mouse IGg (H+L) secondary (Life Technologies, #A11001) and Alexa Fluor 594 goat-anti-mouse IGg (H+L) secondary (Life Technologies, #A11005).

2.2 Live Alkaline Phosphatase Detection Components

1. Basal Media: DMEM/F-12 with GlutaMAX (Life Technologies, #10565-018) (see Note 1).
2. Alkaline Phosphatase Live Stain (500×) in DMSO (Life Technologies, #A14353). Keep stock stored at −20°C for up to 6 months, protected from light. Upon initial thaw, aliquot into small samples and refreeze. Keep away from light and use right away upon dilution in aqueous medias (see Note 3).

2.3 Terminal Alkaline Phosphatase Detection Components

1. Basal Media: DMEM/F-12 (1:1) with GlutaMAX (Life Technologies, #10565-018).
2. Wash solution: D-PBS without $CaCl_2$ and $MgCl_2$ (1×), Liquid (Life Technologies, #14190-250), can be stored and used at room temperature.
3. Fixation solution: 4% (v/v) Paraformaldehyde in PBS (US Biological, #19943). Store at 4°C until use.
4. ELF 97 Endogenous Phosphatase Detection Kit (Life Technologies, #E6601).

2.4 Immunocytochemistry Component

1. Wash solution: D-PBS without $CaCl_2$ and $MgCl_2$ (1×), Liquid (Life Technologies, #14190-250), can be stored and used at room temperature.
2. Fixation solution: 4% (v/v) Paraformaldehyde (PFA) in PBS (US Biologicals #19943). Store at 4°C until use.
3. Blocking and primary antibody incubation solution: 5% (v/v) Normal Goat Serum (Life Technologies #10000C), 1% (v/v) BSA (Life technologies #A10008-01), 0.1% (v/v) Triton X-100(Life technologies #HFH-10) all diluted in D-PBS (Life Technologies #14190-250). All reagents can be mixed as follows to make 100 mL of blocking solution: 5 mL of goat serum, 4 mL of the 25% BSA supplement, and 100 μL of Triton X-100 in 91 mL of D-PBS. Sterile filter to remove contaminants and precipitates. Store at 4°C for up to 4 weeks (see Note 4).
4. Pluripotent antibodies: Rabbit polyclonal primary antibodies raised against Human Oct-4 (Life Technologies #A13998), Sox2 (Life Technologies #481400), or Nanog (Abcam #ab80892).
5. Differentiation antibodies: Mouse monoclonal primary antibodies raised against Human beta-3-Tubulin (b-3-Tub) (Life Technologies #480011), alpha-fetoprotein (AFP) (Life Technologies #180003), and alpha-smooth muscle Actin (SMA) (Life Technologies #180106) (see Note 5).
6. Fluorophore-conjugated secondary antibodies: Alexa Fluor 488 goat-anti-mouse IGg (H+L) secondary (Life Technologies #A11001), Alexa Fluor 594 goat-anti-mouse IGg (H+L) secondary (Life Technologies #A11005), Alexa Fluor 488 goat-anti-rabbit IGg (H+L) secondary (Life Technologies #A11008), and Alexa Fluor 594 goat-anti-rabbit IGg (H+L) secondary antibodies (Life Technologies #A11012).
7. 1× DAPI (4′,6-Diamidino-2-Phenylindole Dihydrochloride): Dissolve 10 mg of DAPI (Life Technologies #D-3571) in 1 mL of Dimethyl Sulfoxide (DMSO) (Sigma #D2650) to create 10,000× stocks. Aliquot stocks and store at −20°C for up to

6 months, protected from light. Prepare working stocks by diluting 1 μL of the stock DAPI in 10 mL of D-PBS (Life Technologies, #14190-250). Working stocks may be kept at 4°C for up to 4 weeks when protected by light. DMSO and DAPI are known mutagens and must be handled with care.

2.5 Flow Cytometry Components

1. D-PBS without $CaCl_2$ and $MgCl_2$ (1×), Liquid (Life Technologies, #14190-250) can be stored and used at room temperature.
2. TrypLE express stable trypsin replacement enzyme (Life Technologies, #12604-013). Keep at 4°C, and pre-equilibrate appropriate aliquot to room temperature prior to use (see Note 6).
3. Pluripotent, fluorophore-conjugated antibodies: mouse monoclonal primary antibodies, raised against Human SSEA4, conjugated to Alexa Fluor 647 or 700 (Life Technologies, #SSEA421 and # SSEA429) or TRA-1-60 (BD Biosciences, #560173), and TRA-1-81 (BD Biosciences, #560174) conjugated to Alexa Fluor 488.

3 Methods

3.1 Immunofluorescence of Live Adherent PSC Cultures Using Pluripotent Surface Markers

This method of staining live PSCs is meant to capture the cells in their native morphology and structure and to observe the normal pattern of expression of the cell surface markers. All medias must be pre-equilibrated to 37 °C before use and all incubations must be performed in the incubator the cells are normally grown under appropriate temperature and atmospheric conditions.

1. Remove the normal PSC growth media from the desired sample well of a 12-well plate of adherent PSC culture. Add 500 μL of DMEM/F-12 to the well and allow the wash solution to be incubated for 3 min at room temperature. Aspirate off the DMEM/F-12 wash from the well and repeat once more (see Note 7).
2. Dilute the primary antibody of choice (SSEA4, TRA-1-60, or TRA-1-81) in DMEM/F-12 to a final dilution of 1:100 (5 μL of antibody in 500 μL of DMEM/F-12) in a micro-centrifuge tube and mix by gently pipetting up and down (see Note 8).
3. Aspirate off the final wash and add the diluted primary antibody solution to the well. Incubate the cultures with the primary antibody at 37 °C for 45 min in the incubator.
4. Following the incubation period, aspirate off the primary antibody solution. Add 500 μL of DMEM/F-12 to the well and allow the wash solution to be incubated for 3 min at room

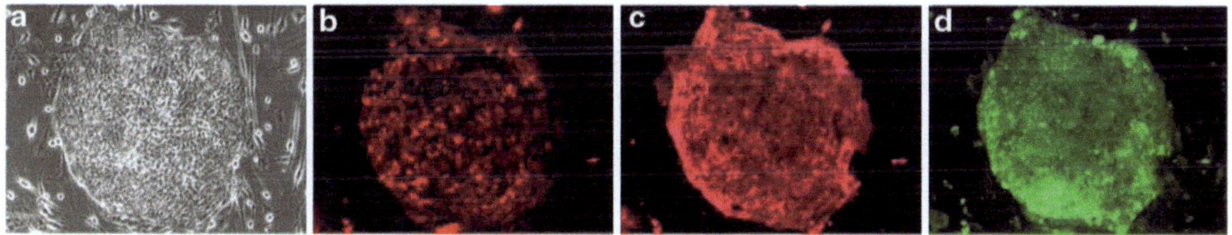

Fig. 1 Fluorescence imaging of pluripotent markers in PSCs. (**a**) Human pluripotent stem cells, grown on feeders, captured at 100× magnification under phase contrast microscopy. (**b**) Immunofluorecent analysis of TRA-1-60 antibody conjugated to Alexa Fluor 594. (**c**) Immunofluorecent analysis of SSEA4 antibody conjugated to Alexa Fluor 647. (**d**) Alkaline Phosphatase Live Stain observed under FITC

temperature. Aspirate off the DMEM/F-12 wash from the well and repeat the wash procedure once more (see Note 9).

5. Dilute the Alexa Fluor 488 or 594 secondary antibody in DMEM/F-12 to a final dilution of 1:500 (1 μL of antibody in 500 μL of DMEM/F-12) in a micro-centrifuge tube and mix gently by pipetting up and down.
6. Aspirate off the final wash and add the diluted secondary antibody solution to the well and incubate the plate at 37°C for 30 min in the incubator.
7. Following the incubation period, aspirate off the secondary antibody solution. Add 500 μL of DMEM/F-12 to the well and allow the wash solution to be incubated for 3 min at room temperature. Aspirate off the DMEM/F-12 wash from the well and repeat the wash procedure twice more.
8. Detect the pluripotent marker using a FITC filter for Alexa Fluor 488 and a Cy3/Rhodamine filter for Alexa Fluor 594 on a fluorescence microscope (see Fig. 1).

3.2 Live Fluorescent Detection of Alkaline Phosphatase Activity on Adherent PSC Cultures

1. Remove the normal PSC growth media from the desired sample well of a 12-well plate of adherent PSC culture. Add 500 μL of DMEM/F-12 to the well and allow the wash solution to be incubated for 3 min at room temperature. Aspirate off the DMEM/F-12 wash from the well and repeat once more (see Note 10).
2. Dilute the 500× Alkaline Phosphatase Live Stain in DMEM/F-12 to a final dilution of 1:500 (1 μL of stain in 500 μL of DMEM/F-12) in a micro-centrifuge tube and mix by gently pipetting up and down (see Note 11).
3. Aspirate off the final wash and add the diluted AP Live Stain solution to the well. Incubate the cultures with the primary antibody at 37°C for 20 min in the incubator (see Note 12).
4. Following the incubation period, aspirate off the AP Live Stain solution. Add 500 μL of DMEM/F-12 to the well and allow

the wash solution to be incubated for 3 min at room temperature. Aspirate off the DMEM/F-12 wash from the well and repeat the wash procedure twice more (see Note 12).

5. Immediately before the visualization of the fluorescent signal, aspirate off the final wash and add 500 μL of DMEM/F-12 to the well of interest (see Note 13).
6. Visualize the AP activity using a FITC filter under fluorescent microscopy (see Fig. 1d).
7. Upon completion of signal visualization and/or manual selection, replace the DMEM/F-12 with normal, pre-warmed hESC/hiPSC culture media, and return culture to a humidified incubator, at 37°C (see Note 14).

3.3 Terminal Detection of Alkaline Phosphatase Activity on Adherent PSC Cultures

This Assay measures AP activity in a terminal manner, where the accumulation of the fluorescent by-product of the reaction is toxic to the cells and hence they may not be propagated further. The method described the technique for PSC growing on feeders. For feeder independent cultures, it is necessary to add a fixation step similar to the immunocytochemistry method in Subheading 3.4 in order to prevent the cells from floating off the plate as they begin to die due to the toxic nature of this assay. All these reactions may be done at room temperature with components that are at room temperature.

1. Remove the normal PSC growth media from the desired sample well of a 12-well plate of adherent PSC culture. Add 500 μL of DMEM/F-12 to the well and allow the wash solution to be incubated for 3 min at room temperature. Aspirate off the DMEM/F-12 wash from the well and repeat once more.
2. Prepare ELF 97 Substrate kit by diluting the substrate (component A) tenfold in detection buffer (Component B). Prepare 200 μL of substrate solution per sample (20 μL of Component A in 180 μL of Component B). (Optional but recommended: Filter the substrate solution through a 0.2 μm filter to remove precipitates.)
3. Remove the wash media from PSC wells and add 200 μL of DMEM/F12 media per well.
4. Add 200 μL of the diluted ELF 97 substrate solution to each sample well and incubate at room temperature, in the dark, for the appropriate time (recommended 20–30 min).
5. Aspirate off the substrate solution from all samples and wash at least once with DMEM/F12.
6. Add 500 μL of fresh DMEM/F12 to each well prior to visualization. AP activity can be visualized using a short pass FITC filter, seen as a green granular accumulation of by-product or by using a long pass Hoechst filter under DAPI settings.

Note: cells cannot be propagated after ELF 97 staining.

3.4 Immunocytochemistry of Adherent PSC Cultures for Detection of Intracellular Markers of Pluripotency and Dif-ferentiation

1. Remove the normal PSC growth media from the desired sample well of a 12-well plate of adherent PSC culture. Add 500 µL of D-PBS to the well and allow the wash solution to be incubated for 3 min at room temperature. Aspirate off the D-PBS wash from the well and repeat once more.
2. Fix the PSC by adding 500 µL of 4% PFA to the desired well and incubate for 20 min at room temperature (RT).
3. Aspirate off the PFA from the well, and add 1 mL of D-PBS to the well and incubate for 3 min at RT. Aspirate and repeat wash step.
4. Permeabilize and block the fixed cells by adding 500 µL of Blocking Solution for 1 h at RT.
5. Prepare the dilution of the particular antibody of interest in cold (4 °C) blocking solution in a micro-centrifuge tube in a 500 µL final volume per well. Use the following dilutions for each pluripotent antigen of interest: Oct-4 (1:1,000), Sox2 (1:1,000), and Nanog (1:200). If staining of EBs or differentiated cells is desired to demonstrate lineage specific differentiation, use the following dilutions for the recommended lineage markers: AFP (1:1,000), SMA (1:100), and beta-3-Tubulin (1:1,000) (see Note 15).
6. Following the blocking step, aspirate off the blocking solution and add the pre-diluted primary antibody. Incubate the primary antibody solution at 4 °C overnight.
7. After the overnight incubation period, aspirate off the primary antibody solution. Add 1 mL of D-PBS to the well and allow the wash solution to be incubated for 3 min at room temperature. Aspirate off the D-PBS wash from the well and repeat the wash procedure twice more (see Note 9).
8. Dilute the Alexa Fluor 488 or 594 secondary antibodies in D-PBS to a final dilution of 1:500 (1 µL of antibody in 500 µL of D-PBS) in a micro-centrifuge tube and mix gently by pipetting up and down.
9. Aspirate off the final wash and add the diluted secondary antibody solution to the well and incubate the plate at room temperature for 30–45 min.
10. Following the incubation period, aspirate off the secondary antibody solution. Add 1 mL of D-PBS to the well and allow the wash solution to be incubated for 5 min at room temperature. Aspirate off the D-PBS wash from the well and repeat the wash procedure once more.
11. Aspirate off the second D-PBS wash and add 500 µL of the 1× DAPI working stock solution and allow to incubate for 10 min at room temperature. Remove the DAPI solution and repeat a wash step with 1 mL of D-PBS. Aspirate off the wash and add

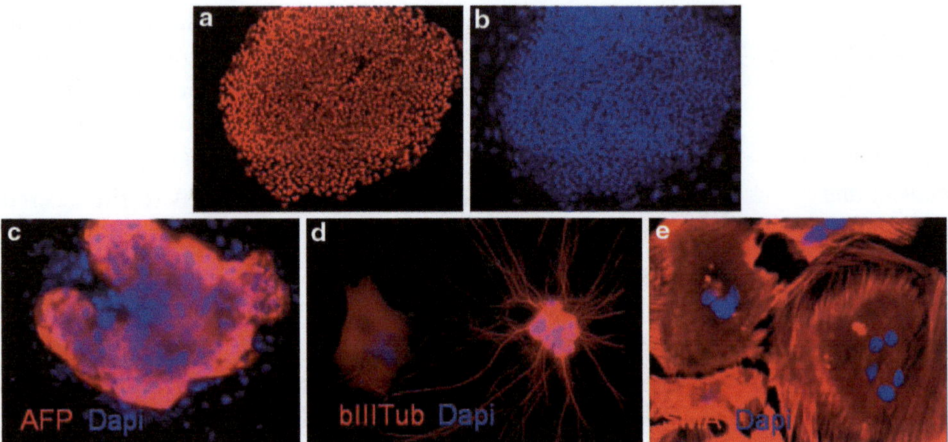

Fig. 2 Immunocytochemical analysis of pluripotent stem cells. (**a**) ICC analysis of intracellular Oct-4 expression of a PSC colony, as observed with Alexa Fluor 594 at 100× magnification. (**b**) ICC analysis of nucleic acid staining with DAPI of a PSC colony, as observed with a Hoechst Filter. (**c**) ICC analysis of PSCs differentiated down the endoderm lineage as visualized with alpha-fetoprotein and Alexa Fluor 594, counterstained with DAPI, 200× magnification. (**d**) ICC analysis of PSCs differentiated down the ectoderm lineage as visualized with beta-3-Tubulin and Alexa Fluor 594, counterstained with DAPI, 100× magnification. (**e**) ICC analysis of PSCs differentiated down the mesoderm lineage as visualized with smooth muscle Actin and Alexa Fluor 594, counterstained with DAPI, 200× magnification

1 mL of D-PBS in order to keep the cells moist until visualization. Keep the samples protected from light by covering the plate in aluminum foil until visualization. Visualization is best performed immediately to prevent photo bleaching, but samples may be stored at 4 °C, protected by light, ensuring the wells do not dry up due to evaporation.

12. Detect the marker of interest using a FITC filter for Alexa Fluor 488 and a Cy3/Rhodamine filter for Alexa Fluor 594 on a fluorescence microscope. Detect the nuclei counter staining of DAPI using a Hoechst filter (see Fig. 2).

3.5 Analysis of PSC Cultures via Flow Cytometry and FACS

1. Remove the normal PSC growth media from the desired plate or dish of PSC to be harvested. Add appropriate amount of D-PBS to the plate or dish and allow the wash solution to be incubated for 3 min at room temperature. Aspirate off the D-PBS wash from the plate.
2. Add the appropriate amount of TrypLE, which has been pre-equilibrated to room temperature, to the dish of interest: 1 mL per 35 mm dish, 2 mL per 60 mm dish. Incubate at room temperature for 3–5 min (see Note 6).
3. Using a 5 mL pipette, triturate the cells to achieve a single cell suspension and transfer to a 15 mL conical tube. Wash the remaining cells on the plate with 2 mL of D-PBS and add to the conical tube. Add an additional 10 mL of D-PBS to the cell

suspension in the conical tube and pipette up and down several times to dilute the TrypLE.

4. Centrifuge the cells at 200 × *g* for 2 min.
5. Carefully aspirate off the supernatant and resuspend the cells in 1 mL of D-PBS. Perform a cell count and dilute your cell suspension to 1×10^6 cells per mL
6. Add 10 μL of the SSEA-4, Tra1-60, or Tra1-81 Alexa Fluor directly conjugated antibody in 1 mL of the single cell suspension. This is the recommended concentration for 1×10^6 cells (see Note 16).
7. Allow the cells to incubate in the primary antibody for 1 h at room temperature with gentle agitation.
8. Following primary antibody incubation, add 10 mL of D-PBS to dilute out any unattached antibody and centrifuge at 200 × *g* for 2 min.
9. Aspirate the supernatant without disturbing the cell pellet. Resuspend the cell pellet with 5 mL of D-PBS and mix by pipetting up and down.
10. Centrifuge the cell suspension at 200 × *g* for 2 min and aspirate supernatant. Repeat the wash step once again.
11. Following centrifugation, aspirate the wash supernatant and resuspend the cell pellet in 500 μL of D-PBS.
12. Gently vortex or mix the samples prior to processing the sample through a FACS Cytometer or the Attune Acoustic Focusing Cytometer. Use the proper laser and filter combination for the fluorophore of interest use. Remember to always run an unstained negative control to adjust the settings to your cell of interest and be able to gate your populations as necessary. Also include Isotype controls where appropriate.

4 Notes

1. For any assay which involves keeping the cells alive and in their normal morphology, use the appropriate basal media (the main constituent of the growth media) that they are most accustomed to. All Media must be pre-warmed/pre-equilibrated prior to usage.
2. Using a 2-step method takes more time, and hence the cells are more stressed. In order to minimize the hands on time and decrease the stress on the cells, use a primary antibody that is conjugated to a fluorophore. Ensure that it is certified free of contaminants. This will also allow you to multiplex different monoclonal markers that are conjugated to different fluorophores in a single sample.

3. AP Live stain is a fluorescent substrate that is light sensitive and hydrolyses readily so aliquot the stock solution into working amounts in amber tubes upon initial thaw. Thaw quickly at room temperature right before use.
4. The blocking solution contains goat serum and BSA and must be sterile filtered to remove impurities and protein aggregates that may impede proper blocking after fixation or cause autofluorescence when visualizing your samples. The addition of Triton X-100 is to aid in membrane permeabilization as this is meant for intracellular immunofluorescence.
5. The differentiation antibodies are used in determining differentiation potential of the PSCs in question. They are meant to be used when the PSCs are forced to differentiate in a non-lineage specific manner as is evident in teratoma formations (in vivo) and EB formation and continued differentiation when dissected and seeded down for 14–21 additional days in the differentiation media of choice. These antibodies are not intended to be used on PSC cultures or early differentiated cells. The same immunocytochemistry procedure is used when using these antibodies.
6. TrypLE is a gentle but effective enzyme to disassociate PSCs into single cell suspensions. It is not meant to be used on PSCs grown on feeders as it will disassociate both the PSCs and feeders equally. This is meant to be used on feeder independent cell culture. If the intent is to analyze or sort PSC grown on feeders, you will have to first harvest PSC colonies with an enzyme that preferentially removes PSCs such as Collagenase IV or Dispase. Alternatively, you may harvest colonies via manual or mechanical dissection. Upon removal of the PSCs from the feeder layer, you may briefly treat the colonies with TrypLE in order to further disassociate the colonies into single cell suspensions. This must be done to prevent clumps from clogging the flow cytometer and to ensure proper analysis of each individual cell of interest.
7. Any manipulation has the potential to introduce contaminants and disrupt cell viability. In order to ensure preservation of PSC health, only use sterile medias that have been pre-warmed to 37 °C. Also when removing any solutions or medias via aspiration, take care not to touch the cells. Tilt the plates so the solutions accumulate on the side wall of the dish and aspirate from there. Also never add solutions directly onto the cells. Tilt the plate and add to the side wall and gently distribute it evenly after addition. All volumes used in this method are based on a single well of a 12-well tissue culture plate, and must be scaled up or down as appropriate for the dimensions of the plate or dish of interest.

8. Never add any antibody or compound directly to any portion of the cell, rather first prepare the desired dilution in a separate tube to ensure thorough distribution of the antibody. This will prevent clumping of the antibody in any one location and will ensure the even distribution throughout the whole culture.
9. Washes are crucial at this point to ensure that all the primary antibody that is attached is specific, and any nonspecific binding of the antibody is removed, as the secondary will attach to all primary molecules and hence may give erroneous/nonspecific positives.
10. The removal of the normal culture media and subsequent washes are crucial to the success of this assay, as KSR, BSA, FBS, or other typical growth media components will interfere with the reaction.
11. Typically using the final dilution of 1× has been proven optimal for hESC and iPSC. You may desire to dilute further if desired to decrease any background fluorescence as the stain is cell permeable and may enter any feeder cells which will contain basal levels of AP activity.
12. Increasing the amount of time of incubation beyond 30 minutes is not necessary as the reaction is kinetic in nature and the cells have been saturated with the substrate within the first 20 minutes. Subsequent washes are essential to remove any excess substrate and any free dye that has been released into the media.
13. As the AP reaction progresses, the fluorescent signal caused by the metabolization of the substrate is seen transiently as is escapes from the cell into the surrounding media. Care must be taken to visualize the PSC immediately after the final wash, as the fluorescence will be completely removed from the cells within 2 h from initial addition of the stain.
14. The live stain is a transient stain that permits the identification of pluripotent cells that have high levels of AP activity, above the basal levels evident in feeders. This is particularly useful in the selection of emerging iPSC in the reprogramming workflow. After visualization and or acquiring desired images or scoring the desired colonies, you may manually pick the scored cells and propagate them further.
15. Antibodies may work well if incubated at room temperature for a minimum of 2 hours. This is not recommended but may be empirically determined by the user as they see fit. The three differentiation antibodies are meant to capture each of the cell types present in the human body that correspond to the three embryonic germ layers seen during development. These may be substituted as desired by the end user for the marker of choice.

16. Directly conjugated dyes allow the rapid processing of cells without compromising cell health and viability. If conjugated antibodies are not available for the marker of interest, you may alternatively perform a 2-step method of antibody reactions as described in Subheading 3.1, but performed in suspension rather than adherent culture.

Acknowledgements

This work was supported by Life Technologies. The products within this publication are for Research Use Only, not intended for animal or human therapeutic or diagnostic use.

References

1. Thomson JA, Itskovitz-Eldor J, Shapiro SS, Waknitz MA, Swiergiel JJ, Marshal VS, Jones JM (1998) Embryonic stem cell lines derived from human blastocysts. Science 282:1145–1147
2. Pera MF, Reubinoff B, Trouson A (2000) Human embryonic stem cells. J Cell Sci 113:5–10
3. Takahashi K, Tanabe T, Ohnuki M, Narita M, Ichisaka T, Tomoda K, Yamanaka S (2007) Induction of pluripotent stem cells from adult human fibroblasts by defined factors. Cell 131:861–872
4. Josephson R, Ording CJ, Liu Y, Shin S, Lakshmipathy U, Toumadje A, Love B, Chesnut JD, Andrews PW, Rao MS, Auernach JM (2007) Qualification of embryonal carcinoma 2102Ep as a reference for human embryonic stem cell research. Stem Cells 25:437–446
5. Chan EM, Ratanasirintrawoot S, Park I, Manos PD, Loh Y, Huo H, Miller JD, Hartung O, Rho J, Ince TA, Daley GQ, Schlaeger TM (2009) Live cell imaging distinguishes bona fide human iPSC cells from partially reprogrammed cells. Nat Biotechnol 27:1033–1037
6. Tavakoli T, Xu X, Derby E, Serebryakova Y, Reid Y, Rao MS, Mattson MP, Ma W (2009) Self-renewal and differentiation capabilities are variable between human embryonic stem cell lines 13, 16 and BG01V. BMC Cell Biol 10:44
7. Singh U, Quintanilla RH, Grecian S, Gee KR, Rao MS, Lakshmipathy U (2012) Novel live alkaline phosphatase substrate for identification of pluripotent stem cells. Stem Cell Rev 8(3):1021–1029
8. Skottman H (2010) Derivation and characterization of three new human embryonic stem cell lines in Finland. In Vitro Cell Dev Biol Anim 46:206–209
9. Itskovitz-Eldor J, Schuldiner M, Karsenti D, Eden A, Yanuka O, Amit M, Soreq H, Benvenisty N (2000) Differentiation of human embryonic stem cells into embryoid bodies comprising the three embryonic germ layers. Mol Med 6(2):88–95

Chapter 15

TaqMan® OpenArray® High-Throughput Transcriptional Analysis of Human Embryonic and Induced Pluripotent Stem Cells

Sunali N. Patel, Yalei Wu, Yun Bao, Ricardo Mancebo, Janice Au-Young, and Elena Grigorenko

Abstract

It is widely accepted that somatic cells can be reprogrammed by a set of transcription factors to become embryonic stem cell-like: These reprogrammed cells, induced pluripotent stem cells (iPSCs), are nearly identical to embryonic stem cells (ESCs), because both have the capacity to self-renew and to form all cellular lineages of the body. Transcriptional differences between ESCs, iPSCs, and fibroblasts can be analyzed by quantitative PCR (qPCR) using TaqMan® Gene Expression assays, a widely used tool for rapid analysis of different cell types. In this chapter, we describe the OpenArray® platform which generates qPCR data from high-throughput instrumentation. We examined the gene signature profiles of ESCs, fibroblasts, and iPSCs with a TaqMan® OpenArray® Human Stem Cell Panel containing 631 TaqMan® Gene Expression assays that represent pathways involved in self-renewal, pluripotency, lineage patterning, transcriptional networks, stem cell differentiation, and development.

Key words Reprogramming, Fibroblast, Induced pluripotent stem cells, iPSC, TaqMan, RT-PCR, Pluripotency, Self-renewal, Embryonic stem cells, Adult stem cells

1 Introduction

In 2006, Takahashi and Yamanaka achieved a major milestone in stem cell research (1) when human and mouse fibroblasts were reprogrammed into iPSCs using a viral vector to introduce four transcription factors: Oct4, Sox2, Klf4, and c-Myc. Recently, a non-integrating, high efficiency method was developed to deliver this set of pluripotency transcription factors resulting in iPSCs with no remnants of a viral genome (2–4), and a kit is now available commercially (CytoTune™-iPS Reprogramming Kit, Life Technologies). In vitro differentiation tests have demonstrated the developmental potential of iPSCs generated by this method, specifically differentiation of cultured cells into three germ layers and expression of cell type specific markers. We describe the transcriptional comparison between

Uma Lakshmipathy and Mohan C. Vemuri (eds.), *Pluripotent Stem Cells: Methods and Protocols*, Methods in Molecular Biology, vol. 997, DOI 10.1007/978-1-62703-348-0_15,

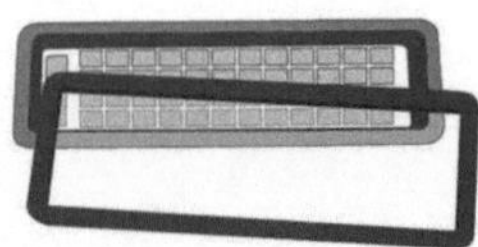

Fig. 1 The OpenArray® System workflow consists of (**a**) mixing sample with master mix in a 384-well plate, (**b**) loading the samples with the AccuFill™ System onto the TaqMan® OpenArray® Human Stem Cell Plate, (**c**) encasing the slide with immersion oil, and then (**d**) running up to four plates on the QuantStudio™ 12K Flex Instrument

fibroblasts, iPSCs, and ESCs to generate gene-specific signatures and pinpoint differences between cell types. We describe a high-throughput gene expression workflow performed on the OpenArray® platform which uses nanoliter fluidics for low-volume solution-phase reactions. Researchers using this technology benefit from low sample input requirements, high-throughput capabilities, and the data quality of PCR-based solution-phase reactions.

OpenArray® technology utilizes a stainless steel slide-sized plate with 3,072 through-holes (5). Each plate is laid out in 48 subarrays, each containing 64 through-holes in an 8×8 array grid—64×48 gives 3,072 through-hold reaction chambers. Each through-hole holds 33 nl of PCR mix through a coating process that makes the inside of the through-hole hydrophilic and the outside hydrophobic. This enables reagents to be held within the through-holes. The OpenArray® platform helps streamline real-time PCR studies that use large numbers of samples, assays, or both. QuantStudio™ 12K Flex OpenArray® plates are the NextGen qPCR solution for genetic variation and gene expression that provide the highest sample throughput for mid-density qPCR analysis: Over 43,000 gene expression reactions and >110,000 genotyping reactions can be run in a single day on the QuantStudio™ 12K Flex Real-Time PCR System without any additional robotics. The QuantStudio™ system can run 40 different configurable formats with throughput up to 5,100 genotyping samples and >760 gene expression samples per day (Fig. 1).

The TaqMan® OpenArray® Human Stem Cell Panel contains 631 gene expression assays, a configuration which provides the capacity to test 16 samples in single and eight samples in duplicate on a single instrument run, equivalent to thirty-two 384-well plates. The stem cell panel contains assays for structural and regulatory protein classes related to pluripotency, early differentiation, and development. These include transcription factors, receptors, kinases, growth factors, signaling molecules, peptide hormones, and proteins essential for morphogenesis, gamete formation, and the immune response. The assays cover a wide variety of pathways such as TGF-β, Wnt signaling, inflammation, angiogenesis, integrin signaling, presenilin, apoptosis, interleukin signaling, and blood coagulation. Additionally, the panel includes a set of endogenous control assays (i.e., for ACTB, B2M, GAPDH, GUSB, HMBS, HPRT1, IPO8, PGK1, POLR2A, PPIA, RPLP0, TBP, TFRC, and YWHAZ).

2 Materials

2.1 Cells, RNA Isolation, and Reverse Transcription

1. BJ human foreskin fibroblasts, human H9 embryonic stem cells, and iPSCs derived from BJ fibroblasts (see Note 1).
2. TRIzol® Reagent (Life Technologies, Carlsbad, CA).
3. High Capacity cDNA Reverse Transcription Kit (Life Technologies).
4. Nuclease-free water.
5. GeneAmp® PCR System 9700 (Life Technologies).

2.2 qPCR

1. OpenArray® 384-well Sample Plates.
2. OpenArray® AccuFill™ System Tips.
3. TaqMan® OpenArray® Real-Time PCR Master Mix (Life Technologies).
4. TaqMan® OpenArray® Human Stem Cell Panel (Life Technologies).
5. OpenArray® Real-Time PCR Accessories Kit (Life Technologies): QuantStudio™ 12K Flex OpenArray® Immersion Fluid with syringe and tips, QuantStudio™ 12K Flex OpenArray® Lids, QuantStudio™ 12K Flex OpenArray® Plugs, and Plate File on CD-ROM which specifies the location of the Stem Cell TaqMan® assays in the OpenArray® nanotiter plates.
6. Corning® 96-Well Microplate Aluminum Seal Tape (Corning Life Sciences).
7. Powder-free nitrile gloves.
8. Disposable transfer pipettes.
9. Pipettes, P10 to P1000.

10. Centrifuge with plate adaptor.
11. OpenArray® Plate Press.
12. QuantStudio™ 12K Flex OpenArray® AccuFill™ System.
13. QuantStudio™ 12K Flex Real-Time PCR System (Life Technologies).

3 Methods

3.1 Identification and Characterization of the iPSC Clones Signature Gene Expression

Today, with many stem cells lines in existence, the question remains as to how reliable and stable they are in terms of their pluripotency and their potential usage in regenerative medicine. Quantitative real-time PCR is one of the well-established approaches to generate pluripotency-specific gene expression signatures. The OpenArray® panel contains 631 assays implicated in pluripotency and self-renewal as well as early differentiation (see Notes 2 and 3).

This panel was used for identifying the gene expression profile of three different cell lines: iPSCs, ESCs, and BJ fibroblast cells.

3.2 Reprogramming of Somatic Cells

1. iPSCs were generated from BJ fibroblasts using the CytoTune™ iPS Reprogramming kit according to the manufacturer's protocol (see Note 1). (http://www.invitrogen.com/cytotune). The kit utilizes Sendai virus particles containing four genes—Oct3/4, Sox2, Klf4, and c-Myc to reprogram somatic cells using a non-integrating, high-efficiency technology.
2. Cells were expanded for over ten passages and confirmed to be pluripotent via marker expression analysis (including live alkaline phosphatase staining) and ability to differentiate into cell types representative of the three germ layers via random embryoid body formation (see Note 1).
3. The clones were also confirmed to be karyotypically normal prior to RNA isolation and subsequent analysis (see Note 1).

3.3 RNA Isolation

1. Cells were harvested in TRIzol® Reagent, and total RNA was isolated according to manufacturer's instructions (http://www.lifetechnologies.com).
2. Samples were treated with amplification grade DNase I for 2 h at 37°C to remove contaminating genomic DNA. Total RNA was precipitated and resuspended in RNase-free water.
3. RNA preps were quantified spectrophotometrically and their purity assessed by electrophoresis on a 1% agarose gel.

3.4 Reverse Transcription Reaction Using RNA Isolated from Cell Culture

1. Thaw the High Capacity cDNA Reverse Transcription Kit reagents and RNA samples on ice.
2. Prepare sufficient RT mix for all the samples according to the manufacturer protocol plus pipetting error overage of 15–20%.

For a single reaction of 20 μl, combine the RT reagents consisting of 2 μl of 10× RT buffer, 1 μl of 100 mM dNTPs, 2 μl of 10× RT random primers, 1 μl of 50 U/μl MultiScribe™ Reverse Transcriptase, and 4 μl Rnase-free water. To 10 μl of RT mix, add 10 μl RNA sample (200 ng/μl) (see Notes 4 and 5).

3. Mix gently and spin the tubes. Incubate in a thermocycler using the following conditions: 22°C for 10 min, 37°C for 2 h, 4°C for 5 min, 75°C for 10 min, and 4°C for 5 min.
4. Spin the tube after completion of the RT step. Place on ice or store at −20°C up to 2 months for later use.

3.5 Plate Overview

1. OpenArray® 384-well plates are used as an intermediate plate prior to loading each PCR mix into the OpenArray® plates with the OpenArray® AccuFill™ System.
2. As shown in Fig. 2, the OpenArray® 384-well Sample Plate can be divided into eight regions, each containing 48 wells. Each 48-well region is used to fill one TaqMan® OpenArray® Human Stem Cell Panel. Each well on the 384-well plate is used to load a subarray of 64 through-holes on the OpenArray® plate. The well dimensions of the OpenArray® 384-well Sample Plates are specifically suited for use with the AccuFill™ System. We do not recommend the use of other microtiter plates.

3.6 TaqMan® OpenArray® Human Stem Cell Panel

1. The TaqMan® OpenArray® Human Stem Cell Panel contains a set of assays in quadruplicate. One set of assays is contained in 12 subarrays. (A subarray consists of 64 through-holes preloaded with TaqMan® assays.)
2. Each TaqMan® OpenArray® Human Stem Cell Panel can accommodate four different RNA-derived cDNA samples (or two samples in duplicates) for a full profiling of 631 genes (see Note 6).

3.7 Preparing Sample Plate for OpenArray® Loading

1. Thaw reverse-transcribed samples, spin the tubes briefly, and place on ice.
2. Prepare PCR master mix by combining 146 μl of 2× TaqMan® OpenArray® Real-Time PCR Master Mix and 76 μl of RNase-free water for total volume of 222 μl.
3. Aliquot 50.2 μl PCR master mix into four tubes, and then add 15.8 μl of reverse-transcribed sample to each tube. Mix gently and spin briefly.
4. Identify the region of the current sample in the 48 well section of the 384-well sample plate, as shown in Fig. 2. Pipette 5 μl of sample/master mix into each well. For each sample: Load 12 subarrays, i.e., pipette sample 1 into wells A1 through D3, sample 2 into A4 through D6, sample 3 into A7 through D9, and sample 4 into A10 through D12 (see Fig. 2).

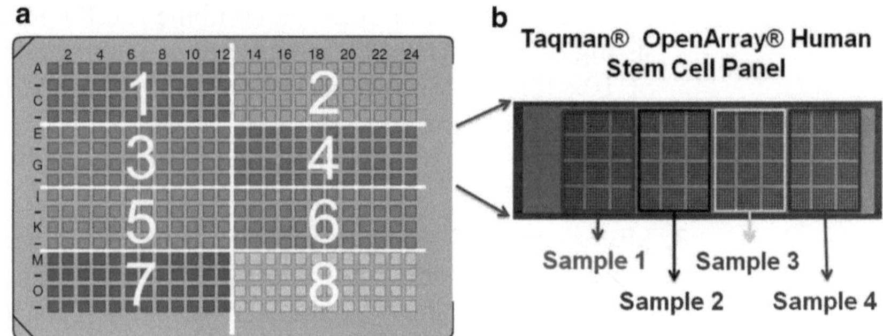

Fig. 2 OpenArray® 384-well Sample Plate and TaqMan® OpenArray® Human Stem Cell Plate sample layout. Schematic shows that (**a**) a single 384-well sample plate accommodates the loading of eight different OpenArray® plates. (**b**) A single TaqMan® OpenArray® Human Stem Cell Panel accommodates samples 1–4, each of which is profiled against 631 pre-spotted assays contained in the rectangle with 12 subarrays

5. Cover the plate with aluminum foil sealing tape. Centrifuge the plate for 1 min at 1,000 rpm to eliminate bubbles. Samples in the plate should be used within an hour of preparation.

3.8 OpenArray® Loading

1. Initialize the OpenArray® AccuFill™ System:
   (a) Double-click the OpenArray® AccuFill™ icon to launch the software and make sure that the enclosure door is closed. The software checks the computer and connections as the system starts. After the software has been launched, it displays a message requesting you to clear the deck and empty the waste bin of used tips. Clear the deck area and run the system test.
   (b) Place the OpenArray® AccuFill™ Systems pipette tip box (contains 384 tips) at the assigned location.
   (c) Remove the lid of pipette tip box. With gloves on, gently run finger across the top of tips to make sure tips are all upright and evenly spaced.
2. Preparing the case:
   (a) Remove OpenArray® plates from −20°C freezer and equilibrate to room temperature for at least 15 min.
   (b) Remove the OpenArray® plate from the mylar pouch and place into appropriate position in AccuFill (see Note 7). Orient the barcode end of the OpenArray® plate to the left, alphanumeric text to the right.
   (c) Remove OpenArray® Lids. Plugs and immersion fluid are included in the OpenArray® Accessory kit.
   (d) Remove the cap from the syringe containing OpenArray® Immersion Fluid, attach the syringe tip to the syringe, and then place the assembly on a clean surface. Immersion fluid must be used within 60 min after being removed from pouch.

3. Loading:
   (a) Insert the 384-well sample plate with the foil cover still in place at the designated area.
   (b) Gently press on the plate until you hear it snap into place.
   (c) Set up the AccuFill software by providing the sample plate and OpenArray® plate's barcode information.
   (d) Verify the location and orientation of OpenArray® plates, state of tips on the deck, and state of the foil on sample plate. When verified, select the check boxes on setup deck window of the software.
   (e) Remove the foil covering the area of the sample plate to be loaded on the OpenArray® plate.
   (f) Close the instrument door and select "Load".
   (g) Once the OpenArray® plate load sequence is done, you will see the Remove OpenArray® plate window.
   (h) Remove the adhesive liner from an OpenArray® Case Lid.
   (i) Using the thumb and index finger of left hand, grasp the OpenArray® plate by the top (nearest the barcode), gently lift the plate from the plate holder, and then load it into the OpenArray® Plate Press. Place the OpenArray® Case Lid (with adhesive liner removed) onto the OpenArray® plate using the alignment pins of the OpenArray® Plate Press for orientation.
   (j) Actuate the OpenArray® Plate Press for 15 sec.
   (k) Disengage the OpenArray® Plate Press and carefully remove the OpenArray® plate, grasping the case by the edges.
   (l) Insert the tip of the syringe into the loading port at the end of the sealed OpenArray® case, and then dispense the fluid in one gentle continuous motion.
   (m) Try to minimize creating air bubbles when dispensing the fluid. Only small air bubbles may be acceptable.
   (n) While holding the OpenArray® plate vertically, seal the loading port with OpenArray® Plug by inserting the plug into the port and twisting the plug clockwise until hand-tight.
   (o) Clean the case with a lint-free tissue that has been thoroughly sprayed with ethanol. To dry the case, wipe the case downward with another clean tissue. Gently handle the case; be careful not to apply pressure on the OpenArray® plate.
   (p) Repeat above steps for additional OpenArray® plates.
   (q) Use the QuantStudio Carrier to transport up to four loaded OpenArray® plates to the QuantStudio 12K Flex Instrument (see Note 8).

3.9 Thermal Cycler Protocol and Imaging

1. Activate the QuantStudio 12K Flex Instrument by touching the screen.
2. Touch the Eject icon at the lower right corner to open the side door and access the plate adapter.
3. Place the QuantStudio Carrier with OpenArray® plates on the plate adapter. Ensure that each plate is properly aligned.
4. Touch the Eject icon again to retract the plate adapter.
5. In the Main Menu screen, touch *Run OpenArray® Plates* (see Note 9).
6. The instrument will retrieve the barcodes and scan for existing experiments with the same barcodes. If no experiments with the same barcode can be found, touch *Source Input* to select a template to use.
7. Touch *Start Run Now* to start the run.
8. Results of OpenArray® experiments can be analyzed with analysis software included with the QuantStudio™ Expression suite software. The software includes the comparative Ct (ddCt) method to rapidly quantitate relative gene expression across a large number of genes and samples (6) (Fig. 3).

4 Notes

1. For a description of iPSC generation, see chapter on CytoTune Sendai Virus for iPSC generation in this volume written by Lieu et al. For further details on iPSC and ESC characterization, refer to chapter on cellular characterization of stem cells by Quintanilla.
2. The GeneAssist™ Pathway Atlas tool is available to search for pathways by genes of interest or to browse the list of over 350 interactive pathways. This free tool provides detailed gene, protein information, and Life Technologies product information. The pathways can be queried by pathway name, gene id, gene name, or gene symbol. Results include a written description of each pathway, gene and transcript information, protein data and disease implication, sequence alignment maps, *Silencer®* siRNA, and TaqMan® GEx products, supporting experimental data, internal product validation data, and products referenced in scientific journals.

   Pathways which are relevant to pluripotency and early development are as follows:

   (a) Embryonic cell differentiation into cardiac lineages

   (b) Human early embryo development

   (c) Human embryonic stem cell pluripotency

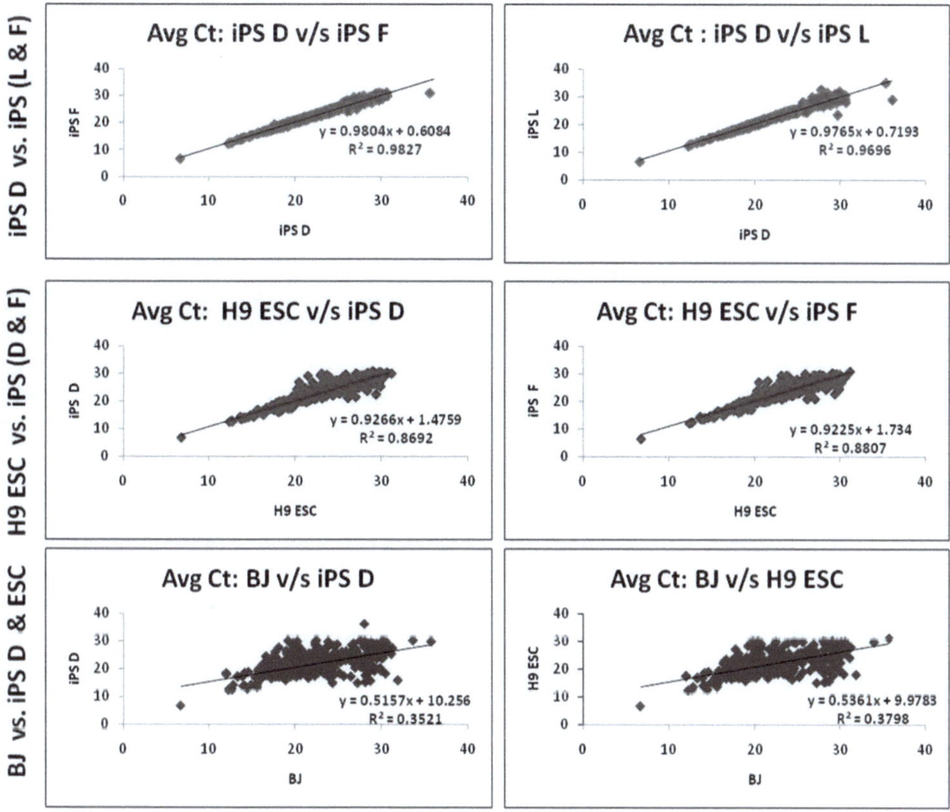

Fig. 3 Pairwise correlation of gene expression levels from 631 transcripts for iPSCs vs. iPSCs (iPS D, F, L clones in Panels **a**, **b**), H9 ESCs vs. iPSCs (Panels **c**, **d**), BJ fibroblasts vs. iPSC clone D (Panel **e**), and BJ fibroblasts vs. H9 ESCs (Panel **f**). Average Ct values are plotted for 631 transcripts, and correlation coefficients (r^2) are shown. The greatest difference in expression was observed for fibroblasts vs. iPSCs or ESCs ($r^2 < 0.38$), across the 631 transcripts. ESCs show greater similarity to iPSCs ($r^2 > 0.86$) while iPSC clones show near-identical transcript expression ($r^2 > 0.96$)

(d) Nanog in mammalian ESC pluripotency

(e) Oct4 in mammalian ESC pluripotency

(f) Transcriptional regulatory network in embryonic stem cell

These pathways from the GeneAssist™ Pathway Atlas Version 1.2 can be viewed at http://www5.appliedbiosystems.com/tools/pathway/

3. The 631 TaqMan® Gene Expression assays were selected from publications (7), as well as GeneAssist™ pathways describing embryonic stem cell and iPSCs. A list of the TaqMan® assay IDs and respective gene symbols on the TaqMan® OpenArray® Human Stem Cell Panel is available at http://www.lifetechnologies.com.

4. More than 1 ng of cDNA per TaqMan assay is recommended in the OpenArray® plate. The Life Technologies protocol

recommends 1–10 ng of cDNA for the same assay in microplates. To achieve this goal, a starting RNA concentration of 200 ng/μl is required. PCR master mix volume is sufficient to load one OpenArray® plate plus an additional 10% volume to account for pipetting. Adjust the volume of the reagents accordingly if more than one OpenArray® plate is used.

5. In case of limited amount of RNA material or low abundant RNA, the preamplification of cDNA with stem cell panel primers can be done before the qPCR step. For this purpose, it is recommended to use TaqMan® PreAmp Master Mix (Life Technologies) in combination with the TaqMan® Stem Cell assay pool. Custom TaqMan assay pools for fixed content Taqman® OA plates are available upon request.
6. The OpenArray® NT cycler is an earlier version of the instrument used for OpenArray® gene expression studies. The TaqMan® OpenArray® Human Stem Cell Panel is designed to run on both instruments; however a special case is required for plates to run on the QuantStudio™ 12K Flex Real-Time PCR System—plates are not interchangeable between platforms.
7. Wear tight-fitting powder-free gloves (i.e., one size smaller than the size you typically wear), to prevent excess glove material from contacting the OpenArray® plate while handling.
8. If you drop a loaded OpenArray plate before thermal cycling, discard it in the appropriate waste container.
9. To start the thermal protocol, several methods can be used. For example, the template file in the QuantStudio Software or the OpenArray® plate setup file has embedded thermal cycling protocols. Please refer to the Applied Biosystems QuantStudio 12K Flex Real-Time PCR system OpenArray® Experiments User Guide to start the thermal cycling program.
10. DataAssist 3.0 Software developed by Life Technologies Corp was used for data analysis. This software is free to download from http://www.lifetechnologies.com/us/en/home/technical-resources/software-downloads/dataassist-software.html. The software allows two different methods of normalization: use of endogenous control(s) or global normalization (8). The use of more than one reference gene increases the accuracy of quantification compared to the use of a single reference gene (9). Strategies for normalization of high-throughput expression profiling experiments generally take advantage of the large amount of data generated and often use all available data points. These strategies range from a straightforward approach based on the mean or median expression value to more complex algorithms such as global normalization method. Global normalization first finds common assays among all

tested samples, and then median Ct values of those assays are used as the normalizer for each sample. Since each data set generated on OpenArray Stem Cell pluripotency panels has 631 data points per sample, we chose global normalization method for the data set presented in Fig. 3.

11. The TaqMan® OpenArray® is for research use only, not for human or animal therapeutic or diagnostic use.

Acknowledgements

The authors thank Rene Quintanilla and Uma Lakshmipathy for providing the cells used in this study, Doug Roberts for assay selection, Vanee Pho for technical guidance and documentation, and Caifu Chen and David Keys for support and helpful discussions.

References

1. Takahashi K, Tanabe K, Ohnuki M, Narita M, Ichisaka T, Tomoda K, Yamanaka S (2006) Induction of pluripotent stem cells from mouse embryonic and adult fibroblast cultures by defined factors. Cell 126:663–676
2. Fusaki N, Ban H, Nishiyama A, Saeki K, Hasegawa M (2009) Efficient induction of transgene-free human pluripotent stem cells using a vector based on Sendai virus, an RNA virus that does not integrate into the host genome. Proc Jpn Acad Ser B Phys Biol Sci 85:348–362
3. Seki T, Yuasa S, Oda M, Egashira T, Yae K, Kusumoto D, Nakata H, Tohyama S, Hashimoto H, Kodaira M, Okada Y, Seimiya H, Fusaki N, Hasegawa M, Fukuda K (2010) Generation of induced pluripotent stem cells from human terminally differentiated circulating T cells. Cell Stem Cell 7:11–14
4. Zhang X, De Los Angeles A, Zhang J (2010) The art of human induced pluripotent stem cells: the past, the present and the future. Open Stem Cell J 2:2–7
5. Morrison T, Hurley J, Garcia J, Yoder K, Katz A, Roberts D, Cho J, Kanigan T, Ilyin SE, Horowitz D, Dixon JM, Brenan C (2006) Nanoliter high throughput quantitative PCR. Nucleic Acids Res 34(18):e123
6. Livak KJ, Schmittgen TD (2001) Analysis of relative gene expression data using real-time quantitative PCR and the 2 (-delta delta C(T)) method. Methods 25(4):402–408
7. Adewumi O, Aflatoonian B, Ahrlund-Richter L et al (2007) Characterization of human embryonic stem cell lines by the International Stem Cell Initiative. Nat Biotechnol 25(7):803–816
8. Mestdagh P, Van Vlierberghe P, De Weer A, Muth O, Westermann F, Speleman F, Vandesompele J (2009) A novel and universal method for miRNA RT-qPCR data normalization. Genome Biol 10:R64
9. Vandesompele J, De Preter K, Pattyn F, Poppe B, Van Roy N, De Paepe A, Speleman F (2002) Accurate normalization of quantitative RT-PCR data by geometric averaging of multiple internal genes. Genome Biol 3:0034-1–0034-11

Chapter 16

Epigenome Analysis of Pluripotent Stem Cells

Christopher L. Ricupero, Mavis R. Swerdel, and Ronald P. Hart

Abstract

Mis-regulation of gene expression due to epigenetic abnormalities has been linked with complex genetic disorders, psychiatric illness, and cancer. In addition, the dynamic epigenetic changes that occur in pluripotent stem cells are believed to impact regulatory networks essential for proper lineage development. Chromatin immunoprecipitation (ChIP) is a technique used to enrich genomic fragments using antibodies against specific chromatin modifications, such as DNA-binding proteins or modified histones. Until recently, many ChIP protocols required large numbers of cells for each immunoprecipitation. This severely limited analysis of rare cell populations or post-mitotic, differentiated cell lines. Here, we describe a low cell number ChIP protocol with next generation sequencing and analysis that has the potential to uncover novel epigenetic regulatory pathways that were previously difficult or impossible to obtain.

Key words Chromatin immunoprecipitation, ChIP-sequencing, Histone modifications, Epigenetics, Pluripotent stem cells

1 Introduction

Epigenetics is the study of non-DNA sequence-related heredity (1). The strictest definition labels epigenetics as any meiotically or mitotically heritable change in gene function that is not explained or results from changes in the underlying primary DNA sequence (2–4). Recently, this definition has been expanded to include epigenetic modifications that may not be heritable, but still result in gene expression changes (2, 3). The modifications can be broadly classified into DNA methylation, covalent modifications of histone tails, and mechanisms involving noncoding RNAs (3–6). The presence of these modifications is of great interest because of the wide range of biological and clinical implications resulting from epigenetic mechanisms. Mis-regulation of gene expression due to epigenetic abnormalities has been linked with complex genetic disorders, psychiatric illness, and cancer (1, 7). In addition, the dynamic epigenetic changes that occur in pluripotent stem cells are believed to impact regulatory networks essential for proper lineage development (8).

Uma Lakshmipathy and Mohan C. Vemuri (eds.), *Pluripotent Stem Cells: Methods and Protocols*, Methods in Molecular Biology, vol. 997, DOI 10.1007/978-1-62703-348-0_16,

Methods to observe the chromatin state in pluripotent stem cells have provided an unparalleled understanding of chromatin status before most lineage restriction pathways are induced. Tracking these marks over time during differentiation into multiple lineages may assist in predicting the phenotypic potential of embryonic stem (ES) or induced pluripotent stem (iPS) cell lines. With the recent revolution of iPS technology where somatic cells are reprogrammed to an embryonic cell-like state, epigenetic analysis is useful to assess reprogramming status. There have been reports of partial reprogramming where the iPS cell epigenetic state of both DNA methylation and/or histone modifications were quite different than ES cells. These can have devastating effects from tumor formation to inappropriate tissue differentiation (9–11). Therefore, effective epigenetic diagnostics can highlight inconsistencies from improper reprogramming that may have been overlooked using morphology and gene expression assays.

Researchers first interrogated isolated gene sequences for various histone modifications using chromatin immunoprecipitation (ChIP) (12, 13). Chromatin consists of DNA bound with histones and nonhistone proteins that promote the proper packaging of DNA by altering its underlying structure and thus influencing gene regulation (7, 14). ChIP is a technique used to isolate and enrich chromatin fragments using antibodies against specific chromatin modifications. These modifications can be DNA-binding proteins, such as transcription factors, or a particular covalent histone modification. Chromatin is sheared into small fragments (200–500 bp) to allow for proper genomic resolution. Enriched regions can be identified using PCR or quantitative PCR (qPCR) with specific primers designed for a selected genomic region. The enrichment level of various modifications is then used to describe and correlate with the transcriptional activity of the interrogated gene. Although this method can be precise, it is limiting because of the specificity needed to probe each genomic region.

To overcome these limitations, this technique has been scaled to produce genome-wide maps of chromatin state. Combining ChIP with microarrays (ChIP-chip) or next generation deep sequencing methods (ChIP-seq) has enabled the global analysis of chromatin state for the entire genome, defined as the epigenome (15–19). The epigenome is the combination of all epigenetic modifications in a particular cell. In contrast to the genome, each organism contains multiple epigenomes that display extensive differences per tissue type, developmental stage, and environmental influences (20). As opposed to ChIP, genome-wide epigenetic analysis is independent of hypotheses, allowing the detection of previously unknown patterns in cancer, development, and disease.

Until recently, many ChIP protocols required millions of cells for each immunoprecipitation. This severely limited analysis of rare cell populations or post-mitotic, differentiated cell lines. This barrier can be overcome by using Invitrogen's MAGnify™ kit.

This method utilizes magnetic beads to recover antibody-bound chromatin fragments, requiring far fewer cells per immunoprecipitation (IP) than previously reported protocols. There are many benefits to using a low cell number ChIP kit such as the ability to IP multiple histone marks or transcription factors using the same biological sample. In addition, it also provides the ability to investigate chromatin profiles from differentiated cultures containing limited quantities. By combining the benefits of low cell number ChIP with next generation sequencing, this protocol can be used to uncover novel epigenetic regulatory pathways that were previously difficult or impossible to obtain.

2 Materials

1. MAGnify™ ChIP Kit Module #1. Store at 4°C.
2. MAGnify™ ChIP Kit Module #2. Store at 4°C.
3. MAGnify™ ChIP Kit Module #3. Store at -20°C.
4. MAGnify™ ChIP Kit Module #4. Store at -20°C.

2.1 Cell Culture

1. Dissociation media: 0.25% Trypsin/EDTA (see Note 1)—Store at -20°C.
2. Trypsin inhibitor—0.25 mg/ml—GibcoBRL #17075-011—Store at -20°C.
3. Formaldehyde—37% Molecular Biology Grade.
4. Glycine—1.25 M—MAGnify™ ChIP Kit Module #1.
5. Phosphate-buffered saline (PBS) (1×).

2.2 Cell Lysis

1. Ice.
2. Lysis buffer—MAGnify™ ChIP Kit Module #4.
3. Protease inhibitors—MAGnify™ ChIP Kit Module #3.
4. Vortex mixer.

2.3 Chromatin Shearing

1. Sonicator—Bioruptor®—Diagenode.
2. 1.5–2% Agarose gel.
3. 1 kbp DNA Ladder—Promega #G571A—Store at 4°C.
4. 250 bp DNA Ladder—Lonza FlashGel DNA Marker #57033—Store at 4°C.
5. DNA Dye as Loading Buffer—EZ-Vision™ Three—AMRESCO #N313-Kit.
6. Ice.
7. 4°C Microcentrifuge.
8. Proteinase K MAGnify™ ChIP Kit Module #1.
9. Heating block.

2.4 Dynabeads Conjugation

1. Primary antibody(s) of interest.
2. Isotype control antibody.
3. Protein A/G Dynabeads MAGnify™ ChIP Kit Module #1.
4. 200 μl 8-strip or individual PCR tubes.
5. DynaMag™-PCR magnet.
6. Rotating mixer.
7. Dilution buffer—MAGnify™ ChIP Kit Module #4.

2.5 Chromatin Dilution/Binding/Washing

1. Dilution buffer—MAGnify™ ChIP Kit Module #1.
2. Protease inhibitors—MAGnify™ ChIP Kit Module #3.
3. IP buffer 1—MAGnify™ ChIP Kit Module #2.
4. IP buffer 2—MAGnify™ ChIP Kit Module #2.

2.6 Reverse Cross-Linking

1. Reverse Cross-linking Buffer—MAGnify™ ChIP Kit Module #1.
2. Proteinase K—MAGnify™ ChIP Kit Module #1.
3. Thermal mixer or thermal cycler.

2.7 DNA Purification and Elution

1. DNA purification magnetic beads—MAGnify™ ChIP Kit Module #1.
2. DNA purification buffer—MAGnify™ ChIP Kit Module #1.
3. DNA wash buffer—MAGnify™ ChIP Kit Module #2.
4. DNA elution buffer—MAGnify™ ChIP Kit Module #2.
5. Thermal mixer or thermal cycler.

3 Methods

3.1 Cell Preparation

1. When culturing cells for ChIP, choose an appropriate tissue culture plate or flask to ensure 1×10^6 to 3×10^6 cells will be harvested per sample.
2. Aspirate the medium and wash cells briefly with 1× PBS.
3. Aspirate the PBS and then add enough trypsinizing reagent (0.25% Trypsin/EDTA) to cover the cells. Immediately rock the plate gently back and forth to mix the trypsinizing reagent throughout the plate (see Note 1).
4. Incubate at 37°C for 2–5 min or until cells detach from their substrate. The timing may depend on the cell type's attachment characteristics and confluency. Once cells have dislodged, add an equal amount of trypsin inhibitor. Transfer the cell suspension into a 15 ml centrifuge tube. Add additional media or 1× PBS and rinse the plate for additional cells in suspension. Transfer the remaining suspension into the centrifuge tube, up to 15 ml and then gently pipette the cell mixture up and down five to ten times.

5. Take a small aliquot and count the cells using a hemocytometer or an automated cell counter. Aliquot a minimum of 1×10^6 to 3×10^6 cells per sample and spin each cell suspension at $200 \times g$ for 5 min at room temperature (see Note 2).
6. Aspirate the supernatant and lightly flick the tube to gently dislodge the cell pellet before resuspending in 500 μl of room temperature 1× PBS. Transfer each sample into 1.5 ml Eppendorf tubes (see Note 3). To cross-link in suspension, add 13.5 μl of 37% formaldehyde to each 500-μl sample, making the final working concentration 1%. Immediately invert each tube approximately five times to mix and incubate for 10 min at room temperature. The time course for optimal cross-linking should be discovered empirically (see Note 4).
7. Add 57 μl of room temperature 1.25 M glycine to quench the cross-linking reaction. Immediately invert each tube approximately five times to mix and incubate for 5 min at room temperature.
8. From this point forward, keep all samples on ice and spin at 4°C. Spin each cross-linked sample at $200 \times g$ for 10 min. Aspirate the supernatant leaving approximately 20–30 μl behind as to not disturb the cell pellet.
9. Lightly flick the cell pellet to gently dislodge it before resuspending in 500 μl cold 1× PBS. Spin at $200 \times g$ for 10 min. Carefully discard the supernatant. Repeat this one more time.
10. Aspirate the supernatant leaving 10–20 μl behind and either flash freeze the cell pellet in liquid nitrogen and store at −80°C or proceed directly to cell lysis (see Note 5).

3.2 Cell Lysis and Shearing

1. Thaw the lysis buffer and protease inhibitors (200×). Mix the lysis buffer by vortexing or pipetting up and down. Prepare a fresh master mix by adding 199 μl lysis buffer to 1 μl protease inhibitor. For each experiment, scale the master mix accordingly and use 50 μl of master mix for every 1×10^6 cells. Resuspend by light vortexing or pipetting, ensuring little to no bubbles. Incubate on ice for at least 5 min. Proceed to chromatin shearing or snap freeze with liquid nitrogen and store at −80°C.
2. Fragmenting the chromatin can be achieved through sonication or enzymatic digestion using Micrococcal nuclease. Although both can result in preferential shearing, sonication is the recommended method when cross-linking with formaldehyde.
3. Shear the lysed chromatin into random fragments using a sonicator with 30 s on/off pulses using the high-power setting on the Bioruptor®. Optimal sonication times should be empirically tested with a sample time course ranging from 10 to 30 pulses, separated in 5 pulse increments. After sonication is complete, confirm that there is minimal foaming and keep the

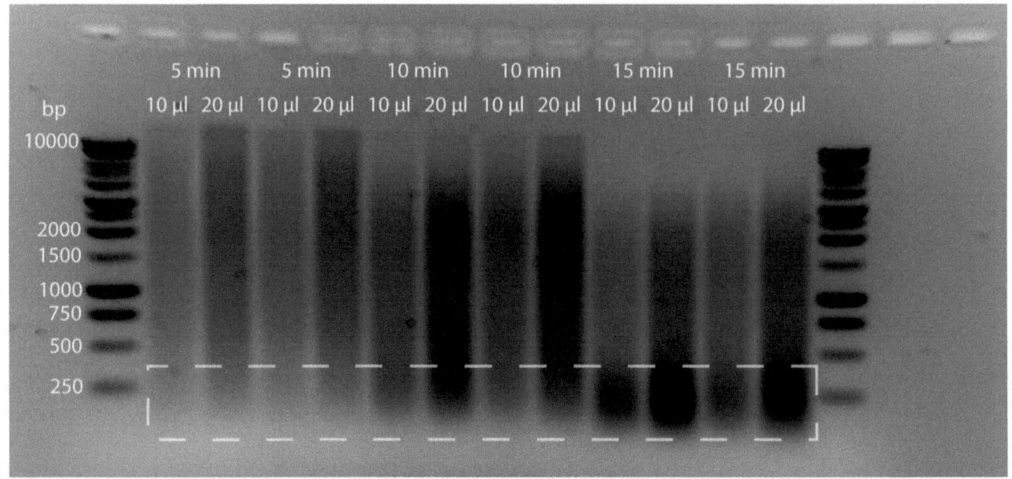

Fig. 1 Optimizing sonicated chromatin sizes. Cross-linked neural precursor cells were lysed and fragmented by sonication using the Bioruptor® with 30 s on/off pulses. Optimal sonication times were empirically tested in duplicate, with a time course ranging from 5 to 15 min pulses. Each sample was then treated with Proteinase K for 20 min at 55°C. Next, 10 μl and 20 μl from each sample were loaded into a 2% agarose gel to verify the fragmented chromatin sizes. The *dotted outline* represents the optimal chromatin size for ChIP-sequencing (~150–300 bp fragments). 15 min of sonication was optimal as these samples contained the majority of sheared chromatin within the desired fragmented range

samples on ice (see Note 6). If there is a substantial amount of foaming, discard the sample.

4. Spin the sonicated samples at >16,000 × *g* at 4°C for 5 min to pellet the cell debris. Transfer the supernatant containing the sonicated chromatin to a new 1.5 ml Eppendorf tube.
5. The sonicated fragment sizes should be verified by running a small aliquot on a 2% agarose gel. First, take an aliquot of approximately 10 μl of sonicated chromatin and a non-sonicated control. Next, add 1 μl of Proteinase K and incubate at 55°C for 20 min. Run both samples on a 2% agarose gel with the appropriate DNA ladders to verify sizes. The sonicated chromatin should appear as a smear (Fig. 1). To achieve proper resolution for ChIP-sequencing, the fragmented chromatin sizes should be in the range of 150–300 base pair (bp) due to the need for additional adaptor sequences later in the ChIP-sequencing protocol.

3.3 Chromatin Immunoprecipitation

1. Primary antibodies need to be coupled to the protein A/G Dynabeads before chromatin immunoprecipitation. Thaw the dilution buffer and place on ice along with the Dynabeads, ChIP magnet, sample tubes, and all primary antibodies.
2. Resuspend the protein A/G Dynabeads by lightly pipetting up and down. Do not vortex as this can damage the beads.

3. Wash the Dynabeads. Add 100 μl of cold dilution buffer to each individual or 8-strip 0.2 PCR tubes. Next add 10 μl of the protein A/G Dynabeads and gently pipette 5–10 times (see Note 7).
4. Insert each tube or strip into the DynaMag™-PCR magnet and wait 30 s to 1 min for the beads to pellet against the side of the tube. While the tubes are still in the magnet, carefully remove the liquid and discard without disturbing the pellet. Remove the tubes or strip from the magnet and once again add 100 μl of cold dilution buffer to each tube and gently pipette five to ten times.
5. Add the primary antibody of interest to each sample tube. The amount of antibody used should be determined empirically and based on the manufacturer's recommendations. Generally, 1–10 μg is a standard starting range. It is also important to include a negative control antibody, (i.e.—a mouse or rabbit IgG) based on the isotope of the primary antibody of interest. Do not pipette up and down, instead gently flick and invert the tubes five to ten times to fully resuspend the bead/antibody mix. Place the tubes in a rotator and mix end over end at 4°C for at least 1 h.
6. Dilute the chromatin during the antibody/bead conjugation. Thaw the sheared chromatin and set on ice. Prepare the dilution buffer/protease inhibitor (200×) master mix similar to step 3.2.1 and keep on ice. The final dilution volume for each sample is 100 μl and the ratio of sheared chromatin to dilution buffer depends on the number of cells used per immunoprecipitation (IP). For example, the concentration of sheared chromatin was 1×10^6 cells per 50 μl of lysis buffer. Therefore, to immunoprecipitate 0.2×10^6 cells, add 10 μl of sheared chromatin to 90 μl of the dilution buffer/protease inhibitor mix. Make a sample master mix based on the number of IPs per sample plus one extra 100 μl dilution per sample to use as an input control. Mix the sample thoroughly and aliquot 10 μl into a new 0.2 ml PCR tube to use as the input control for each sample (see Note 8).
7. Once the primary antibody/bead coupling is complete, remove any liquid in the caps by gently and briefly spinning the sample tubes from step #5 on a tabletop centrifuge capable of holding 0.2 ml PCR tubes. Place the tubes into the DynaMag™-PCR magnet and wait 30 s to 1 min for the bead pellet to form. Carefully discard the liquid. The beads now contain the conjugated antibodies needed for immunoprecipitation. Add 100 μl of the diluted chromatin master mix, from step #6, to their respective antibody/bead PCR tube. Do not pipette, instead cap the tubes, gently flick and invert five to ten times to fully resuspend the bead complexes. Rotate the tubes as previously described at 4°C for at least 2 h to overnight.

8. Before proceeding with the wash steps after chromatin binding, the Proteinase K, reverse cross-linking buffer, DNA purification beads, and DNA purification buffers should all be at room temperature.
9. After the chromatin has been bound to the beads, gently spin all sample tubes briefly and place the tubes into the DynaMag™-PCR magnet and wait 30 s to 1 min for the bead pellet to form. Remove the liquid carefully without disturbing the pellet. Add 100 μl of cold IP Buffer 1 to each sample tube and gently flick and invert briefly to resuspend the beads. Rotate the tubes at 4°C for approximately 5 min. Repeat these steps two more times.
10. Continue the washes by repeating step #9 with IP Buffer 2 in place of IP Buffer 1, for a total of two washes.
11. Prepare a reverse cross-linking master mix by adding 1 μl of Proteinase K and 53 μl of reverse cross-linking buffer for each sample. For example, eight samples would require 8 μl of Proteinase K plus 424 μl of reverse cross-linking buffer. When making the reverse cross-linking master mix, prepare a separate master mix for the input samples that were set aside in step #6. For these samples, add 1 μl of Proteinase K and 43 μl of the reverse cross-linking buffer.
12. Following the last wash with IP Buffer 2, put the tubes into the DynaMag-PCR magnet, wait for the bead pellet, and carefully discard the liquid without disturbing the pellet. Remove the tubes from the magnet and add 54 μl of the reverse cross-linking master mix to each sample tube containing beads. Separately, for each of the 10 μl diluted input samples, add 1 μl of Proteinase K to 43 μl of the reverse cross-linking buffer and scale accordingly. Incubate all sample tubes at 55°C for 15 min in a thermal shaker or other available heat source.
13. Following the reverse cross-linking steps, the immunoprecipitated DNA has now been released from the bead complexes and the liquid now contains the sample DNA, and therefore should not be discarded. Place each tube into the DynaMag™-PCR magnet and wait for the bead pellet to form. Carefully transfer the liquid without disturbing the pellet to a new 0.2 ml PCR tube. The tubes containing the used magnetic beads can be discarded. Incubate all of the newly transferred sample tubes along with the input samples at 65°C for another 15 min.
14. Briefly resuspend the DNA purification magnetic beads by light vortexing or pipetting up and down five to ten times. Prepare a DNA purification master mix by adding 50 μl of DNA purification buffer to 20 μl of DNA purification magnetic beads for all samples, including input controls. Pipette up and down to mix thoroughly. The DNA purification magnetic beads and buffer should be at room temperature before use.

15. After the reverse cross-linking incubation, briefly spin to collect any liquid. Next, set the tubes on ice or let cool to room temperature for 5 min.
16. Add 70 μl of the DNA purification master mix to each sample tube and input controls. Gently mix by pipetting up and down five to ten times. Let stand for at least 5 min at room temperature. Place each tube into the DynaMag™-PCR magnet and wait for the pellet to form. Remove the liquid carefully without disturbing the pellet (see Note 9).
17. Remove the tubes from the magnet and add 150 μl of DNA wash buffer and mix five to ten times. Place each tube into the magnet and repeat the wash step one more time by removing the liquid and adding 150 μl of DNA wash buffer.
18. Add the tubes back into the magnet and wait for the bead pellet to form. Discard the liquid, leaving a few μl behind making sure not to disturb the bead pellet. Remove the tubes from the magnet and add 60 μl of DNA elution buffer to all tubes. Pipette gently five to ten times to mix thoroughly. Add all tubes to a thermal shaker or other heat source and incubate for 20 min at 55°C.
19. Following the incubation, briefly spin to collect any liquid, put the tubes into the DynaMag™-PCR magnet, and wait for the bead pellet to form. The liquid now contains the purified sample. Gently transfer the liquid to a new 0.2 ml PCR tube without disrupting the bead pellet. Once transferred, the tubes containing the DNA purification beads can then be discarded (see Note 10).
20. The purified DNA can now be temporarily stored at 4°C if immediately analyzed by qPCR or can be stored at −20°C.

3.4 ChIP qPCR

1. Before proceeding towards ChIP-sequencing, it is beneficial to run a few real time quantitative PCR (qPCR) quality control assays using the optional 10 μl of each eluted purified DNA that was set aside previously (see Note 10), to confirm that reasonable chromatin enrichment was achieved.
2. Design positive and negative primers to regions of the genome where immunoprecipitated enrichment is both expected and not predicted. Amplicons should range from 80 to 150 bp since the optimal range of sheared chromatin was 150–300 bp for ChIP-sequencing. Primer amplification efficiency should first be verified by running a qPCR reaction with tenfold serial dilutions using either input DNA or genomic DNA as standards. A standard curve with a slope of −3.1 to −3.6 is an acceptable efficiency range. This method will produce a relative quantity of each sample's enrichment compared to its own input.

3.5 ChIP-sequencing

1. Following the elution of the purified ChIP-enriched DNA, fragment libraries will need to be constructed. Please refer to the SOLiD™ ChIP-seq Kit Guide for a more-detailed protocol.

However, for the purpose of generating barcoded libraries for multiplexing ChiP samples, one should refer to the SOLiD™ ChIP-seq Library Preparation with Barcodes protocol for additional details. We generally prepare 12–20 barcoded ChIP-seq libraries in parallel for each SOLiD slide to obtain the required sequencing depth (>20×10^6 aligned reads per library).

2. A starting quantity of 1–10 ng is recommended for library construction. If the enriched ChIP sample is less than 10 ng/μl, quantify the DNA using Invitrogen's Qubit® fluorometer. The Qubit® dsDNA HS Assay is sensitive and accurate for double-stranded DNA and has a linear detection range of 0.2–100 ng.
3. Proceed to library construction. Briefly, the DNA fragments are end repaired, polished, followed by ligation of adaptors to each end of DNA. If you will be multiplexing using barcodes, the P1 adaptor is replaced by the Multiplex Library P1 adaptor and the P2 adaptor replaced by the Barcode-0XX.
4. Once the barcoded DNA is purified, each sample is nick translated and amplified. If barcoding is used, the Library PCR Primer 1 and PCR Primer 2 will be replaced by Multiplex Library PCR-1 and Multiplex Library PCR-2. During the amplification step, the 250-μl amplified sample is placed on ice after ten cycles. To determine the optimal number of amplification cycles, remove 20 μl from the 250 μl sample and divide into five tubes of 4 μl each. These separate samples are further amplified at two cycle increments from 10 to 18 cycles and then run on a Lonza FlashGel™ to determine the minimum number needed for amplification. In some cases, where there is lower than 1 ng of enriched chromatin, it may be necessary to increase the amplification cycle increments up to 20 cycles (Fig. 2). Run the remaining 230 μl the appropriate number of cycles that was determined from the Lonza FlashGel™ results. The optimal number of PCR cycles produces a barely detectable quantity of product in the desired size range.
5. The amplified DNA is then purified and quantified by qPCR. Once complete, the fragment library can be stored in elution buffer at −20°C or you may proceed to the next step by referring to the detailed Applied Biosystems SOLiD™ System Templated Bead Preparation Guide for emulsion PCR and sequencing.
6. After sequencing, the reads and quality strings should be aligned to the reference species genome. Bowtie (21) is one example that can be used to identify the single, best-quality match location, selecting the option to produce output in SAM format. Results should be converted to sorted and indexed BAM format using Samtools (22) (see Note 11).
7. Duplicate reads aligning with identical genomic positions are likely to be the product of overamplification during

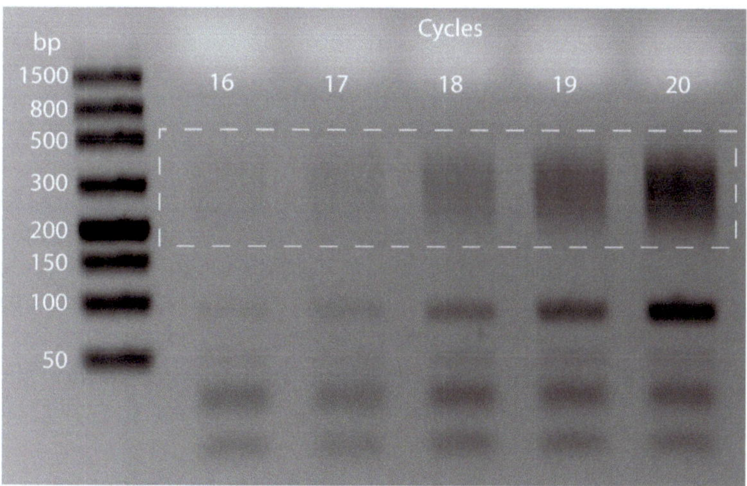

Fig. 2 Optimizing ChIP-sequencing sample amplification. Following DNA purification, each sample was nick translated and amplified. During the amplification step, the 250 μl amplified sample was placed on ice after 10 cycles. To determine the optimal number of amplification cycles, 20 μl was removed from the 250 μl sample and divided into five tubes of 4 μl each. These separate samples were further amplified at two cycle increments from 16 to 20 cycles and then run on a Lonza FlashGel™ to determine the minimum number needed for amplification. The *dotted outline* represents the fragmented ChIP-sequenced chromatin after the ligation of adaptors to each end of the DNA. For this example, 17 cycles was chosen as the minimum number of cycles needed for sufficient amplification

library construction. The Java package Picard (22) is useful for removing exact duplicate reads.

8. There are many peak calling programs using various algorithms to highlight chromatin enrichment (23). Find Peaks 4.0 (24) is one example program that compares each sequenced sample against the input control. Using the peak calling algorithm generates both bed position files and wiggle peak tracks, which can then be visualized using a genome browser. For example, using the UCSC Genome Browser (25, 26), each peak file can be designated its own track within a user-defined browser session and then be simultaneously viewed by stacking each sample track in a defined order.
9. ChIP-sequencing has the potential to generate large numbers of peaks throughout the genome. A first step in managing this wealth of data is to annotate each peak based on a set of criteria. We recommend using the ChIPpeakAnno algorithm (27), a plug-in from the Bioconductor package within the R statistical programming environment. The algorithm is designed to annotate enriched peaks based on the nearest gene. Various filtering options are available based on peak distance and overlapping between samples to generate meaningful data. In addition, Gene Ontology (GO) terms that are associated with the nearest genes can be generated.

4 Notes

1. Depending on the sensitivity of your cells, a gentler dissociative enzyme may be used such as TrypLE™. We have also observed that if you are planning to harvest cells for separate RNA analysis, using TrypLE™ yielded higher quality RNA. If using TrypLE™, there is no need for trypsin inhibitor, just dilute with culture media. Cross-linking can also be performed on adherent cells directly in a dish as an alternative to cross-linking the cells in suspension.
2. The number of cells per immunoprecipitation should be equal; therefore it is important to count the cells as accurately as possible.
3. This tube transfer step will make it easier for cross-linking and for spinning multiple samples in a 4°C centrifuge in later steps.
4. Too much cross-linking can result in the masking of epitopes, resulting in less than optimal immunoprecipitation, while not enough cross-linking can inefficiently bind the protein of interest to the chromatin.
5. Sometimes it is necessary to store cross-linked cell pellets before cell lysis and sonication (for example, when multiple time points are necessary). However, we do not recommend the long-term storage of frozen cross-linked cell pellets that have not been lysed as this may negatively affect the yield of chromatin immunoprecipitation.
6. It is important to always keep the samples cold during the sonication process because excess heat can reverse the cross-linking. Add ice cold water to the sonicator water bath every 5 min to ensure the samples remain cool. In addition, ensure that there is no foaming during sonication because this could also disrupt the cross-linking. An additional option is to also save 10 μl of non-sonicated chromatin. This can be run on a 2% agarose gel to verify the size by running alongside the sonicated chromatin samples.
7. The use of a multichannel pipette is highly recommended for the duration of this protocol for both consistency and efficiency.
8. Non-immunoprecipitated input controls are critical for evaluating enrichment. It is not always easy to accurately count cells when harvesting for ChIP. Therefore, obtaining an input control for each sample can be used as a normalizing factor when analyzing enrichment.
9. This pellet may be larger than previous pellets with the protein A/G beads so it is acceptable to leave a few microliters of liquid behind in order to not disturb the purification beads as these will now contain the bound DNA fragments.

10. It is suggested that 10 μl from the eluted 60 μl samples be set aside for later ChIP-qPCR analysis before or after ChIP-sequencing.
11. Bowtie, Samtools, Picard, and Findpeaks can all be downloaded and run on Windows, Mac, or Linux computers. While a computer cluster is useful for processing many sequencing projects at once, a reasonably powerful desktop computer is adequate for running one at a time. Generally the limiting factor is storage space more than processing power. For occasional ChIP-seq projects, a RAID-capable network attached storage device is adequate. For greater usage, storage area network capacity should be used.

Acknowledgements

This work is supported by grants from NIH (1R21MH085088 and 1RC1CA147187). C.L.R. was an NSF IGERT fellow (DGE 0801620).

References

1. Feinberg AP (2008) Epigenetics at the epicenter of modern medicine. JAMA 299:1345–1350
2. Bird A (2007) Perceptions of epigenetics. Nature 447:396–398
3. Hamby ME, Coskun V, Sun YE (2008) Transcriptional regulation of neuronal differentiation: the epigenetic layer of complexity. Biochim Biophys Acta 1779:432–437
4. Wu H, Sun YE (2006) Epigenetic regulation of stem cell differentiation. Pediatr Res 59:21R–25R
5. Bird A (2002) DNA methylation patterns and epigenetic memory. Genes Dev 16:6–21
6. Goldberg AD, Allis CD, Bernstein E (2007) Epigenetics: a landscape takes shape. Cell 128:635–638
7. Mehler MF (2008) Epigenetics and the nervous system. Ann Neurol 64:602–617
8. Bernstein BE, Mikkelsen TS, Xie X et al (2006) A bivalent chromatin structure marks key developmental genes in embryonic stem cells. Cell 125:315–326
9. Takahashi K, Yamanaka S (2006) Induction of pluripotent stem cells from mouse embryonic and adult fibroblast cultures by defined factors. Cell 126:663–676
10. Stadtfeld M, Maherali N, Breault DT et al (2008) Defining molecular cornerstones during fibroblast to iPS cell reprogramming in mouse. Cell Stem Cell 2:230–240
11. Yamanaka S (2009) A fresh look at iPS cells. Cell 137:13–17
12. Solomon MJ, Larsen PL, Varshavsky A (1988) Mapping protein-DNA interactions in vivo with formaldehyde: evidence that histone H4 is retained on a highly transcribed gene. Cell 53:937–947
13. Hebbes TR, Thorne AW, Crane-Robinson C (1988) A direct link between core histone acetylation and transcriptionally active chromatin. EMBO J 7:1395–1402
14. Kouzarides T (2007) Chromatin modifications and their function. Cell 128:693–705
15. Ren B, Robert F, Wyrick JJ et al (2000) Genome-wide location and function of DNA binding proteins. Science 290:2306–2309
16. Lieb JD, Liu X, Botstein D et al (2001) Promoter-specific binding of Rap1 revealed by genome-wide maps of protein-DNA association. Nat Genet 28:327–334
17. Johnson DS, Mortazavi A, Myers RM et al (2007) Genome-wide mapping of in vivo protein-DNA interactions. Science 316:1497–1502
18. Robertson G, Hirst M, Bainbridge M et al (2007) Genome-wide profiles of STAT1 DNA association using chromatin immunoprecipitation and massively parallel sequencing. Nat Methods 4:651–657
19. Barski A, Cuddapah S, Cui K et al (2007) High-resolution profiling of histone methylations in the human genome. Cell 129:823–837
20. Schones DE, Zhao K (2008) Genome-wide approaches to studying chromatin modifications. Nat Rev Genet 9:179–191

21. Langmead B, Trapnell C, Pop M et al (2009) Ultrafast and memory-efficient alignment of short DNA sequences to the human genome. Genome Biol 10:R25
22. Li H, Handsaker B, Wysoker A et al (2009) The sequence alignment/map format and SAMtools. Bioinformatics 25:2078–2079
23. Park PJ (2009) ChIP-seq: advantages and challenges of a maturing technology. Nat Rev Genet 10:669–680
24. Fejes AP, Robertson G, Bilenky M et al (2008) FindPeaks 3.1: a tool for identifying areas of enrichment from massively parallel short-read sequencing technology. Bioinformatics 24:1729–1730
25. Mangan ME, Williams JM, Kuhn RM et al (2009) The UCSC genome browser: what every molecular biologist should know. In: Ausubel FM et al (eds) Current protocols in molecular biology. Chapter 19, Unit 19 19
26. Karolchik D, Hinrichs AS, Kent WJ (2011) The UCSC Genome Browser. In: Haines JL et al (eds) Current protocols in human genetics. Chapter 18, Unit18 16
27. Zhu LJ, Gazin C, Lawson ND et al (2010) ChIPpeakAnno: a Bioconductor package to annotate ChIP-seq and ChIP-chip data. BMC Bioinformatics 11:237

Chapter 17

Single Cell Gene Expression Analysis of Pluripotent Stem Cells

Ronald V. Abruzzese and Richard A. Fekete

Abstract

Analyzing gene expression profiles from cells en masse provides an average profile for the population which may obscure differences in individual cells. Using an optimized workflow for qRT-PCR, gene expression profiles of undifferentiated pluripotent stem cells reveal distinct gene expression profiles for individual cells, and a large expression level range of almost every gene. Importantly, this technique allows for the identification and characterization of small subpopulations.

Key words Single cell, Gene expression, qRT-PCR, FACS, SuperScript III, Preamplification, Assays on demand, PreAmp master mix

1 Introduction

The analysis of RNA and proteins from single cells has long been performed using flow cytometry and microscopy (1, 2). However, the number of analytes that can be analyzed at one time using these techniques is limited to the capacity of the instrument (3). This has led to the analysis of RNA from single cells using tools such as qRT-PCR (4), gene expression microarrays (5), and next-generation sequencing (NGS; 6). Each of these techniques has its own set of limitations and usually requires a customized method for sample preparation. qRT-PCR techniques are well studied; however, reactions are usually limited to a predetermined number of genes. NGS is less studied and protocols have not been optimized, however, allows analysis of the entire transcriptome from a single cell. The main difficulty in any of these methods of analysis is that each cell contains a very small amount of material, between 10 and 30 pg of RNA. This has led to either the miniaturization of the analysis reactions (7) or incorporation of some type of amplification method such as linear amplification (5) or preamplification (8). The acquisition of single cells has also been

Uma Lakshmipathy and Mohan C. Vemuri (eds.), *Pluripotent Stem Cells: Methods and Protocols*, Methods in Molecular Biology, vol. 997, DOI 10.1007/978-1-62703-348-0_17, © Springer Science+Business Media New York 2013

investigated using a variety of methods such as fluorescently activated cell sorting (FACS; 9), micromanipulation (10), laser capture microdissection (LCM; 11), and more recently microfluidic devices (8). The steps after obtaining single cells, but before amplification and analysis, are lysis and stabilization and a number of methods have been investigated for this. Use of the buffers used in reverse transcription is commonly used to lyse cells; however, divalent cations such as magnesium in these buffers can cause chemical cleavage of RNA, particularly at elevated temperatures. Other methods such as freezing and thawing in water are also commonly used; however, these methods do not inactivate cellular RNases which can cause RNA degradation and a change in the cellular RNA profile. A common error made when analyzing small amounts of material, as found in single cells, is that samples are pipetted and transferred from tube to tube. It has been found that with each of these transfers, material is lost on pipette tips and tubes. Every effort should be made to not pipette samples until the limited material has been amplified.

By performing expression analysis at the single cell level, profiles associated with specific differentiated states and cell types can be obtained. Using FACS we sort thousands of single cells, stabilize their profiles, preamplify 100 selected genes, and analyze these by qPCR using the Single Cell-to-Ct™ kit from Ambion®. This gives the relative copy number for each transcript in a single cell and allows cells to be classified based on expression profiles. In this way we are able to characterize the heterogeneity within a cell population.

2 Materials

2.1 Cell Isolation and Staining

1. 2×10^6 cells in 1×10^6 cell/mL aliquots.
2. 1× PBS chilled to 4°C, diluted from 10× PBS (Ambion).
3. LIVE/DEAD® Viability/Cytotoxicity Assay Kit (Invitrogen).
4. General lab materials and equipment.

2.2 Cell Lysis, Reverse Transcription, and Preamplification

1. Single Cell-to-Ct™ Kit (Ambion).
2. Tris-EDTA pH 8.0 (Ambion).
3. MicroAmp Optical 96-Well Reaction Plates (Applied Biosystems).
4. MicroAmp Clear Adhesive Film (Applied Biosystems).
5. TaqMan® Gene Expression Assays (Applied Biosystems).
6. GeneAmp® PCR System 9700 Thermal Cycler (Applied Biosystems).
7. General lab materials and equipment: nuclease-free microcentrifuge tubes, aerosol-resistant pipette tips, pipettes, centrifuge, and disposable gloves.

2.3 qPCR Reagents

1. TaqMan® Gene Expression Master Mix (Applied Biosystems).
2. Nuclease-Free Water (Ambion).
3. Applied Biosystems 7900HT Fast Real-Time PCR Instrument and User Guide.
4. ABI PRISM® 96-Well or 384-well Clear Optical Reaction plate with barcode (Applied Biosystems).
5. ABI PRISM® Optical Adhesive Covers (Applied Biosystems).

3 Methods

3.1 Cell Isolation and Staining

1. There are a number of ways to isolate single cells (see Subheading 1). This chapter will focus on a protocol to isolate single cells using a FACSAria machine (see Note 1).
2. The following protocol uses the LIVE/DEAD® Viability/Cytotoxicity Assay Kit to stain the cells before FACS.
3. Allow all reagents to come to room temperature.
4. Make an 80-fold dilution of calcein AM (Component A) in DMSO to make a 50 μM working solution (i.e., add 2 mL of Component A to 158 mL DMSO). The working solution should be used within 1 day.
5. Prepare a 1 mL suspension of cells with 0.1 to 5×10^6 cells/mL for each assay. Cells may be in culture medium or buffer.
6. Add 2 μL of 50 μM calcein AM working solution and 4 μL of the 2 mM ethidium homodimer-1 stock to each milliliter of cells. Mix the sample.
7. Incubate the cells for 15–20 min at room temperature, protected from light.
8. As soon as possible after the incubation period (within 1–2 h), analyze the stained cells by flow cytometry using 488 nm excitation and measuring green fluorescence emission for calcein (i.e., 530/30 bandpass) and red fluorescence emission for ethidium homodimer-1 (i.e., 610/20 bandpass). Gate on cells to exclude debris. Using single color stained cells, perform standard compensation. The population should separate into two groups: live cells will show green fluorescence and dead cells will show red fluorescence (see Note 2).

3.2 Cell Lysis, Reverse Transcription, and Preamplification

1. Add 1 μL Single Cell DNase I to 9 μL single cell lysis solution for each single cell to be sorted.

Component	Lysis mix (μL)
Single Cell DNase I	1
Single cell lysis solution	9
Final volume	10

2. Aliquot 10 μL of the lysis/DNase I solution to the appropriate number of wells on a 96-well PCR plate.
3. Load 10^6 cells (in 1 mL 1× PBS) into the FACS machine and sort the single cells directly into the 10 μL lysis/DNase I solution.
4. Incubate at room temperature for 5 min, no mixing required (see Note 3).
5. Add 1 μL single cell stop solution to each sample. Incubate at room temp for 2 min (no mixing required), and place samples on ice (see Note 4).
6. Prepare sufficient RT reaction mix for all samples, then add 4.5 μL to each lysed cell sample:

Component	RT reaction mix (μL)
Single cell VILO™ RT buffer	3
Single cell SuperScript® RT mix	1.5
Final volume RT mix	4.5

7. Add RT reaction mix directly to lysate (do not transfer tubes) to make a 15.5 μL RT reaction, and mix thoroughly. Cover the plate using a MicroAmp Clear Adhesive Film, and incubate at 25°C for 10 min, 42°C for 60 min, and 85°C for 5 min (see Note 4).
8. Pool the TaqMan® Gene Expression Assays (20×) for your targets of interest, then dilute the pooled assays using 1× TE, pH 8.0 so that each assay is at a final concentration of 0.2× (see Note 5).
9. Prepare sufficient PreAmp reaction mix, then add 11 μL to each reverse-transcribed sample:

Component	PreAmp reaction mix (μL)
5× PreAmp Mix	5
0.2× pooled assay mix	6
Final volume of PA reaction mix	11

10. Add single cell PreAmp reagents directly to RT reaction (do not transfer tubes) to make a 26.5 μL PreAmp reaction, and mix thoroughly.
11. Cover the plate using a MicroAmp Clear Adhesive Film. Centrifuge briefly to collect the contents at the bottom of the wells.
12. Place the 96-well PCR plate containing the preamplification samples in a thermal cycler and run using the following cycling conditions:

	Stage	Reps	Temp (°C)	Time
Enzyme activation (hold)	1	1	95	10 min
PCR (cycle)	2	14	95	15 s
			60	4 min

13. Upon completion, immediately remove the plate from the thermocycler and place on ice or store at −20°C (see Note 6).
14. Dilute the PreAmp products 1:20 with 1× TE pH 8.0 before using in real-time PCR (see Note 7).

3.3 qPCR

1. Prepare the real-time PCR plate using 10 μL of each diluted preamplification product in a 96-well PCR plate using the table below (see Note 8):

Component	25 μL/well
TaqMan® Gene Expression MMix (2×)	25 μL
Diluted Pre-Amp product	10 μL
TaqMan® Gene Expression Assay (20×)	2.5 μL
Nuclease-free water	12.5 μL
Final volume	50 μL

2. Cover the plate using an ABI PRISM® Optical Adhesive Cover. Centrifuge briefly to collect the contents at the bottom of the wells.
3. Run the qPCR samples using the real-time PCR instrument cycling conditions shown below:

	Stage	Reps	Temp (°C)	Time
UDG incubation (hold)	1	1	50	2 min
Enzyme activation (hold)	2	1	95	10 min
PCR (cycle)	3	40	95	5 s
			60	1 min

3.4 Data Analysis

1. Use an automatic baseline and set the threshold to 0.2. Review the amplification plots and remove outliers. Omit samples that are undetectable for all assays tested. We suggest comparing the expression profile of the target genes for each individual cell to every other cell. Expect to see large variation in target gene(s) expression across cells. If your sample size is large enough you may be able to detect subpopulations of cells with similar expression profiles (see Note 9).

4 Notes

1. FACSAria I Flow Cytometer (BD Biosciences -San Jose, CA) running FACSDiva analysis software (version 4.1.2) and equipped with a dual laser setup (488 nm blue argon and 633 nm HeNe lasers), seven photomultiplier tubes with seven sets of optical filters, and an Automatic Cell Deposition Unit (ACDU) to allow for sorting onto multi-well plates. Cells were gated by side scatter vs. forward scatter (P1), then side scatter-height vs. side scatter-width (P2), and finally forward scatter-height vs. forward scatter-width (P3) to eliminate possible cell doublets. Cells were then sorted on the basis of Calcein AM alone or in combination with EthHD-A fluorescence.
2. LIVE/DEAD® Viability/Cytotoxicity Assay Kit (Invitrogen Cat# 649124) which provides a two-color fluorescence cell viability assay. The kit contains the polyanionic dye Calcein AM, which is well retained within live cells, and Ethidium Homodimer-A (EthHD-A, availability, cell-impermeant high-affinity nucleic acid stain that is weakly fluorescent until it binds to DNA). Cells are collected as 10^6 cells/mL aliquots and need at least 2×10^6 cell aliquots. One aliquot is split into four 0.25 mL aliquots. The first aliquot is left unstained. The second aliquot is stained with the live stain. The third aliquot is stained with the Dead stain. The fourth aliquot is stained with both the Live and Dead stain. These aliquots will be used to perform standard compensation of the instrument. The remaining 1 mL cell sample stained for both live and dead cells will be sorted onto the 96-well plate. Cells were sorted on 96-well plates as follows: 0 (4 wells), 1 (84), 10 (4), and 100 (4) cell samples/plate.
3. This is a stopping point for the protocol. The cell lysis sample is stable at room temperature for up to 30 min.
4. This is a stopping point for the protocol. Samples can be freeze–thawed up to five times with no harmful effects.
5. For example, to pool 50 assays, combine 10 μL of each assay, and then add 500 μL 1× TE buffer for a total of 1 mL. Up to 100 targets can be added to the TaqMan® Gene Expression Assay pool; however, all need to be at 0.2×. Do not include the 18S TaqMan® Gene Expression Assay in the pool due to the high expression level of this gene. Do include other control assays, such as the ACTB or XenoRNA Gene Expression Assays included with the TaqMan Cells-to-Ct Control Kit.
6. This is a stopping point for the protocol. The preamplification product can be stored at −20°C; minimize freeze–thaw cycles
7. The presence of TaqMan® fluorescent probes in the preamplification product necessitates diluting the samples 1:20

with TE buffer to reduce background fluorescence during the qPCR. The entire sample does not need to be diluted. The diluted preamplification product can be stored at −20°C. We recommend storing as aliquots to minimize freeze–thaw cycles.

8. The table shows the reaction setup using the recommended amount of diluted preamplification product in each PCR, 20%. If desired, up to 45% of the PCR volume can be diluted preamplification product; adjust the quantity of nuclease-free water accordingly. The total sample volume will vary depending on the format chosen:
   (a) 384-well is 20 μL
   (b) TaqMan® array is 900 μL
9. Due to variations in starting sample quantity, sample quality, and variable PCR efficiency, one of the most widely used methods for expression analysis is relative quantification. This method involves comparison of target molecules, which may vary due to different treatments or tissue types, to a reference target which does not vary. However, when working with single cell samples we have found that trying to compare target molecules with control or reference molecules is problematic. Gene expression of every gene can vary widely within each single cell, which may be due to microenvironments and transcriptional noise (12). When working with large numbers of cells or tissues, this noise is averaged out for each gene allowing results to be normalized to control genes. In single cells normalization actually increases the spread of calculated expression levels due to differences in the amount of each gene within the same cell. Similarly, these normalized values are not the same within each cell and vary depending on the genes compared. This gives different "profiles" for every pair of genes in each cell. Hence, qPCR data for each cell should be expressed at C_t or C_q per cell, since only one cell was added to each reaction.

References

1. Givan AL (2001) Flow cytometry: first principles, 2nd edn. Wiley, Indianapolis, IN
2. Lichtman JW, Conchello JA (2005) Fluorescence microscopy. Nat Methods 2:910–919
3. DeRose PC, Resch-Genger U (2010) Recommendations for fluorescence instrument qualification: the new ASTM Standard Guide. Anal Chem 82:2129–2133
4. Ståhlberg A, Bengtsson M (2010) Single-cell gene expression profiling using reverse transcription quantitative real-time PCR. Methods 50:282–288
5. Ginsberg SD (2005) RNA amplification strategies for small sample populations. Methods 37:229–237
6. Tang F, Barbacioru C, Wang Y, Nordman E, Lee C, Xu N, Wang X, Bodeau J, Tuch BB, Siddiqui A, Lao K, Surani MA (2009) mRNA-Seq whole-transcriptome analysis of a single cell. Nat Methods 6:377–382
7. White AK, VanInsberghe M, Petriv OI, Hamidi M, Sikorski D, Marra MA, Piret J, Aparicio S, Hansen CL (2011) High-throughput microfluidic single-cell RT-qPCR. Proc Natl Acad Sci USA 108:13999–14004
8. Klatsky PC, Wessel GM, Carson SA (2010) Detection and quantification of mRNA in single human polar bodies: a minimally invasive test of gene expression during oogenesis. Mol Hum Reprod 16:938–943

9. Davies D (2009) Book review: "Flow Cytometry: A Basic Introduction" by Michael G. Ormerod. Cytometry Part A 75A: n/a
10. Barer R, Saunders-Singer AE (1948) A new single-control micromanipulator. Q J Microsc Sci 89:439–447
11. Fink L, Bohle RM (2005) Laser microdissection and RNA analysis. Methods Mol Biol 293: 167–185
12. Elowitz M, Levine AJ, Siggia ED, Swain PS (2002) Stochastic gene expression in a single cell. Science 297:1183–1186

Chapter 18

Profiling Stem Cells Using Quantitative PCR Protein Assays

David Ruff and Pauline T. Lieu

Abstract

Reprogramming human somatic cells to induced pluripotent stem cells is an important avenue in biological research. Advances in the profiling of human stem cells have identified important pluripotency maintenance factors. The presence and relative expression levels of these essential markers are commonly used to define the pluripotency status and potential of reprogrammed stem cells. We reprogram human dermal fibroblasts with four transcription factors, OCT3/4, SOX2, KLF4, and cMYC delivered by viral vectors. We describe a real-time quantitative PCR methodology to quantify the levels of key protein factors and examine the kinetics during reprogramming as well as comparing protein expression in different iPS clones. This report describes three applications of TaqMan® Protein Assays for reprogramming studies: (1) monitoring of reprogramming proteins over the induction time course, (2) characterization of pluripotent cells by protein expression profiles, and (3) identification of potential iPSC colonies in high-throughput screening protocols. This approach is fast, simple, sensitive and generates a pluripotency scorecard for reprogrammed stem cells.

Key words Induced pluripotent, Reprogrammed stem cells, Protein expression, Gene expression, Real-time PCR, Lentivirus vector, Immunoflourescence

1 Introduction

Human and mouse somatic cells can be reprogrammed to induce pluripotency by cellular expression of four key transcription factors, OCT3/4, SOX2, KLF4, and MYC (1–3). The simultaneous ectopic expression of a combination of proteins is readily accomplished by use of viral vectors. The intracellular levels of each of the transcription factors required for efficient reprogramming is largely unknown. Also, many of the molecular pathway alterations that are engaged in this process remain to be elucidated. Here, we employ TaqMan® Protein Assays to measure transcription factor expression levels in cells undergoing reprogramming. This method is a rapid, straight-forward, selective and sensitive approach for protein quantification. The assay usually needs less than 100 cell equivalents or 10 ng total cellular protein. The procedure uses the proximity

Uma Lakshmipathy and Mohan C. Vemuri (eds.), *Pluripotent Stem Cells: Methods and Protocols*, Methods in Molecular Biology, vol. 997, DOI 10.1007/978-1-62703-348-0_18, © Springer Science+Business Media New York 2013

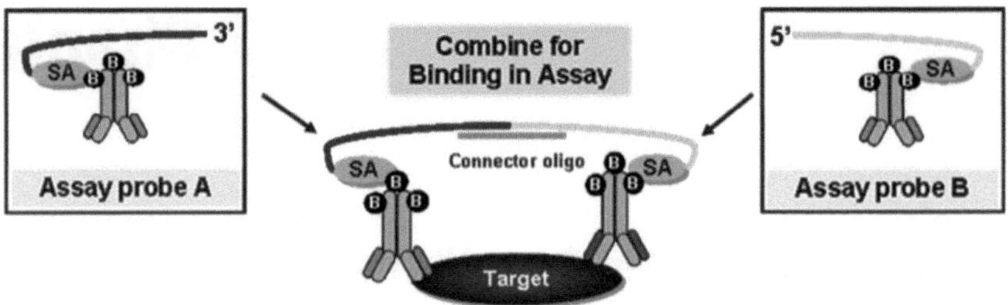

Fig. 1 TaqMan® Protein Assays Overview

ligation assay (PLA™) for real-time PCR detection (4). This approach relies on two oligonucleotide-conjugated antibodies to bind to antigen in unfractionated cell lysates (Fig. 1).

The antigen–antibody interactions create a close proximity of the two oligonucleotides. A complimentary sequence connector oligonucleotide hybridizes to the ends of the two proximal oligonucleotides. Addition of a DNA ligase enzyme allows for the completion of the proximity ligation event. The full-length product serves as an amplicon template for real-time quantitative PCR (qPCR) detection. By using the same starting cell culture for both protein and mRNA qPCR measurements, direct gene to protein correlations can be investigated (5). This is a highly useful tool since many studies report poor concordance between gene and protein expression due to a cadre of influences from posttranscriptional, posttranslational, mRNA/protein turnover rates and microRNA impacts. We also describe use of cellular genomic DNA as an endogenous control to normalize assay inputs more accurately to cell inputs. This is particularly useful when screening sample limited specimens such as stem cell colonies. We utilize the 18S ribosomal genomic DNA sequence to produce Cq values for data normalization by the delta Cq method (6). Combining these qPCR tools into parallel experimental workflows, correlation of protein and mRNA levels of these transcription factors during reprogramming has been previously described (7).

The assay workflow is simple: unfractionated cell lysate is combined with two TaqMan® Protein Assay probes, and after a 1 h binding incubation step, a connector oligonucleotide and DNA ligase enzyme complete the reaction. Unbound TaqMan® Protein Assay probes produce very low background ligation templates while bound probes produce large numbers of templates. The key reagent is the TaqMan® Protein Assay probe. TaqMan® Protein Assay probes are made by combining biotinylated-antibody with streptavidin-oligonucleotide conjugates. Two oligonucleotide sequences are used distinguish the TaqMan® Protein Assay probe A and Assay probe B. Monoclonal antibody pairs can be used to make Assay probes A and B. Alternatively, a single polyclonal antibody can be split into two tubes and each acquires either the A or B oligonucleotide sequence. (B = biotinylated moieties on antibody, SA = streptavidin, 3′ = 3′end-free oligonucleotide, 5′ = 5′end-free oligonucleotide).

2 Materials

All are from Life Technologies™ unless indicated otherwise.

2.1 Protein Assay and Real-Time PCR

1. Protein Expression Sample Preparation Kit (includes TaqMan® Protein Assay resuspension solution) 4405443.
2. TaqMan® Protein Assays Open Kit 4453745.
3. TaqMan® Protein Assays Core Reagents Kit 4448591.
4. Human KLF4 Biotinylated Affinity Purified PAb, Goat IgG antibody (BAF3640 R&D Systems, Minneapolis, MN, USA).
5. OCT3/4, SOX2, LIN28 TaqMan® Protein Assays 4405489, 4405495, 4405477.
6. 18S TaqMan® Ribosomal RNA Control Reagents 4308329 (for 18S gDNA assay).
7. MicroAmp® Optical 96-Well Reaction Plate 403012.
8. MicroAmp® Optical Adhesive Film 4311971.
9. TaqMan® Fast Universal PCR Master Mix (2×) for 18S assay 4352042.
10. MicroAmp® Fast Optical 96-Well Reaction Plate 4346906.
11. Real-time PCR system: Applied Biosystems StepOnePlus™ real-time PCR system 4376600.

2.2 Media Components and Reagents

1. Viral vectors expressing OCT3/4, SOX2, KLF4, and MYC.
2. Neonatal human dermal fibroblasts catalog #C-004-5C.
3. D-MEM with GlutaMAX™-I (high glucose) 10569-010.
4. KnockOut™ D-MEM/F-12 10660-012.
5. Fetal Bovine Serum (FBS), ES-Cell Qualified 16141-079.
6. KnockOut™ Serum Replacement 10828-028.
7. β-Mercaptoethanol (1,000×), liquid 21985-023.
8. FGF-basic, AA 1-155 Recombinant Human PHG0264.
9. GlutaMAX™-I Supplement 35050-061.
10. Trypsin/EDTA Solution 25300-054.
11. TrypLE™ Select Cell Dissociation Reagent 12563-011.
12. Geltrex™ hESC-qualified Reduced Growth Factor Basement Membrane Matrix A14133-02.
13. StemPro® hESC SFM A1000701.

2.3 Prepare Human iPSC Medium to Culture on Feeders (Inactivated MEFs)

To prepare 100 mL of human iPSC medium, aseptically mix the components listed below. Human iPSC medium can be stored at 2–8°C for up to 1 week.

KnockOut™ D-MEM/F-12	89 mL
KnockOut™ SerumReplacement—20%	10 mL
MEM nonessential amino acids solution	1 mL
GlutaMAX™-I Supplement 100×	1 mL
β-Mercaptoethanol 55 mM	182 μL
Penicillin–streptomycin (optional)	1 mL
Basic FGF 10 μg/mL	40 μL

Prepare the iPSC medium without bFGF, and then supplement with fresh bFGF when the medium is used.

2.4 Prepare Feeder Free Medium with StemPro® hESC SFM

For 100 mL of human feeder free medium, aseptically mix the components listed below. This medium can be stored at 2–8°C for up to 1 week.

D-MEM with GlutaMAX™-I	90.8 mL
StemPro hESC SFM growth supplement	2 mL
β-Mercaptoethanol 55 mM	182 μL
Basic FGF 10 μg/mL	80 μL
BSA 25%	7.2 mL

Prepare the feeder free medium without bFGF, and then supplement with fresh bFGF when the medium is used.

3 Methods

3.1 Time Course Study of Fibroblasts Transduced with Viruses Carrying OCT3/4, SOX2, and KLF4

1. Two days before transduction, plate human fibroblast cells onto two 6-well plates at the appropriate density of 100,000 cells per well.
2. Transduce fibroblasts with virus of choice carrying OCT3/4, SOX2, and KLF4 at the MOI of 3–5.
3. Next day, replace virus with fibroblast medium.
4. To examine protein levels during reprogramming, 4–6 days post transduction, seeds cells at 50,000 per well in 24-well plates using 0.5 mL of TrypLE™ Select reagent or 0.05% trypsin/EDTA following the procedure recommended by the manufacturer and incubate at room temperature (see Note 1). When the cells have rounded up (1–3 min later), add 2 mL of fibroblast medium into each well, and collect the cells in a 15-mL conical centrifuge tube. Centrifuge the cells at 200 × *g* for 4 min, aspirate the medium, and resuspend the cells in an

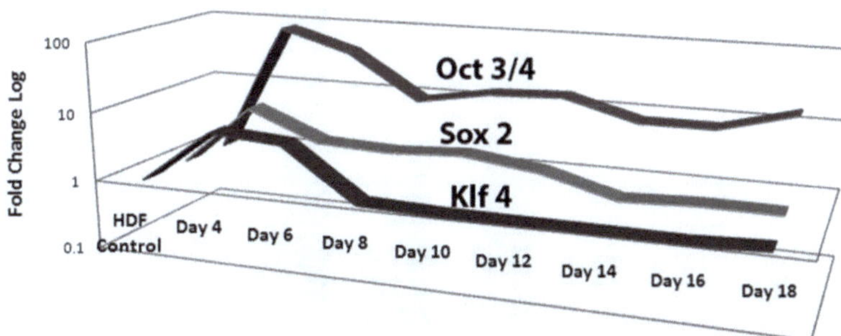

Fig. 2 Kinetics and levels of KLF4, SOX2, and OCT3/4 protein expression in human dermal fibroblast-fetal transduced with lentivirus. At day 4 post-transduction, cells were harvested and transferred to iMEFs (irradiated mouse embryonic fibroblasts) or feeder-free conditions. Every other day, cells on feeder-free plates were harvested for protein expression assays. TaqMan® Protein Assays time course and levels of KLF4, SOX2, and OCT3/4 expression transduced by lentivirus from day 4 to day 18. Samples are normalized against the endogenous reference 18S ribosomal genomic DNA and plotted (log10 scale) relative fold change to the expression level of un-transduced human dermal fibroblasts (HDF control)

appropriate amount of fibroblast medium. Plate 50,000 cells per well in a 24-well plate on feeder free condition with StemPro® hESC SFM.

5. Every 48 h, one well was sacrificed for each TaqMan® Protein Assay time point.
6. Follow the Taqman® Protein Assays below (Subheadings 3.4–3.11).
7. Example of data from a time course experiment is displayed in Fig. 2.

3.2 Identification of iPS Clones by TaqMan® Protein Assays

Follow reprogramming method of choice to generate induced pluripotent stem cells. Once colonies emerge, apply TaqMan® Protein Assays to screen for individual iPSC colonies for protein expression of the pluripotent markers, OCT3/4, SOX2, and LIN28.

1. Transfer a single colony manually from the reprogramming experiment to one well of a 6-well plate. Let colonies grow in iPSC medium for 4–5 days.
2. Pick 1–2 colony manually into 1.5 mL microfuge tube, treat with 0.05% trypsin/EDTA following the procedure recommended by the manufacturer and incubate at room temperature.
3. Count total cell number by standard method.

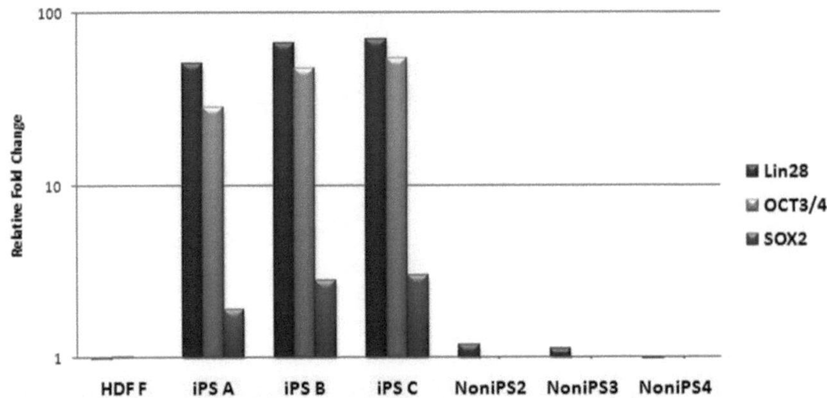

Fig. 3 TaqMan® Protein Assays to screen for potential iPSC (iPSA, B, and C) versus Non-iPSC (NoniPS2,3, and 4) colonies. All six lines stained positive for SSEA4+ and were then subjected to Taqman® Protein Assays for OCT3/4, SOX2, and LIN28. Samples are normalized against the endogenous reference 18S ribosomal genomic DNA and plotted (log10 scale) relative fold change to the expression level of un-transduced human dermal fibroblasts (HDF F)

4. Wash cells with culture medium and microcentrifuge to remove the trypsin solution.
5. Resuspend cells in the appropriate volume of TaqMan® Protein Assay resuspension solution to obtain 250 cells per μL.
6. Follow the Taqman® Protein Assays below (Subheadings 3.4–3.11).
7. Example of iPS clone screening results is shown in Fig. 3.

3.3 Characterization of iPS Clones by TaqMan® Protein Assays

Follow reprogramming method of choice to generate induced pluripotent stem cells. Once colonies emerge, apply TaqMan® Protein Assays to characterize individual iPSC line for protein expression of the pluripotent markers, OCT3/4, SOX2, and LIN28.

1. Pick 1–2 colony manually into 1.5 mL microfuge tube, treat with 0.05% trypsin/EDTA following the procedure recommended by the manufacturer and incubate at room temperature.
2. Count total cell number by standard method.
3. Wash cells with culture medium and microcentrifuge to remove the trypsin solution.
4. Resuspend cells in the appropriate volume of TaqMan® Protein Assay resuspension solution to obtain 250 cells per μL.
5. Follow the Taqman® Protein Assays below (Subheadings 3.4–3.11).
6. Example of iPS clone characterization data is illustrated in Fig. 4.

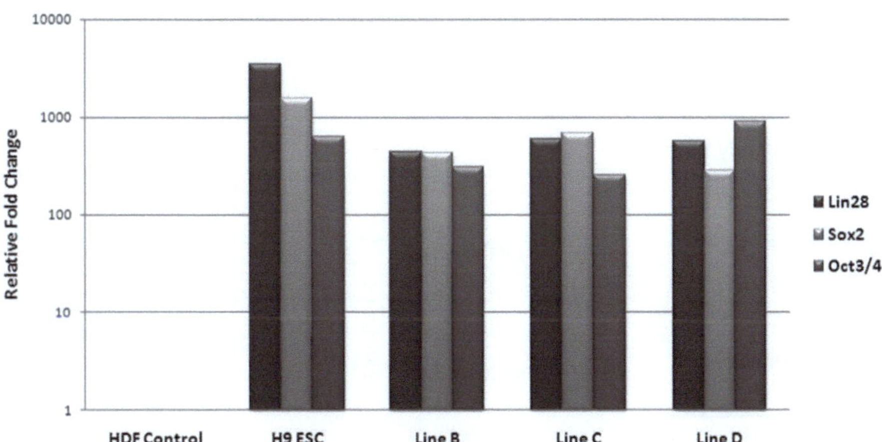

Fig. 4 Comparison of levels of pluripotent markers in WA09 ESC and three different iPSC lines by TaqMan® Protein Assays. Results show relative quantitative data of OCT3/4, SOX2, and LIN28. Samples are normalized against the endogenous reference 18S ribosomal genomic DNA and plotted (log10 scale) relative fold change to the expression level of un-transduced human dermal fibroblasts (HDF Control)

3.4 TaqMan® Protein Assays: Example of Cell Line Culture for Lysate

We use SOX2 TaqMan® Protein Assay as an example with six cell lysate samples and one No Protein Control (NPC), for a total of seven samples. Reagents and buffers are supplied in TaqMan® Protein Assay kits. The workflow overview is outlined in Fig. 5.

1. A total of 50,000 cells was transferred to a 1.5 mL microfuge tube. The cells were centrifuged at 2,000 × g for 10 min at 4°C.
2. The cell pellet was directly resuspended in 100 μL of TaqMan® Protein Assay resuspension solution. After complete resuspension, a 100 μL aliquot of TaqMan® Protein Assay lysis reagent was added and the mixtures thoroughly mixed by gently vortexing for 15 s. The final cell equivalent concentration is 250 cells per μL. This lysate is stable at 4°C for 2–3 weeks, store frozen at −20°C for longer periods.
3. To conduct TaqMan® Protein Assays, a 2 μL aliquot of the lysate is used directly in the probe binding step.

3.5 TaqMan® Protein Assays: Binding Reaction (SOX2 Example)

1. Preparation of the SOX2 TaqMan® Protein Assay proximity probe working solution.
2. Each sample will be assayed in triplicate—make an excess volume of proximity probe to allow for pipetting margin. Seven samples in triplicate will be 21 wells. Prepare 60 μL of proximity probe (sufficient for 30 binding reactions):

Pipet Proximity Probe Dilution Buffer to a microfuge tube	54 μL
Pipet SOX2 proximity probe A	3 μL
Pipet SOX2 proximity probe B	3 μL

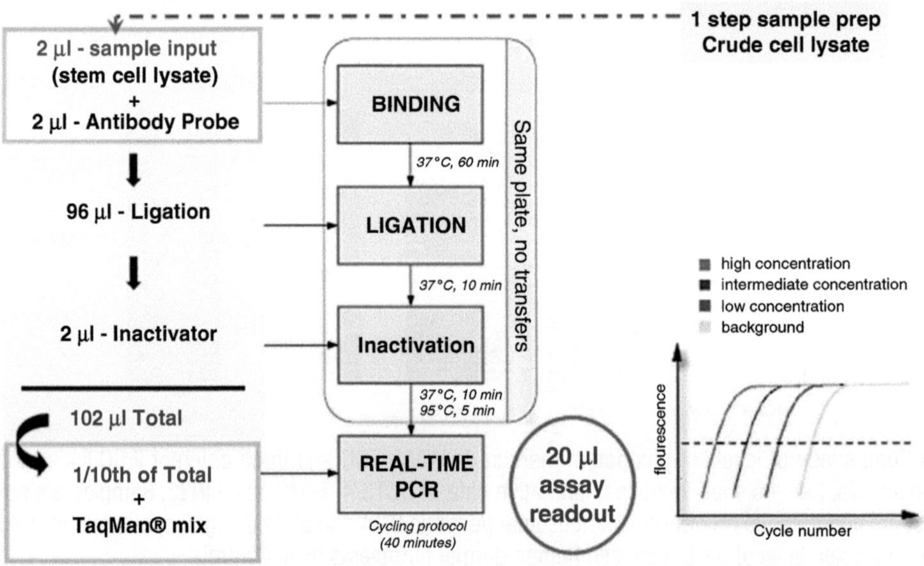

Fig. 5 TaqMan® Protein Assays Workflow overview. After generation of the cell lysate, the PLA workflow is divided into four parts: (1) binding reaction, (2) ligation reaction, (3) ligase inactivation, and (4) Real-time quantitative PCR. Work steps are described in detail for each procedure section. All work steps are sequentially ordered in each section

3. Gently mix the solution by pipetting up and down three times, be careful to avoid introducing bubbles. Spin down briefly to collect contents.
4. This should be done just prior to the experiment. Store working solution on ice.

3.6 Binding Plate Setup

1. The binding step is carried out in a 96-well PCR plate (MicroAmp® Optical 96-Well Reaction Plate and use MicroAmp® Optical Adhesive Film). Use of multiple channel pipettor facilitates the workflow. The 96-well plate should be on ice during the liquid delivery steps (this reduces evaporation). The proximity probe working solution is first delivered into the bottom of each assay well.
2. TaqMan® Protein Assay Liquid handling guidelines: When pipeting the proximity probes and lysate, be certain each pipet tip delivers exactly 2 μL—any volumetric variation will generate significant noise in the TaqMan® Protein Assay Cq values. The liquids should be dispensed to the very bottom of each

well—no mixing. IMPORTANT: Do not manually mix the 2 μL aliquots in the binding plate wells—instead centrifuge the PCR plate. The reactants will diffuse rapidly and sufficiently in the small volume.

3. A handy approach is to aliquot 18 μL of SOX2 proximity probe working solution into three adjacent PCR tray wells. Then use a multiple channel pipettor with three tips to deliver 2 μL×7 deliveries to fill the binding reaction plate in a 3×7 layout grid.
4. Next, the cell lysates are delivered and combined with the proximity probe. Each sample should be assayed in triplicate. Use a multiple channel pipettor with seven tips to deliver the lysate samples (and No cell control) three times to the binding reaction plate. Use new pipet tips for each pipetting step. The plate is sealed with an adhesive cover, briefly centrifuged, and the binding step incubation commenced at 37°C.

3.7 Binding Step Procedure

1. Incubate the binding reaction plate for 1 h at 37°C.

3.8 Ligation Reaction

1. Dilute DNA Ligase with Ligase Dilution Buffer: Transfer 1.0 μL of DNA Ligase to 500 μL DNA ligase dilution buffer (provided in kit). This diluted Ligase should be made fresh for each experiment.
2. Dilute Ligation buffer to 1× and add diluted ligase as follows (make sufficient ligase reaction mix for 30 reactions):

Component	1× Reaction volume (μL)	30× Reaction volume (μL)
20× Ligation buffer	5	150
H_2O	90	2,700
Diluted ligase	1	30
Total	96	2,880

3. Add 96 μL ligation mix to each binding reaction. Reseal plate and incubate the ligation reaction in 37°C for 10 min.
4. Ligation mix step liquid handling guidelines: recommend using reservoir and multichannel pipettor and new tips for each pipetting step. Also mix with binding products once by pipetting up and down.

3.9 Ligase Inactivation Reaction

1. Terminate the ligation reaction by adding 2 μL of Inactivation protease solution (diluted 1:1,000 in 1× PBS just prior to use: Add 999 μL of room temperature PBS to dilute 1 μL of concentrated protease from the supplied microamp tube).

Place the 1 mL of diluted protease solution into a reagent reservoir and use a 12-channel pipettor to deliver 2 μL to each well. Additional mixing is not required—sufficient Inactivation protease activity diffuses throughout the sample in the 10 min timeframe.

2. Reseal the plate and incubate at 37°C for 10 min followed by 5 min at 95°C to heat kill the protease.
3. Protease-treated TaqMan® Protein Assay products can be stored at 4°C for up to 2 days.

3.10 Real-Time PCR Assay Plate Setup

The TaqMan® Protein Assay reaction products are usually transferred into a real-time PCR assay immediately after the protease heat-kill step. The simplest PCR plate setup approach is to use a multiple channel pipettor to transfer the same sample layout grid pattern from the binding reaction plate to the real-time PCR assay plate. TaqMan® Protein Assay uses 20 μL reaction volumes for the qPCR detection.

3.11 Real-Time PCR Reaction and Readout

Prepare qPCR reaction master mix as follows for the SOX2 TaqMan® Protein Assay products (since there are 21 samples, make excess master mix—for 25 reactions):

Component	1× Reaction volume (μL)	25× Reaction volume (μL)
20× Primers and Probe	1	25
Fast UMM	10	250
Total	11	275

1. Aliquot 11 μL of the PCR reaction mix and then add 9 μL of the protease-treated-ligation product in a 96-well real-time PCR reaction plate (keep plate on ice to minimize evaporation during reagent transfer). For Fast qPCR cycling, use the MicroAmp® Fast Optical 96-Well Reaction Plate.
2. Seal plate with an optical blanket and centrifuge at 164 × *g* for 2 s.
3. Readout on a 7900 Fast or StepOnePlus™ instrument. Fast cycling conditions are recommended, but comparable results can be obtained with standard conditions. For StepOnePlus™, select cDNA as the template. Use the FAM dye layer with no quencher setting. Analyze the plate run using a threshold setting of 0.2 with automatic baseline.
4. The ΔCq values were calculated for each protein sample. To enable this approach, in parallel follow the steps in Subheadings 3.10 and above in 3.11, except use the TaqMan Ribosomal primer and probes in place of the TaqMan® Protein

Assay primer/probe mix. The 18S dye layer on the qPCR system should be set to VIC. The 18S gDNA TaqMan® Protein Assay Cq values were subtracted from the corresponding TaqMan® Protein Assay sample. The ΔCq method was used for calculations to determine the fold difference.

5. Because the same cell inputs for TaqMan® Protein Assay measurements were also subjected in parallel to 18S genomic ribosomal DNA assays to obtain endogenous control Cq, the ΔCq method is easily applied. The 18S Cq value normalizes for any variations in cell counting. The ΔCq between the 18S gDNA reference and the protein assay for each cell lysate sample was calculated. The day 0 cell sample was designated as the calibrator normalizer control for each experiment. To determine the fold difference using the ΔΔCq method, the slope of each protein target assay was obtained by performing TaqMan® Protein Assays with a 1:4 serially diluted lysate using 500, 125, 31 and 7.8 cells (see Note 2) (6). TaqMan® Protein Assay efficiency is calculated based on the equation: $= 10^{(-1/\text{slope})} - 1$. The 18S gDNA reference assay efficiency is 1.0 (100%). Fold differences between the different samples was determined with the following equation, fold difference = $((1 + \text{protein assay efficiency})^{\Delta\Delta Cq})$. See Note 3 for example calculation.

4 Notes

1. Because the cells can be very sensitive to trypsin at this point, minimize trypsin exposure time and incubate the cells at room temperature.
2. TaqMan® Protein Assay efficiency calculation. This method requires a TaqMan® Protein Assay Cq standard curve for each Assay antibody-probe set (determine KLF4, LIN28, OCT3/4, and SOX2 individually). A convenient approach is to dilute a stem cell lysate serially 1:4 from 500, 125, 31, and 7.8 cells per assay well (in triplicate). A linear trendline can be readily obtained by plotting average Cq values vs. log of cell input. The TaqMan® Protein Assay slopes are typically between −2.5 and −3.0. To calculate the TaqMan® Protein Assay efficiency, use this equation: $= 10^{(-1/\text{slope})} - 1$. If the slope is −2.5, then the observed assay efficiency $= 10^{(-1/-2.5)} = 1.51$. While many qPCR assays show an efficiency of 100% (1.0), TaqMan® Protein Assay often has a higher than 1.0 apparent slope efficiency due to several issues including antibody binding and ligation reaction cooperativity effects.
3. ΔΔCq method calculation example (for SOX2). Using the above example of slope −2.5. Assume the calibrator untransduced sample at roughly 500 cell input has a SOX2 TaqMan® Protein

Assay Cq of 30.0 and 18S of 17.0. A transduced cell input at about 500 cells has SOX2 TaqMan® Protein Assay Cq of 26 and 18S of 18.0. The calibrator ΔCq = 30.0 − 17.0 = 13.0. The 500 cell input ΔCq = 26.0–18.0 = 8.0. To calculate the $\Delta\Delta$Cq, 13.0–8.0 = 5.0. Fold difference in SOX2 expression between the calibrator and the transduced cell = $2.51^{(5)}$ = 100-fold.

Acknowledgment

This work was supported by Life Technologies Corporation.

References

1. Takahashi K, Yamanaka S (2006) Induction of pluripotent stem cells from mouse embryonic and adult fibroblast cultures by defined factors. Cell 25:663–676
2. Lowry WE, Richter L, Yachechko R, Pyle AD, Tchieu J, Sridharan R, Clark AT, Plath K (2008) Generation of human induced pluripotent stem cells from dermal fibroblasts. Proc Natl Acad Sci USA 105:2883–2888
3. Liu Y, Thyagarajan B, Lakshmipathy U, Xue H, Lieu P, Fontes A, MacArthur CC, Scheyhing K, Rao MS, Chesnut JD (2009) Generation of platform human embryonic stem cell lines that allow efficient targeting at a predetermined genomic location. Stem Cells Dev 18: 1459–1472
4. Fredriksson S, Gullberg M, Jarvius J, Olsson C, Pietras K, Gústafsdóttir SM, Ostman A, Landegren U (2002) Protein detection using proximity-dependent DNA ligation assays. Nat Biotechnol 20:473–477
5. Swartzman E, Shannon M, Lieu P, Chen SM, Mooney C, Wei E, Kuykendall J, Tan R, Settineri T, Egry L, Ruff D (2010) Expanding applications of protein analysis using proximity ligation and qPCR. Methods 50:S23–S26
6. Livak KJ, Schmittgen TD (2001) Analysis of relative gene expression data using real-time quantitative PCR and the 2(-Delta Delta C(T)). Methods 25:402–408
7. Ruff D, MacArthur C, Tran H, Bergseid J, Tian J, Shannon M, Chen SM, Fontes A, Laurent L, Swartzman E, Taliana A, Rao M, Lieu P (2012) Applications of quantitative polymerase chain reaction protein assays during reprogramming. Stem Cells Dev 21(4):530–538

Part IV

Pluripotent Stem Cell Modification and Applications

Chapter 19

Labeled Stem Cells as Disease Models and in Drug Discovery

Catharina Ellerström, Raimund Strehl, and Johan Hyllner

Abstract

Human pluripotent stem cells provide unique possibilities for in vitro studies of human cells in basic research, disease modeling as well as in industrial applications. By introducing relevant genome engineering technology, and thereby creating, for example, reporter cell lines, one will facilitate and improve safety pharmacology, toxicity testing, and can help the scientists to better understand pathological processes in humans. This review discusses how the merger of these two fields, human pluripotent stem cells and genome engineering, form extremely powerful tools and how they have been implemented already within the scientific community.

In sharp contrast to immortalized human cell lines, which are both easy to expand and very simple to transfect, the genetically modified pluripotent stem cell line can be directed to a specific cell lineage and provide the user with highly relevant information. We highlight some of the challenges the field had to solve and how new technology advancements has removed the early bottlenecks.

Key words Human, Pluripotent stem cells, Genome engineering, Drug discovery, Disease models

1 Introduction

1.1 Human Pluripotent Stem Cells

We are approaching one and a half decades since the first human embryonic stem (hES) cells were derived (1) and indeed, it has been a tremendous development within the stem cell field. Not only have we gained insights into undifferentiated human pluripotent stem (hPS) cell characteristics, but we have also gained a lot of knowledge around how to direct lineage specific differentiation. No longer are our culture conditions based on adapted protocol from the murine counterpart; instead defined media, adjusted to the need and growth characteristics of human pluripotent stem cells have been developed (2, 3). We have learned how to passage the pluripotent cells as single cells (4, 5), a prerequisite for many industrial applications. Finally, the methods of reprogramming human somatic cells into human induced pluripotent stem (hiPS) cells (6–8) have certainly revolutionized the stage of human pluripotent stem cells.

Uma Lakshmipathy and Mohan C. Vemuri (eds.), *Pluripotent Stem Cells: Methods and Protocols*, Methods in Molecular Biology, vol. 997, DOI 10.1007/978-1-62703-348-0_19, © Springer Science+Business Media New York 2013

Not too far back in time the industry had to rely on nonhuman systems, or on immortalized human cell lines, both for in vivo and in vitro studies. The majorities of the in vivo systems, are, or have been, based on animal studies. Not surprisingly, there are many significant differences between humans and animals with respect to patterns of development, expression of differentiation markers, as well as expression profiles of cytokines, cell cycle, and cell death-regulating genes. There are definitely also ethical issues related to these types of tests and they are not cost-effective, but rather the opposite. Many human systems, such as transformed and immortalized cell lines may not come with a similar ethical burden, but there are issues both with the biological and clinical relevance instead, and this in turn has a severe effect on the costs.

Human pluripotent stem cell derived models have the potential to revolutionize the fields of safety testing and drug discovery and to increase the human relevance and the predictive value of in vitro models, but major endeavors in human pluripotent stem cell related research will be required. Now, the cell culture technical bottlenecks associated with human pluripotent stem cells have been eliminated, and the arrival of hiPS cells has made the stem cell community bigger. The pace of discovery consequently increases and there are no doubts any longer, human pluripotent based assays and disease models are here to stay.

1.2 Specialized Cells Used Within Drug Discovery

It is no news that pharmaceutical companies today are facing increasing developmental costs in combination with a decreased success rate in bringing novel and safe compounds to the market (9). In general, it takes between 10 and 15 years to develop a new drug and the costs associated with the process add up to approximately one billion USD. The two major reasons for late-stage attrition during drug development are lack of safety and efficacy (10). Obviously it is very much in the pharmaceutical industries' interest to identify and employ improved in vitro models based on readily available, physiologically relevant human cells. These new systems must provide a rapid throughput and a high predictive rate, with few false-positive results. Furthermore, the new cell-based assays need to be of high and robust quality not to jeopardize the data output. Human pluripotent stem cells can provide the users with a reliable and reproducible source of cells, but how about the cells differentiated thereof? Cardiotoxicity is highly relevant for the pharmaceutical industries as candidate compounds can, and do often, have unintended effects on the potassium current that may result in QT-prolongation and potentially lead to fatal arrhythmia (10). The currently employed in vitro models are based on different animal-derived primary cardiomyocytes and therefore unfortunately, in many cases it is not before the initial clinical trials that the test compounds de facto come in contact with human cardiomyocytes. As the field of hPS cell derived cardiomyocytes is a very active research field, several different groups have published protocols for how to direct hPS cells to cardiomyocyte cells (for review see Vidarsson, (11)). The general conclusion is that morphology, gene expression, ion channel functionality as well as response to pharmacological compounds of these hPS cell derived cardiomyocytes show great similarities in particular with embryonic and fetal cardiomyocytes. No doubt, soon we will have access to hPS cell derived cardiomyocytes with a more adult phenotype. As the availability of quick and easy end-points simplifies the readout procedure substantially and allows larger scale experiments to be carried out, the utility of reporter lines is obvious when screening for compound effects on hPS cell derived cardiomyocytes on a larger scale. To take it one step further, several of the recently published enrichment studies illustrate the advantage of using the transgenic approach based on cardiac specific drug resistance selection (12–14) Significantly increased purity of cardiomyocytes was achieved, by utilizing a bicistronic vector expressing EGFP gene for direct visualization and drug resistance gene for positive selection (>90% purity) under the control of the MYH6 promoter (13). In addition, a negative selection strategy employing a proliferation-base suicide system to eliminate noncardiac cells was evaluated in parallel (13).

Another cell type, highly relevant for drug discovery is the hepatocyte. So far, the majority of hepatic toxicity studies have been based on either animal studies or on, for example, human, primary hepatocytes. The disadvantage of the first choice has in principle already been discussed above, and for the latter it has unfortunately been shown that the key enzymes involved in metabolizing known and potential compounds, as well as several of the transporter functions, rapidly decrease and are lost when cells are transferred to an in vitro situation (15). Obviously, there is an urgent demand for new and better model systems that more reliably predict effects of the compound candidates.

Today scientists can guide the hPS cells in their differentiation into populations of hepatocyte-like cells, but there is still room for optimization of the protocols, and homogeneity within the culture as well as level of differentiation needs to be improved. Nevertheless hPS cell derived hepatocytes have already now proven useful for toxicity evaluations. Yildirimman et al. have recently published an elegant report (16) in which hepatocyte-like cells derived from human embryonic stem cells were used to provide a human relevant in vitro test system to evaluate carcinogenic hazard of chemicals. The results demonstrate, for the first time, the potential of human embryonic stem cell-derived hepatic cells as an in vitro model for hazard assessment of chemical carcinogenesis, although it should be noted that more compounds are needed to test the robustness of the assay. Also this field has taken advantage of genetic engineering and reporter cell lines. One of the earliest studies used differentiated hES cells that had been enriched after transduction with a lentiviral vector containing the green fluorescent protein (GFP) gene driven by the alpha1-antitrypsin promoter (17). The GFP gene was expressed only in committed hepatocyte progenitors and hepatocytes and could therefore be used for purification of the desired cell type.

1.3 Disease Models

In vitro cultures, based on immortalized cell strains from diseased atients has been an invaluable resource for medical research for half a century, starting with the amazing HeLA cells (18). These systems are human but, as discussed previously, largely limited to tumor cell lines or transformed derivatives of native tissues, with a minor relevance to the in vivo situation. Therefore it was no big surprise that the field of disease modeling early on realized the huge potential of human pluripotent stem cells. Initially, all derivations were made from surplus blastocysts from conventional IVF treatments, but rather soon, as the initial technical hurdles had been passed, one began to create ES cell lines from disease-affected blastocysts specifically. These hES cell lines, derived with specific genotypes related to severe human diseases, such as Charcot-Marie-Tooth disease, Fragile X Syndrome Hemophilia B, Huntington's disease, and Marfan syndrome were generated from blastocysts that were determined to carry the disease through preimplantation genetic diagnosis (PGD). Consequently, the disease spectrum is therefore restricted to very severe and monogenetic disorders.

Since, 2007 the advent of hiPS cell technology has opened up exciting new avenues for generating disease-specific cell lines and modeling human diseases (for review, see Robinton and Daley (19)). No longer is the generation of patient/disease specific pluripotent stem cell lines restricted to PGD material, which significantly broadens the spectrum of potential disease models. Already 2008, dedicated scientists at Harvard announced that they had produced more than 20 different hiPS cell lines from patients with diseases such as adenosine deaminase deficiency-related severe combined immunodeficiency (ADA-SCID), Type I Diabetes, Duchenne and Becker muscular dystrophy, Huntington's disease, and Parkinson's (20). The study published by Brennand et al. (21) was particularly highlighted by the media for good reasons. The researchers isolated skin cells from a young man diagnosed with schizophrenia. The cells were successfully reprogrammed into hiPS cells and this disease model represents the first relevant in vitro model for such a condition. Such models will constitute very important research tools as the biological basis of mental diseases, such as schizophrenia and bipolar disorders still is poorly understood (21). Recently, a human cellular model of Alzheimer's disease has also been described. As many adults with Down's syndrome develop an early onset of Alzheimer's disease, scientists have generated hiPS cell lines from these patients. They found that cortical neurons generated from hiPS cells (and hES cells) from Down syndrome patients developed AD pathologies over months in culture, rather than years in vivo (22). Disease-specific pluripotent stem cell lines such

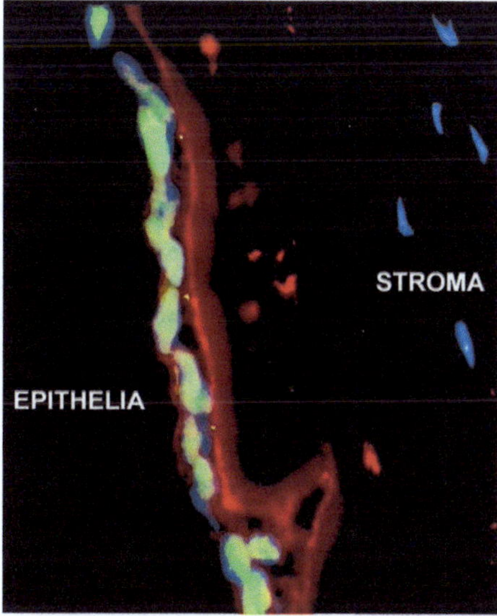

Fig. 1 A genetically modified human embryonic stem cell line, which express GFP under the constitutively expressed promoter EF1-alpha was used in a clinically oriented study in where donated and damaged human cornea received GFP positive hESC derived cells in vitro (43). The GFP-tagged cells made it possible to discern the interaction between donor and recipients cells in real time. Moreover, the GFP positive cells facilitated the previously tedious histological preparations during the more in depth morphological analysis: GFP positive cells (*green*), CK3 expression (*red*), Nuclear staining with DAPI (*blue*)

as the ones described above will provide an unprecedented opportunity to better understand pathological events on a cellular level as well as an excellent assay system for relevant and safer drug development.

There is no doubt that genetically engineered stem cells and their derivatives will facilitate disease modeling (see Fig. 1), drug screening and compound testing, and thereby provide us with as a better understanding of the underlying mechanisms of severe diseases. They are expected to lead to significantly increased throughput at reduced labor expenditure providing us with more accurate pharmaceutical drugs and with fewer severe side effects. For the pharma industry this potentially means fewer of the much dreaded and catastrophic late stage attritions of novel drugs.

2 Overview

As the characteristics of pluripotent stem cells provide a unique resource for genetic modification and subsequent experimental application, it was no surprise that the scientists behind this method were awarded the Nobel Prize in medicine year 2007. As the pluripotent stem cells, in contrast to all other normal cells, can theoretically be expanded indefinitely there is ample time for clone selection, expansion, banking, and the necessary quality control. This will provide scientists, both in academia and industry, with well-characterized stem cell lines as a robust platform for their studies. That circumvents the issues with batch to batch variation which comprises one of the major problems when primary cells are used. In contrast to the immortalized human cell lines, which are both easy to expand and often very simple to transfect, the genetically modified pluripotent stem cell line can be directed to a specific cell lineage and provide the user with highly relevant information.

The predecessor of human pluripotent stem cells, the mouse pluripotent stem field has since long been taking full advantage of the powerful combination of pluripotent stem cells and genetic engineering. Modified murine pluripotent stem cells have been employed as model system for tissue development as well as for disease modeling and several sophisticated methods have been developed and employed.

In contrast to the mouse pluripotent stem cell field, genetic modification of human pluripotent stem cells has been accompanied with challenges due to technical bottlenecks (e.g., ref. (23, 24)). Significant differences between the mouse and human pluripotent stem cells have limited the success rate when trying to translate genetic modification protocols from one species to another. Many of the difficulties have been associated with suboptimal culture conditions, with heterogeneous cell populations due to spontaneous differentiation, with lack of cell supply to thoroughly optimize the transfection event, with lack of survival of the hPS cells as individual cells. These difficulties have made hPS cells less amenable to genetic manipulation.

Today, robust culture protocols for feeder free culture of hPS cells are available (2, 3). Researchers are not limited anymore by tedious mechanical or cluster passaging procedures but can enzymatically passage and transfect hPS cells as single cell suspensions (4, 5) similar but yet not identical to mES cells. No doubt, the development of pluripotent reporter cell lines expressing, for example, GFP under the control of a selected and specific promoter would facilitate a whole range of experimental setups by providing a rapid readout that indicates the state of cells, but there are challenges yet to be resolved regarding robust and reproducible genetic modification of hPS cells. So far, viral transduction has been the easiest and fastest road to choose for genetic manipulation of hPS cells (e.g., (17, 25–27)). No doubt, at first glance, viral transduction, based on, for example, Lentiviral technologies may sound very attractive. They offer among other advantages a high efficiency of gene delivery. However, viral methods come with limitations, such as insertional mutagenesis by integrating viral vectors. Only rather small transgenes can be employed, which may have a negative impact on the goal to achieve high and stable expression. Further, the potential risk for gene disruption due to random integration further decreases their potential benefit, as lentiviral vectors clearly reveal an integration profile that favors transcriptionally active genes (28).

Recently, there have been several alternative systems that have emerged as nonviral alternatives such as CPhi31 Integrase (29), *Sleeping Beauty* Transposase (28), Zinc-finger nucleases (30, 31), and TALENs (32). The majority of these methods does not depend on random integration and have already proven useful. This suggests that even the previous challenges associated with genetic engineering of human pluripotent stem cells might soon be behind us, thus allowing us to take full advantage of genetic engineering in this exciting new field.

One alternative to achieve an increased frequency of nonviral integration is to employ DNA transposon, like the *Sleeping Beauty*. Evolutionary, the *Sleeping Beauty* Transposon originates from element that was able to colonize several fish genomes millions of years ago (33) but represents today a synthetic transposable element that has been modified to better serve its new purpose—enable robust and stable gene transfer on demand. The *Sleeping Beauty* transposon system is plasmid-based system, made of two components: (a) A transposon containing a gene-expression cassette of interest and (b) a source of transposase enzyme which mediates stable integration and reliable long-term expression of the gene of interest (28, 34). By transposing the expression cassette from a plasmid into the genome, continuous transcription of the transgene can be achieved (34). The efficiency of stable gene transfer employing this system is impressive—it ranges from 1 to 10% of the transfected cells when human blood cells are used (34). The *Sleeping Beauty* system integrates more or less with a random integration pattern, but may be less prone to integrate into transcribed genes or transcriptional regulatory regions compared to, for example, the lentivirus. These tend to target genes and their upstream regulatory regions (34). One limitation with this system is the cargo size since it appears to have an upper limited around 10 kb.

Indeed, the *Sleeping Beauty* has successfully been used in hPS cells. Wilber et al. (28) published 2007 a study in where the authors demonstrate that this system provides an efficient method to achieve stable genetic modifications of human ES cells.

Robust and stable expression of transgenes in human pluripotent stem cells can be obtained in several ways. Integration in hES cells using Jump-In technology provides an attractive system and takes advantages of semi-targeted integration (29). Initially it is a two-step procedure. In the first step, the aim is to create a R4 platform line. PhiC31 integrase mediated recombination is employed to stably integrate the DNA sequences, containing the R4 integrase target sequence together with a selection cassette, at specific

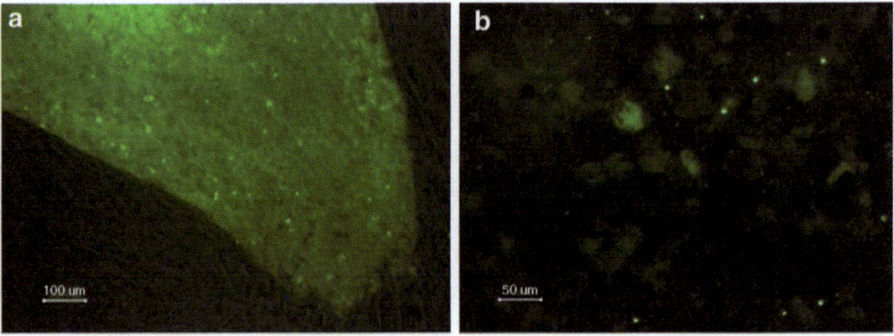

Fig. 2 The Jump-In technology provides the possibility to with a two-step procedure create isogenic lines, but with different promoters driving the GFP expression. In panel (**a**) a retargeted hES cells express GFP under constitutively expressed promoter. Panel (**b**) reveals hESC derived cells expressing GFP under an endodermal specific promoter

locations called *attP* sites. Selected and characterized clones, now containing the R4 sequence, should be banked as they now are an excellent platform for the second step, retargeting, mediated by R4 integrase. At this step, the DNA sequence of interest is site-specifically integrated into the platform line at the R4*attP* target site. A new round of selection is then initialized, this time for obvious reason a different selection substance is employed. The system greatly simplifies the creation of transgenic lines expressing the gene of interest since it is designed to eliminate clonal selection at the retargeting stage (35). Only successful retargeting will result in an expression of the selections cassette for this second step and consequently, no false positive clones can be obtained, minimizing the work since one can choose to pool all the surviving cells (35). Figure 2 reveals two different reporter lines created with the Jump-In technology, one expressing a constitutive active promoter, while the other expresses the reporter in hepatocyte-like cells only as the GFP gene expression is driven by an endodermal specific promoter. On the downside this approach is not always precise (35) and it has been associated with chromosomal rearrangements when studied in, for example, human fibroblasts (36).

Gene targeting has several benefits—and frankly, targeted genetic engineering of hPS cells is a prerequisite for taking full advantage of the merge of these two technologies, hPS cells and genetic engineering. Homologous recombination is a powerful technology which has been used extensively within the mES cell field, but again, it has proven more difficult in human pluripotent stem cells. Zwaka et al. reported 2003 the first homologous recombination in hES cells in where they successfully target HPRT1, the gene encoding hypoxanthine phosphoribosyltransferase-1 (HPRT1), and POU5F1, the gene encoding Oct-4 (23). Recent advances in improved culture conditions have had a definitive positive impact and consequently the numbers of publications based on homologous recombination in hPS cells is steadily increasing (24, 31, 37, 38). The technology builds on already existing mechanisms in the cells, where the targeting vectors are made so that they have homologous sequences flanking the regions of interest. Using this approach, the targeting vector is integrated into the specific gene of interest during cell cycle, but in practice this method demands extensive screening of resulting clones in order to identify the correctly integrated ones. Successful homologous recombination technology was used to insert GFP into the SOX17 locus, and thereby created a valuable research tool for endoderm oriented research (39). In order to achieve this, 200 resistant clones had to be screened in order to identify one with a homologous event and correct targeting. In general, the (low) targeting efficiency is related to factors such as accessibility of target site in the hPS cells, origin and design of target vectors, but there are other so far undefined factors too.

The possibility to engineer rare cutting nucleases, first described 1996 for Zinc finger nucleases (40) has opened up new possibilities. These precisely designed nucleases mediate site-specific genome modifications and may increase the probability of homologous recombination while significantly reducing the workload (41, 42). Unfortunately the Zinc finger nucleases are fairly expensive to manufacture, preventing many users to take advantage of them. Recently, a new player has entered the scene, the TALE nucleases, originally identified in plant pathogens to subvert host genome regulatory networks (32). They provide an attractive new alternative strategy, and engineered TALE nucleases have successfully been used to genetically engineer human PSCs (32). No doubt, more studies are required, but if these technologies meet up to their expectations and proves reliable and robust in PSC, a whole range new application for advanced disease modeling and drug therapy becomes doable.

Regardless of which gene targeting approach that is chosen, the foreign DNA must be efficiently delivered into the nucleus. There are several methods for introducing foreign DNA into hPS cells but they can in principle be divided into three groups: (a) Viral based technology, (b) transfection via physical methods, and (c) chemical based transfection. The viral methods are not discussed further as they not are within the scope for this review.

A common method to introduce foreign DNA into hPS cells is based on a physical method, electroporation. By exposing the hPS cells to a high voltage pulse (or multiple), transient pores are made in the cell membrane of the hPS cells and as a consequence of this, the membrane potential is increased which allows the negatively charged foreign DNA to move into the cell and hopefully into the nucleus. Early transfection protocols rendered extremely poor survival of the hPS cells, which either died due to the harsh electroporation conditions or due to suboptimal post-transfection culture conditions. Luckily, there is range of improvements that has been made, both concerning the electroporation devices, electroporation buffers and how to increase the pampering of the freshly transfected hPS cells. To avoid the phenomena of "boiled cells" one reduced the gap size of the two electrodes, decreased the surface area of the electrodes and thereby lowered the generated heat that often destroyed the cells. Further, most of the newly developed machines today allow more parameter adjustments such as pulse width, number of pulses and voltage. A classic electroporation device, such as GenPulser from Bio-Rad or BTX830 from Harvard apparatus, together with an optimized electroporation buffer will generate appropriate survival and transfection efficiency. This type of machine can easily transfect ten million cells at a time. A little device, with proven high efficiency and survival when employed on hPS cells is the Neon (former Microporator), from Life Technologies. Just like the BTX or the Genepulser, it allows the operator to control the length, voltage and number of pulses. In contrast to most other system, the transfection is performed in a pipette tip, allowing either 10 or 100 μl of cells and DNA. Consequently the cells can quickly return to the incubator, something that understandably is highly beneficial for their survival. On the downside one is forced to use the provided proprietary transfection buffers—on the upside, one can use the same buffer for all types of cells. The electroporation system used most successfully with hPS cells are probably is probably the AMAXA nucleofection system. Here everything come included, recommended settings as well as proprietary buffers that are exclusively adjusted for say hPS cells (or one of around 60 other cell types).

Generally all above mentioned technologies require the hPS cells to be in single cell suspension at the time of transfection. The combination of two factors, hPS cells dissociated to single cells and the electroporation conditions per se, puts the cells into a very challenging situation and any unfavorable handling by the operator jeopardizes the result as the cell survival rate will decrease dramatically. Minimizing the handling time is one factor. By using pampering culture condition with highly supporting feeders after the transfection (e.g., human foreskin fibroblasts) higher survival can be obtained. Finally, by adding a Rho Kinase Inhibitor to the single cell suspensions, a significant increase in cell survival will be achieved (5).

An alternative to electroporation would be to use chemical based transfection reagents. This process does not require any transfection device but is solely based on the use of chemical reagents. The majority of the chemical based reagents such as Lipofectamin, Fugene HD, Genejammer, GeneJuice, and ExGen 500 are added together with the plasmid directly to the medium of the hPS cells within the cell culture dish. This eliminates the harsh environment of an electroporation, and consequently more cells survive the transfection procedure. Yet the majority of the studies performed with chemical transfection report a low efficiency.

However, recently Yao et al. (41) have published a report describing homologous recombination in hPS cells, employing chemical transfection technology combined with a defined and feeder free culture system. One could assume that these improved new, more robust feeder free culture systems would also have a positive effect on these methods.

3 Perspectives

Human pluripotent stem cells, hES cells and hiPS cells, both provide unparalleled possibilities for in vitro studies of human cells in basic research, disease modeling as well as in industrial applications. By introducing relevant genetic engineering technology, such as labeled cells and reporter cell lines, one will facilitate and improve safety pharmacology, toxicity testing, and can help the scientists to better understand pathological processes in mankind.

Many of the difficulties associated with genetic engineering in hPS cells have originated from suboptimal culture conditions. Heterogeneous cell populations, cluster passaging, limited supply of cells and lack of a pampering post-transfection environment all have contributed to the difficulties. However, these roadblocks have been removed as we today have much improved culture systems that allow robust up-scaling, single cell dissociation and higher survival of the cells post-transfection. The next important step is to merge the stem cell technology with the emerging new tools for genetic modifications, such as Jump In and *Sleeping Beauty*. The event of homologous recombination should be greatly enhanced as engineered Zinc fingers or TALENs identify and cleave the targeted site with high precision.

In the future, genetic engineering of human pluripotent stem cells will allow us to precisely and accurately design the tools required for research and industrial application. Even though a number of challenges still have to be overcome in order to make this technology an everyday commodity, the possibilities have already been clearly demonstrated.

Acknowledgments

The authors are grateful to Drs. Bhaskar Thyagarajan, Jon Chesnut, and Mahendra Rao for fruitful collaboration and for providing the tools for the Jump In technology. The authors would also like to thank Drs. Charles Hanson, Ulf Stenevi, and Thorir Hardarson for the providing the picture with human corna, and Tina Nilsson, Jenny Johannisson, Dorra El Hajjam, Ingrid Rydström, and Jenny Lindqvist for the execution and characterization of the reporter lines shown in the figures of this chapter.

References

1. Thomson JA, Itskovitz-Eldor J, Shapiro SS, Waknitz MA, Swiergiel JJ, Marshall VS, Jones JM (1998) Embryonic stem cell lines derived from human blastocysts. Science 282:1145–1147
2. Ludwig TE, Levenstein ME, Jones JM, Berggren WT, Mitchen ER, Frane JL, Crandall LJ, Daigh CA, Conard KR, Piekarczyk MS, Llanas RA, Thomson JA (2006) Derivation of human embryonic stem cells in defined conditions. Nat Biotechnol 24:185–187
3. Wang L, Schulz TC, Sherrer ES, Dauphin DS, Shin S, Nelson AM, Ware CB, Zhan M, Song CZ, Chen X, Brimble SN, McLean A, Galeano MJ, Uhl EW, D'Amour KA, Chesnut JD, Rao MS, Blau CA, Robins AJ (2007) Self-renewal of human embryonic stem cells requires insulin-like growth factor-1 receptor and ERBB2 receptor signaling. Blood 110:4111–4119
4. Ellerström C, Strehl R, Noaksson K, Hyllner J, Semb H (2007) Facilitated expansion of human embryonic stem cells by single cell enzymatic dissociation. Stem Cells 25:1690–1696
5. Watanabe K, Ueno M, Kamiya D, Nishiyayama A, Matsumura M, Wataya T, Takahashi JB, Nishikawa S, Nishikawa S, Muguruma K, Sasai Y (2007) A ROCK inhibitor permits survival of dissociated human embryonic stem cells. Nat Biotechnol 25:681–686
6. Takahashi K, Yamanka S (2006) Induction of pluripotent stem cells from mouse embryonic and adult fibroblast cultures by defined factors. Cell 12:663–676
7. Takahashi K, Tanabe K, Ohnuki M, Narita M, Ichisaka T, Tomoda K, Yamanaka S (2007) Induction of pluripotent stem cells from adult human fibroblasts by defined factors. Cell 131:861–872
8. Yu J, Vodyanik MA, Smuga-Otto K, Antosiewiz-Bourget J, Frane JL, Tian S, Nie J, Jonsdottir GA, Routti V, Stewart R, Sluvkin II, Thomson JA (2007) Induced pluripotent stem cell lines derived from human somatic cells. Science 21:1917–1920
9. Jensen J, Hyllner J, Björquist P (2009) Human embryonic stem cell technologies and drug discovery. J Cell Physiol 219:513–519
10. Sartipy P, Björquist P (2011) Human pluripotent stem cell-based models for cardiac and hepatic toxicity assessment. Stem Cells 29:744–748
11. Vidarsson H, Hyllner J, Sartipy P (2010) Differentiation of human embryonic stem cells to cardiomyocytes for in vitro and in vivo applications. Stem Cell Rev 6:108–120
12. Guddati AK, Kessler JA (2010) Novel method of single antibiotic selection of cells containing multiple stage-specific promoter-fluorophore plasmids. J Biomol Tech 2:18–24
13. Anderson D, Self T, Mellor IR, Goh G, Hill SJ, Denning C (2007) Transgenic enrichment of cardiomyocytes from human embryonic stem cells. Mol Ther 15:2027–2036
14. Xu XQ, Zweigerdt R, Soo SY, Ngoh ZX, Tham SC, Wang ST, Graichen R, Davidson B, Colman A, Sun W (2008) Highly enriched cardiomyocytes from human embryonic stem cells. Cytotherapy 10:376–389
15. Rodríguez-Antona C, Donato MT, Boobis A, Edwards RJ, Watts PS, Castell JV, Gómez-Lechón MJ (2002) Cytochrome P450 expression in human hepatocytes and hepatoma cell lines: molecular mechanisms that determine lower expression in cultured cells. Xenobiotica 32:505–520
16. Yildirimman R, Brolén G, Vilardell M, Eriksson G, Synnergren J, Gmuender H, Kamburov A, Ingelman-Sundberg M, Castell J, Lahoz A, Kleinjans J, van Delft J, Björquist P, Herwig R (2011) Human embryonic stem cell derived hepatocyte-like cells as a tool for in vitro hazard assessment of chemical carcinogenicity. Toxicol Sci 124:278–290
17. Duan Y, Catana A, Meng Y, Yamamoto N, He S, Gupta S, Gambhir SS, Zern MA (2007) Differentiation and enrichment of hepatocyte-like cells from human embryonic stem cells in vitro and in vivo. Stem Cells 25:3058–3068
18. Scherer WF, Syverton JT, Gey GO (1953) Studies on the propagation in vitro of poliomyelitis viruses IV. Viral multiplication in a stable strain of human malignant epithelial cells (strain HeLa) derived from an epidermoid carcinoma of the cervix. J Exp Med 97:695–710
19. Robinton DA, Daley GQ (2012) The promise of induced pluripotent stem cells in research and therapy. Nature 481:295–305
20. Park IH, Arora N, Huo H, Maherali N, Ahfeldt T, Shimamura A, Lensch MW, Cowan C, Hochedlinger K, Daley GQ (2008) Disease-specific induced pluripotent stem cells. Cell 134:877–886
21. Brennand KJ, Simone A, Jou J, Gelboin-Burkhart C, Tran N, Sangar S, Li Y, Chen G, Yu D, McCarthy S, Sebat J, Gage FH (2011) Modelling schizophrenia using human induced pluripotent stem cells. Nature 473:221–225
22. Shi Y, Kirwan P, Smith J, MacLean G, Orkin SH, Livesey FJ (2012) A human stem cell model of early Alzheimer's disease pathology in Down syndrome. Sci Transl Med 4:124–129
23. Zwaka TP, Thomson JA (2003) Homologous recombination in human embryonic stem cells. Nat Biotechnol 21:319–321
24. Braam SR, Denning C, van der Brink S, Kats P, Hochstenbach R, Passier R, Mummery CL (2008) Improved genetic manipulations of

human embryonic stem cells. Nat Methods 5:389–392

25. Gerrard L, Zhao D, Clark AJ, Cui W (2005) Stably transfected human embryonic stem cell clones express OCT4-specific green fluorescent protein and maintain self-renewal and pluripotency. Stem Cells 23:124–133
26. Liu YP, Dovzhenko OV, Garthwaite MA, Dambaeva SV, Durning M, Pollastrini LM, Golos TG (2004) Maintenance of pluripotency in human embryonic stem cells stably overexpressing enhanced green fluorescent protein. Stem Cells Dev 13:636–645
27. Zhou BY, Ye Z, Chen G, Gao ZP, Zhang YA, Cheng L (2007) Inducible and reversible transgene expression in human stem cells after efficient and stable gene transfer. Stem Cells 25:779–789
28. Wilber A, Linehan JL, Tian X, Woll PS, Morris JK, Belur LR, McIvor RS, Kaufman DS (2007) Efficient and stable transgene expression in human embryonic stem cells using transposon-mediated gene transfer. Stem Cells 25:2919–2927
29. Thyagarajan B, Liu Y, Shin S, Lakshmipathy U, Scheyhing K, Xue H, Ellerström C, Strehl R, Hyllner J, Rao MS, Chesnut JD (2008) Creation of engineered human embryonic stem cell lines using phiC31 integrase. Stem Cells 26:119–126
30. Hockemeyer D, Soldner F, Beard C, Gao Q, Mitalipova M, DeKelver RC, Katibah GE, Amora R, Boydston EA, Zeitler B, Meng X, Miller JC, Zhang L, Rebar EJ, Gregory PD, Urnov FD, Jaenisch R (2009) Efficient targeting of expressed and silent genes in human ESCs and iPSCs using zinc-finger nucleases. Nat Biotechnol 27:851–857
31. Zou J, Maeder ML, Mali P, Pruett-Miller SM, Thibodeau-Beganny S, Chou BK, Chen G, Ye Z, Park IH, Daley GQ, Porteus MH, Joung JK, Cheng L (2009) Gene targeting of a disease-related gene in human induced pluripotent stem and embryonic stem cells. Cell Stem Cell 5:97–110
32. Hockemeyer D, Wang H, Kiani S, Lai CS, Gao Q, Cassady JP, Cost GJ, Zhang L, Santiago Y, Miller JC, Zeitler B, Cherone JM, Meng X, Hinkley SJ, Rebar EJ, Gregory PD, Urnov FD, Jaenisch R (2011) Genetic engineering of human pluripotent cells using TALE nucleases. Nat Biotechnol 29:731–734
33. Ivics Z, Hackett PB, Plasterk RH, Izsvák Z (1997) Molecular reconstruction of Sleeping Beauty, a Tc1-like transposon from fish, and its transposition in human cells. Cell 91:501–510
34. Aronovich EL, McIvor RS, Hackett PB (2011) The Sleeping Beauty transposon system: a non-viral vector for gene therapy. Hum Mol Genet 15:14–20
35. Thyagarajan B (2011) Bacteriophage integrases for site specific integration. In: Lakshmipathy L, Thyagarajan B (eds) Primary and stem cells. Gene transfer technologies and applications, Wiley. pp 199–210
36. Liu J, Jeppesen I, Nielsen K, Jensen TG (2006) Phi c31 integrase induces chromosomal aberrations in primary human fibroblasts. Gene Ther 15:1188–1190
37. Davis RP, Ng ES, Costa M, Mossman AK, Sourris K, Elefanty AG, Stanely EG (2008) Targeting a GFP reporter gene to the MIXL1 locus of human embryonic stem cells identifies human primitive streak-like cells and enables isolation of primitive hematopoietic precursors. Blood 111:1876–1884
38. Irion S, Luche H, Gadue P, Fehling HJ, Kennedy M, Keller G (2007) Identification and targeting of the ROSA26 locus in human embryonic stem cells. Nat Biotechnol 25:1477–1482
39. Wang P, Rodriguez RT, Wang J, Ghodasara A, Kim SK (2011) Targeting SOX17 in human embryonic stem cells creates unique strategies for isolating and analyzing developing endoderm. Cell Stem Cell 8:335–346
40. Gonzalez B, Schwimmer LJ, Fuller RP, Ye Y, Asawapornmongkol L, Barbas CF 3rd (2010) Modular system for the construction of zinc-finger libraries and proteins. Nat Protoc 4:791–810
41. Soldner F, Laganiére J, Cheng AW, Hockemeye D, Gao Q, Alagappan R, Khurana V, Golbe LI, Myers RH, Lindquist S, Zhang L, Guschin D, Fong LK, Vu BJ, Meng X, Urnov FD, Rebar EJ, Gregory PD, Zhang HS, Jaenisch R (2011) Generation of isogenic pluripotent stem cells differing exclusively at two early onset Parkinson point mutations. Cell 146:318–331
42. Yao Y, Nashun B, Zhou T, Qin I, Qin L, Zhao S, Xu J, Esteban MA, Chen X (2012) Generation of CD34+ cells from CCR5-disrupted human embryonic and induced pluripotent stem cells. Human Gene Ther 23:238–242
43. Hanson C, Hardarson T, Ellerström C, Nordberg M, Caisander G, Rao M, Hyllner J, Stenevi U (2012) Transplantation of human embryonic stem cells onto a partially wounded human cornea in vitro. Acta Ophtalmol Jan 26. (Epub ahead of print)

Chapter 20

Cloning Technologies

Andrew Fontes

Abstract

One major obstacle in realizing the potential behind human embryonic stem cells (hESC) is the availability of efficient and reliable engineering methods. Such methods require cloning technologies that can be applied to a variety of platforms and can serve multiple functions. In the last two decades cloning technologies have become more efficient, widening the bottleneck in creating engineered hESC lines. Using TOPO® TA cloning kits, genes can be efficiently amplified and inserted into target vectors with minimal manipulation and purification. For more complex cloning procedures we introduce the Multisite Gateway® system. This is a cloning platform based on integrase technology that allows for the generation of complex multicistronic gene configurations that can transverse a variety of platforms with ease. These technologies allow the end user to quickly and efficiently select clones, as well as combine multiple genetic elements of interest between platform technologies in a high-throughput manner, providing scientists with a toolbox to create tools to dissect stem cell biology.

Key words Gateway®, TOPO®, hESC, Cloning, Engineering

1 Introduction

Modification of human embryonic stem cells (hESC) allows for the generation of useful tools for basic research, clinical applications, and the dissection of developmental biology. The generation of engineered hESC lines relies on useful technologies to construct and transfer complicated gene configurations into cells. Gene cloning technologies are becoming increasingly more high throughput and efficient (1). Previous methods to recombine a complicated series of genes employed time consuming restriction endonuclease reactions, slowing down the entire workflow. With the growing need for screening tools of engineered lines, a variety of technologies have provided useful platforms for efficient cloning methods. Each tool serves a specific function to enable the user to reach a desired downstream application.

Uma Lakshmipathy and Mohan C. Vemuri (eds.), *Pluripotent Stem Cells: Methods and Protocols*, Methods in Molecular Biology, vol. 997, DOI 10.1007/978-1-62703-348-0_20, © Springer Science+Business Media New York 2013

The TOPO® TA Cloning system, most commonly referred to as TOPO®, eliminates the bottleneck of selecting your gene of interest and cloning it into a plasmid of choice for further manipulation. The technology allows for directional insertion to a vector from the original PCR without further adaptation. The only requirements of the reaction are the use of a Taq polymerase for the PCR amplification and a vector that has been covalently bonded with the topoisomerase enzyme. This eliminates the need for ligation enzymes, gel purification, and the addition of sequence specific ends on your gene of interest. The reaction occurs at room temperature in as little as 5 min, generating plasmid clones with high efficiency for immediate transformation and expansion (2). This rapid technology is ideal for the generation a library of clones in a specific vector backbone.

Many researchers require the engineering of hESC with more complex multicistronic gene configurations. These methods demand high throughput protocols for plasmid generation that can apply to a variety of platform technologies. Here we describe the Multisite Gateway® technology that utilizes lambda phage integrase enzymes to recombine multiple fragments with directionality and high accuracy (2). Each gene is amplified using Gateway® specific primers that contain recombination sites that correspond to the gene orientation and order as seen in Fig. 1.

Amplified cassettes flanked with att sites are first recombined into donor (pDONR™) vectors carrying the corresponding recombination sites in a reaction termed a BP reaction (Fig. 2). Once the gene is inserted, the vector becomes an Entry clone. The Entry clone serves as an intermediate vector and can be stored in a library and used for multiple cloning experiments. In a LR reaction, Entry clones are recombined with destination (pDEST™) vectors creating Expression vectors. Depending on the gene configuration this reaction can become complex with up to five elements including the plasmid as shown in Fig. 1. Each destination vector, represented in Fig. 2b, contains genes owned by a platform technology of choice. This enables the user to transition the same Entry clones into a variety of platform technologies with ease as shown in Fig. 2 (3, 4).

2 Materials

All reagents are available from Life Technologies (www.lifetechnologies.com).

1. Mutlisite Gateway® Pro Plus (Includes chemically competent cells, vectors, and Clonase™ enzymes) (Cat No. 12537-100).
2. BP Clonase™ II Enzyme Mix (Cat No. 11789-020).
3. LR Clonase™ II Plus Enzyme Mix (Cat No. 12538-120).
4. TOPO® TA Cloning® Kit (Cat No. K4500-40).

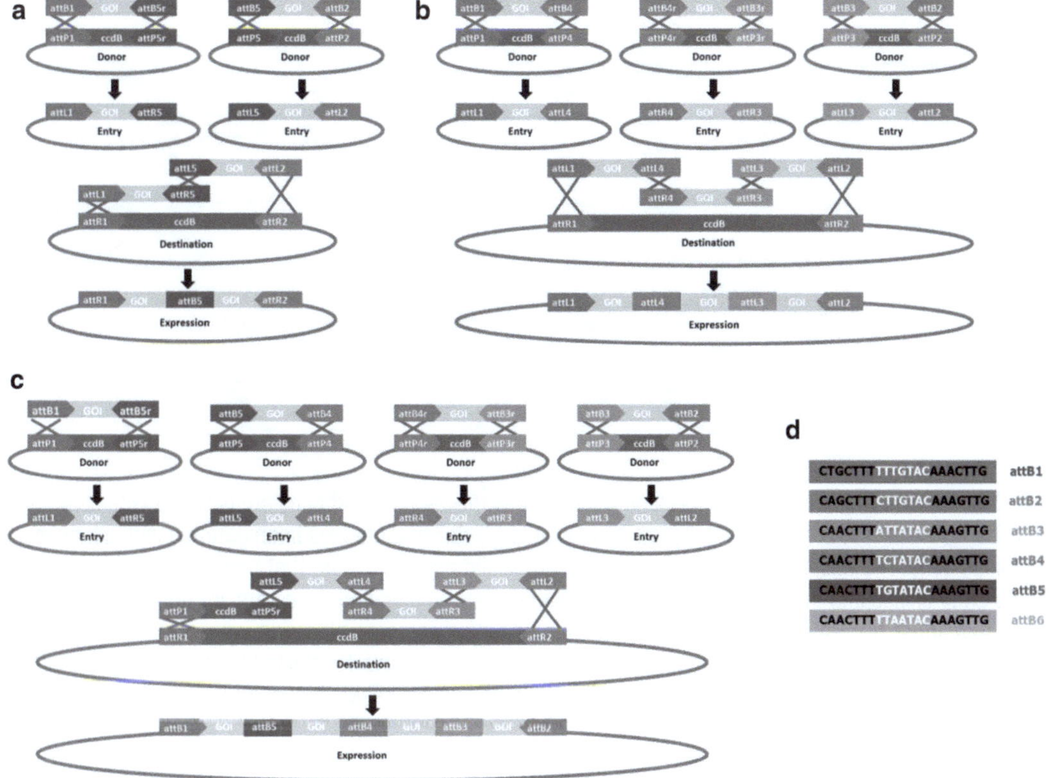

Fig. 1 Multiple primers allow for a variety of configurations based on the complexity of your cloning needs. Gateway® primer sequences rely on sequence homology and are essential for the construction of the final expression vector. Depending if you have a 2(**a**), 3(**b**), or 4(**c**) gene configurations, you will use variable att sites. Additional attB6 site (**d**) can be utilized for custom configurations. The gateway primers (**d**) contain att sites that vary at the single base pair level, shown in *white*. The specificity of each att site allows virtually zero undesirable cross recombination events. The addition of an "r" to the att B and attP sequences signifies the directionality of the sequence and results in the creation of an R site rather than the L site

5. One Shot® Top 10 chemically competent *Escherichia coli* (Cat No. C3030-03).
6. Platinum® *Taq* DNA Polymerase High Fidelity (Cat No. 11304011).
7. *Pfx* 50 DNA Polymerase (Cat No. 12355012).
8. PureLink™ Quick Plasmid Miniprep Kit (Cat No. K210010).
9. Purelink™ PCR Purification Kit (Cat No. K3100-02).
10. PureLink™ Quick Gel Extraction Kit (Cat No. K2100-25).
11. PureLink™ Hipure Plasmid Filter Midiprep Kit (Cat No. K210014).
12. S.O.C. Medium Cat No. (Cat No. 155044034).
13. M13 Forward (−20) Sequencing Primer (Cat No. N520-02).
14. M13 Reverse Sequencing Primer (Cat No. N530-02).

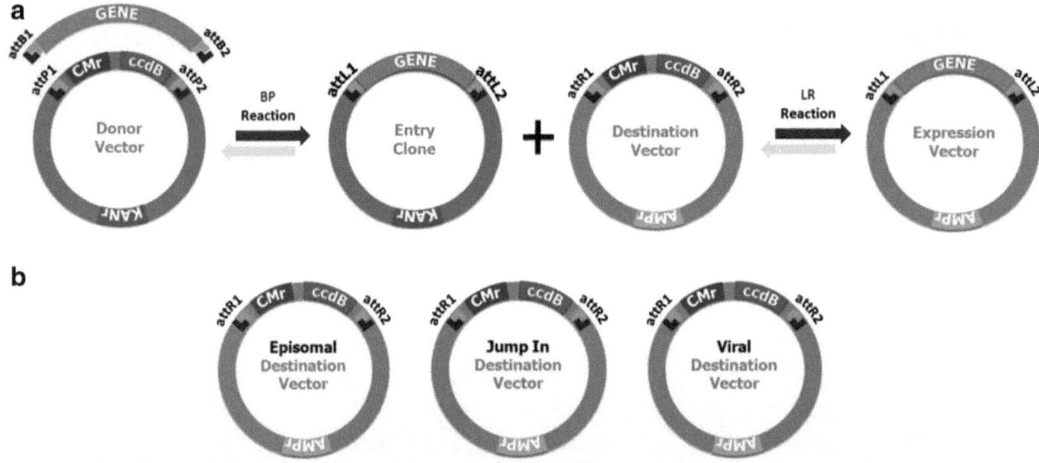

Fig. 2 To improve the efficiency of this cloning technology the vector backbones are outfitted with a variety of selection tools (**a**). For selection and expansion of clones the donor vector has been outfitted with a kanamycin antibiotic gene. The destination vector contains an ampicillin gene to ensure that there is no carryover of any of the Entry clones into the final expression pool. Both donor and destination vectors are outfitted with a ccdB gene that must be removed during recombination to select for only plasmids where recombination has occurred (remove any transformants). The Gateway® process allows for flexibility in choosing a platform technology (**b**). A bank of Entry clones can be utilized in any vector based genomic platform (**b**). hESC on: On feeders and feeder free

15. Dpn I (Cat No. 15242-019).
16. Ampicillin 20 ml (Cat No. 11593-019).
17. Kanamycin Sulfate 100 ml (Cat No. 15160-054).
18. Gateway® Vector Conversion System (Cat No. 11828-029).

3 Methods

3.1 TOPO TA

3.1.1 Primer Design

1. Design appropriate PCR primers at the five prime and three prime end of your gene of interest (see Note 1).
2. Perform a standard PCR using a Taq polymerase using the setup provided in Table 1. Include cycling parameters appropriate for your specific primers and template. We recommend adding a 7–30 min extension period at 72°C after final cycle to ensure full length product (see Table 1 below).
3. Verify PCR using gel electrophoresis to identify a single clear band at the correct molecular weight (see Note 2).

3.1.2 TOPO® TA Reaction

1. Set up the following reaction mixture for the TOPO® TA reaction Table 2.
2. Mix the reaction by gently flicking the tube and incubate for 5 min at room temperature (see Note 3).
3. Place the reaction on ice or store at –20°C.

Table 1
DNA template volume and DNAase/RNAase-free water volume will vary depending on concentration of DNA template

Reagent	Amount
DNA template	10–100 ng
10× PCR buffer	5 μl
50 mM dNTPs	0.5 μl
Primers	1 μM each
DNAase/RNAase-free water	To final volume of 49 μl
Taq polymerase	1 unit/μl (1 μl)

The total fill volume of the full reaction should be 50 μl
-When amplifying your gene of interest from genomic DNA, use higher concentration
-To determine the optimal TOPO vector to best suit your cloning needs, refer to www.lifetechnologies.com

Table 2
PCR product volume and DNAase/RNAase-free water volume will vary depending on concentration of PCR product

Reagent	Volume (μl)
PCR product	0.5–4
Water	Add to total volume 5 μl
TOPO vector	1

The total fill volume of the full reaction should be 6 μl
-The TOPO reaction listed is for chemically competent *E. coli.* If you are using electro competent cells refer to TOPO manual

3.1.3 Transformation (see Note 4)

1. Thaw One Shot® Top 10 chemically competent *E. coli* cells on ice.
2. Add 2 μl of the reaction mix to the chemically competent cells. Mix by gently flicking the tube.
3. Incubate on ice for 30 min.
4. Heat-shock the cells without shaking for 30 s in a 42°C water bath.
5. Immediately transfer to ice for 1 min.
6. Add 250 μl of S.O.C. medium to the mixture.

Table 3
PCR product volume and TE buffer volume will vary depending on concentration of PCR product

Components	Volume (μl)
attB PCR product (15–150 ng)	1–7
pDONR™ vector (150 ng/μl)	1
1× TE buffer, pH 8.0	to 8

The total fill volume of the full reaction should be 8 μl

7. Shake the mixture horizontally at 37°C and 200 rpm for 1 h.
8. Dispense 20 and 200 μl of the transformation onto separate selective agar plates (prewarmed to 37°C). Dispense transformation evenly using either sterile beads or a sterile bacterial spreading tool (see Note 5).

3.1.4 Analyzing TOPO Reaction

1. Manually pick six positive colonies and inoculate 3 ml LB medium with the appropriate antibiotic overnight.
2. Plasmid DNA can be isolated manually or using an isolation kit such as the PureLink® Quick Plasmid Miniprep Kit.
3. Analyze the plasmids using restriction analysis. It is recommended to cut the plasmid once inside the vector backbone and once in your insert to accurately determine if you have the correct insert size and orientation (see Note 6).

3.2 Multisite Gateway

3.2.1 Primer Design

1. Using Fig. 1 as a reference, design primers with appropriate flanking sites (see Note 7).
2. Perform a standard PCR of your gene of interest using Table 1 as a guideline for DNA and primer concentrations. For detailed amplification protocol, refer to the manual for the DNA polymerase (see Note 8). Use cycling parameters adjusted to your primers and template. Include a 7 min extension at 72°C after the final cycle to ensure full length product.
3. Determine the quality of your PCR product by running 5–10 μl of your PCR on an agarose gel (see Note 9).
4. Purify your PCR using the Purelink™ PCR Purification Kit (see Note 10).

3.2.2 Performing BP Reaction (see Note 11)

1. Combine the components of the BP reaction according to the table below in Table 3.
2. Thaw BP Clonase™ II on ice for 2 min.
3. Briefly vortex the vial twice before each use.
4. Add 2 μl of BP Clonase™ II enzyme mix and gently mix the reaction mix by flicking the tube.

Table 4
Entry clone volume and TE buffer volume will vary depending on concentration of entry clone

Components	Volume (µl)
Entry clones (10 fmol each)	1–7
Destination vector (20 fmol)	1
1× TE buffer, pH 8.0	to 8

Conversion of femtomoles (fmol) to nanograms = $(x\ \text{fmol})(N)\left(\frac{660\ \text{fg}}{\text{fmol}}\right)\left(\frac{1\ \text{ng}}{10^{6}\ \text{fg}}\right)$

x is the number of femtomoles, N is the size of the DNA in base pairs

The total fill volume of the full reaction should be 8 µl. The equation represents a means of calculating femtomoles (fmol) when starting with nanograms (ng)

5. Incubate BP reaction at 25°C for 1 h (see Note 12).
6. Add 1 µl of Proteinase K solution to stop the BP reaction.
7. Transform the BP reaction using a general transformation protocol (Subheading 3.1.3). For the propagation and selection of Entry clones, the addition of kanamycin (50 µg/ml) is required, see Fig. 2.

3.2.3 Analyzing BP Reaction

1. Manually pick 6 positive colonies and inoculate 3 ml LB medium with the appropriate antibiotic overnight.
2. Plasmid DNA can be isolated manually or using an isolation kit such as the PureLink® Quick Plasmid Miniprep Kit.
3. Analyze the plasmids using restriction analysis. It is recommended to cut the plasmid once inside the vector backbone and once in your insert to accurately determine if you have the correct insert size and orientation (see Note 13).

3.2.4 Performing LR Reaction

1. Combine the components of the LR reaction according to the chart below in Table 4 (see Note 14).
2. Thaw LR Clonase™ II on ice for 2 min.
3. Briefly vortex the vial twice before each use.
4. Add 2 µl of LR Clonase II enzyme mix and gently mix the reaction mix by flicking the tube.
5. Incubate LR reaction at 25°C overnight (see Note 15).
6. Add 1 µl of Proteinase K solution (2 µg/µl) for 10 min at 37°C to stop the reaction.
7. Transform the LR reaction according to a general transformation protocol (Subheading 3.1.3). For the propagation and selection of Expression clones, the addition of Ampicillin (100 µg/ml) is required, see Fig. 2.

3.2.5 Analyzing LR Reaction

1. Manually pick 6 positive colonies and inoculate 3 ml LB medium with the appropriate antibiotic overnight.
2. Plasmid DNA can be isolated manually or using an isolation kit such as the PureLink® Quick Plasmid Miniprep Kit.
3. Analyze the plasmids using restriction analysis. It is recommended to cut the plasmid once inside the vector and once in your insert to accurately determine if you have the correct insert size and orientation (see Note 13).

4 Notes

1. The directionality of TOPO® TA requires Thymine and Adenine overhang. With the use of a Taq polymerase, it is not necessary to add this to your primers. Traditional primer design will work for the TOPO® reaction.
2. If multiple bands appear on the agarose gel, it is necessary to gel purify the band of the correct size using the PureLink™ Quick Gel Extraction Kit.
3. If the reaction is unsuccessful, it may become necessary to leave the reaction longer than 5 min.
4. The transformation protocol provides general heat shock transformation guidelines for chemically competent bacteria. Please refer to the manual for your TOPO® or Gateway® vector of choice for details of colony selection. For large or complicated vector configurations, slower growing *E. coli* may provide more optimal conditions. Refer to Life Technologies website for additional chemically competent cells.
5. Depending on your vector of choice you may either have ampicillin, kanamycin, or both as a selection tool for your vector. Blue/white screening is optional on additional TOPO® vectors. Refer to TOPO® vector manual for plating instructions for blue/white selection.
6. TOPO® vectors have been outfitted with M13 forward and reverse primer binding sites surrounding the TOPO® adapted ends. This provides users with the option of sequencing their gene of interest without further primer design.
7. Forward and reverse primers should have four Guanine nucleotides at the five prime end, followed by the appropriate att site, and 25 bp of the five prime end of the gene of interest.
8. For PCR products less than 5–6 kb for use in protein expression, use Pfx 50 DNA Polymerase. To generate PCR products for use in other applications (e.g., functional analysis) or from genomic DNA, use Platinum® Taq DNA Polymerase High Fidelity.

9. If your DNA template contains a kanamycin selection cassette, we recommend treating the PCR with Dpn I restriction endonuclease. This will digest the plasmid and eliminate background of your BP reaction transformation.
10. We recommend purifying your PCR product to remove attB primers and primer-dimers. Excess primers and primer-dimers may recombine with the Entry clone and create background transformants. If multiple bands are present from the PCR, gel extraction of the correct size band is required using your method of choice or the PureLink™ Quick Gel Extraction Kit. Further clean-up can be performed using the PEG Cleanup protocol outlined in the MultiSite Gateway® Manual, or a PureLink™ PCR Purification Kit.
11. We recommend including a positive control for the BP reaction that is included in the Gateway® kits. Detailed instructions for using the control vectors are provided in the Multisite Gateway® Manual.
12. For larger and more complicated inserts it may be necessary to carry out the reaction for more than 1 h.
13. Gateway® pDONR™ and pDEST™ vectors have been outfitted with M13 forward and reverse primer binding sites surrounding the att recombination sites. This provides users with the option of sequencing their gene of interest and recombination sites without further primer design.
14. It is recommended to use higher quality DNA for LR reaction. This quality DNA can be obtained using PureLink™ Hipure™ plasmid midiprep kit.
15. When performing gateway reactions, you will see decreasing efficiency with increasing complexity/size.

References

1. Katzen F (2007) Gateway recombinational cloning: a biological operating system. Expert Opin Drug Discov 2:571–589
2. Shuman S (1994) Novel approach to molecular cloning and polynucleotide synthesis using vaccinia DNA topoisomerase. J Biol Chem 269: 32678–32684
3. Hartley JL, Temple GF, Brasch MA (2000) DNA cloning using in vitro site-specific recombination. Genome Res 10:1788–1795
4. Sasaki Y, Sone T, Yoshida S, Yahata K, Hotta J, Chesnut J, Honda T, Imamoto F (2004) Evidence for high specificity and efficiency of multiple recombination signals in mixed DNA cloning by the Multisite Gateway System. J Biotechnol 107: 223–243

Chapter 21

Stable Transfection Using Episomal Vectors to Create Modified Human Embryonic Stem Cells

Ying Liu, Kate Judd, and Uma Lakshmipathy

Abstract

Gene delivery into stem cells can be achieved using viral and nonviral methods. Nonviral methods are more appealing and the use of episomal vectors that do not integrate into the genome enables expression of transgene that are not subject to genomic loci effects that could affect expression levels. Here we describe in detail transfection and stable pooled clone creation of human embryonic stem cells with episomal vectors.

Key words Gene transfer, Neon, Electroporation, Embryonic stem cells, Episomal, Nonintegrating

1 Introduction

Several viral and nonviral methods are available for the transfer of ectopic DNA into embryonic stem cells (1–6). The most commonly used method is the transfer of naked DNA into cells via liposome-based methods or via electroporation (7–10). Electroporation is an instrument-based method that uses high-voltage electric shock to introduce DNA into cells for transient or stable expression. The high voltage disrupts areas of the phospholipid bilayer of the cell membrane resulting in the formation of temporary aqueous pores. The electric potential across the membrane simultaneously rises allowing the charged DNA molecule to be driven across the membrane through the pores in a directional manner. As DNA flows through the pores, the cell membrane discharges, the phospholipid bilayer reassemble to quickly close the temporary pores. Major disadvantage of using electroporation-based method is that use of wrong electric parameters can result in cell damage and death. The transport of material across the cell membrane during the electropermeability phase is nonspecific and requires pristine media and DNA quality (11).

Uma Lakshmipathy and Mohan C. Vemuri (eds.), *Pluripotent Stem Cells: Methods and Protocols*, Methods in Molecular Biology, vol. 997, DOI 10.1007/978-1-62703-348-0_21, © Springer Science+Business Media New York 2013

Under the right conditions, the disrupted membranes quickly reseal leaving the cells intact after delivery of the DNA into the cell. The need for optimal electric parameters and use of large number of cells has led to several modified electroporation-based methods that try to minimize the disadvantages while retaining the high efficiency of gene transfer into cells (12–16). Neon™ transfection system is a next generation electroporation-based transfection system. It differs from the traditional electroporation systems in using a pipette tip instead of the standard cuvette as the electroporation chamber. The electrode in the pipette tip is designed to produce a uniform electric filed to result in cell toxicity with higher transfection efficiency of diverse cell types including primary and stem cells. The high efficiency, low toxicity and simplicity of the Neon™ system had enabled its use as a routine tool for creating pools of stable expressing clones (16) or for isolating single clones with the transgene inserted site-specifically inserted into the genomic locus of choice (12–15).

The Epstein-Barr virus-based episomal vectors provide an easy and rapid method to create pooled stable clones for downstream analysis. The EBNA1 and OriP elements contained in the plasmid allows replication and episomal maintenance of the plasmid without integration into the host genome thereby minimizing chromosomal effects associated with genomic integration methods. The parent pEBNA-DEST vector is Multisite Gateway enabled, allowing assembly of single or multiple genes of choice via MultiSite Gateway Cloning, and the Hygromycin resistance cassette can be utilized to select for stable clones (16). This chapter describes in detail the procedure for transfecting episomal vectors for rapid creation of pooled human ESC clones.

2 Materials

All reagents are from Life Technologies (www.lifetechnologies.com) unless otherwise stated.

2.1 Reagents and Medium

2.1.1 MEF Media (for 100 ml Complete Medium)

DMEM	89 ml
FBS, ESC Qualified	10 ml
Nonessential amino acids	1 ml
β-Mercaptoethanol	182 μl

2.1.2 ESC Media (for 100 ml Complete Medium)

DMEM/F12	79 ml
KSR	20 ml
Nonessential amino acids	1 ml
Basic FGF (10 μg/ml)	40 μl
β-Mercaptoethanol	182 μl

2.1.3 Preparation of MEF Conditioned Medium

1. Plate 9.4×10^6 mitomycin C treated MEF in a T175 Flask coated with 0.1% gelatin in MEF medium.
2. The following day, replace the MEF medium with 90 ml ESC medium.
3. Collect MEF-CM from the flasks after 24 h of conditioning for up to 7 days.
4. Filtered CM can be stored at −20° until use.
5. At the time of use, freshly supplement conditioned media with extra Glutamax (2 mM) and bFGF (4 ng/ml).

2.2 Preparation of Dishes

2.2.1 Preparation of Geltrex-Coated Dishes

1. Thaw 5 ml bottle of Geltrex™ hESC Qualified at 2–8°C overnight.
2. Dilute 1:1 with cold DMEM/F-12 to prepare 1 ml aliquots in tubes chilled on ice. These can be frozen at −20°C or used immediately.
3. Dilute 1:100 with cold DMEM on ice. Note: An optimal dilution of Geltrex™ hESC Qualified may need to be determined for each cell line. Try various dilutions from 1:30 to 1:100.
4. Cover the whole surface of each culture dish with the Geltrex™ hESC Qualified solution.
5. Incubate in a 37°C, 5% CO_2 incubator for 1 h. Note: Dishes can now be used or stored at 2–8°C for up to a week. Do not allow dishes to dry.
6. Pipette out diluted Geltrex™ hESC Qualified from culture container and discard. Cells can be passaged directly into MEF conditioned media onto Geltrex™ hESC Qualified coated culture containers. Note: It is not required to rinse off Geltrex™ hESC Qualified from the culture container after removal.

2.2.2 Preparation of MEF Dishes

1. One to two days before initiating or passing hESC culture, prepare MEFs on the desired TC dishes/plates.
2. Add Attachment Factor to each culture dish, 1.5 ml/10 cm^2 of surface area. Place the dishes in the 37°C incubator for 1 h.
3. Following the attachment factor treatment of the TC dishes, thaw MEFS (1×10^6 cells/vial) as described in Subheading 3. Gently aspirate off attachment factor solution from dishes and seed desired MEFS as per described on Table 1. Allow at least 24 h for cell attachment at 37°C prior to use of MEFs.
4. The plates are now ready for seeding hESC. MEF dishes can be used up to 3–4 days after seeding. Optional: If time permits, 24 h prior to seeding ESC (day 2), replace MEF media with ESC media.

Table 1
Density of inactivated MEFs for hESC coculture

Culture area (cm^2)	MEFs to seed for hESC	Optimum MEF media volume (mL)
10	2.5×10^5	2
20	5×10^5	5
60	1.5×10^6	10

3 Methods

Table 2 summarizes the workflow and media used during for the generation of stable episomal ESC cells.

3.1 Expansion of Human ESC Culture for Transfection

1. Prior to any gene delivery method, culture hESC s in 60 mm Geltrex-coated dish under feeder-free culture conditions, to eliminate feeder fibroblasts cells (see Note 1).
2. Add 2 ml of 10 mg/ml Collagenase IV (Cat#17104) solution to each 60 mm dish of ESC on MEF feeders and incubate for 3 min. Adjust volume of Collagenase IV for various dish sizes, i.e., 35 mm dishes need 1 ml Collagenase IV.
3. Incubate for 30–60 min at 37°C in incubator (see Note 2).
4. Stop incubation when the edges of the colonies are starting to pull away from the plate.
5. After incubation, remove Collagenase IV from dish and rinse the dish with 3 ml DPBS.
6. Gently dislodge the cells with a 10-ml pipet to collect most of the cells and transfer cells into a 15-ml tube.
7. Centrifuge the cells at $200 \times g$ for 2–4 min at room temperature.
8. Discard supernatant; gently tap the tube to loosen the cell pellet from the bottom of the tube.
9. Dissociate cells into small clusters (50–500 cells) by gently pipetting (see Note 3).
10. Seed the cells onto Geltrex or CellStart-coated plates. The final volume of medium depends on the plates used.
11. Return the plate to the incubator. Be sure to shake the plate left to right and back to front to obtain even distribution of cells.

Table 2
Overview of workflow

Step	Duration	Day	Passage#	Dish	Growth conditions	Comments
Expand cells feeder free	1 week	1	46	1 (6 cm)	MEFs in DMEM+KSR	
		4	47	2 (6 cm)	Feeder-free in MEF CM	
		7	48	4 (6 cm)	Feeder-free in MEF CM	
Transfect and drug select	2–3 weeks	10	49	1 (6 cm)	Hyg MEFs in DMEM+KSR	Neon transfection pooled clones
		14	50	1 (6 cm)	HygR MEFs, DMEM+KSR+Hyg	
Expand	2–3 weeks	40	51	2 (6 cm)	HygR MEFs, DMEM+KSR+Hyg	Expand freeze 1–2 vials
		44	52	1 (6 cm)	HygR MEFs, DMEM+KSR+Hyg	
		48	53	2 (6 cm)	HygR MEFs, DMEM+KSR+Hyg	
		52	54	4 (6 cm)	HygR MEFs, DMEM+KSR+Hyg	
Characterize	2–3 weeks	56	55	8 (6 cm)	HygR MEFs, DMEM+KSR+Hyg	Bank, Karyotype, characterize

3.2 Preparation of DNA for Transfection

1. Assemble promoter and reporter/genes of interest into pEBNA DEST episomal vector using MultiSite Gateway cloning (see Chapter 20).
2. Prepare sufficient amounts of plasmid DNA using an endotoxin-free DNA purification method (see Note 4).
3. Determine the concentration of DNA and the ratio of absorbance at 260 and 280 nm to assess purity (see Note 5).
4. Digest the isolated plasmid DNA to confirm the presence of gene of interest and if using the construct for the first time, test for expression of the gene of interest in a cell line such as HEK293.

3.3 Harvesting Cells for Transfection

1. When cells are at over 80% confluent, harvest ESC grown under feeder-free conditions for transfection.
2. Wash the plated hESC cells once with PBS (−/−) and add 1 ml of TrypLE to the cells. After 2 min, add 1 ml of medium and gently pipette to resuspend cells using a 5 ml serological pipette (see Note 6).
3. Transfer cells to 15 ml conical tube and add ESC medium up to 5 ml.

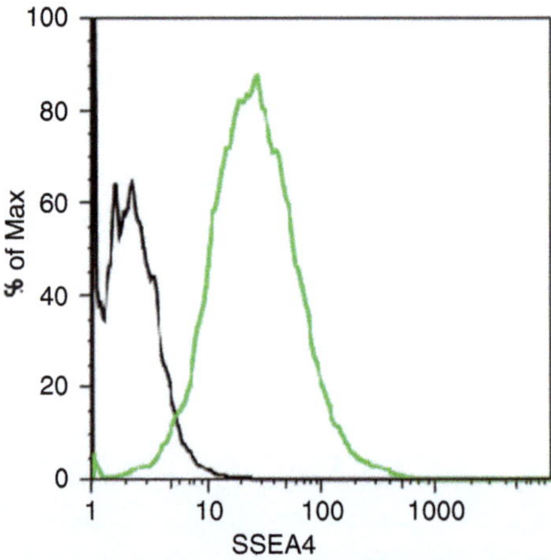

Fig. 1 Assessment of cell quality prior to transfection. ESC cultured under feeder-free conditions harvested for electroporation is also tested for pluripotent marker expression, SSEA4, using FACS analysis (*Green*) and percentage of positive cells relative to the unstained control (*Black*) determined

4. Spin cells at 200 × *g* for 2 min at room temperature.
5. Aspirate medium and cells are ready for assessment of viability and transfection (Note: cells with over 90% viable cells are recommended for further transfection).
6. Cell viability can be assessed using Trypan blue or Live/dead cell Vitality Assay Kit.
7. Set aside a small aliquot of the cells used for transfection to confirm that cells used for the experiment are pluripotent based on expression of the surface marker SSEA4 (Fig. 1).

3.4 Electroporation: Neon System (See Note 7)

1. Turn on the Neon™ Transfection system device (Cat#MPK5000) and adjust parameter to 850 V, 30 ms, 1 pulse.
2. Rinse hESC grown in feeder-free conditions with 2 ml PBS (−/−) once.
3. Add 1 ml TrypLE/60 mm dish and incubate for 3–5 min. After incubation, use a 5 ml pipet to gently dislodge cells.
4. Transfer cells to a 15 ml conical tube, wash the dish with at least 2 ml PBS, and transfer to the same tube.
5. Save a small aliquot of cells for cell counting and for assessing viability using Live/Dead Viability assay kit or by Trypan Blue.
6. Centrifuge the remaining volume of cells at 200× *g* 4 min.
7. Remove supernatant and resuspend cells using Neon Resuspension buffer at a concentration of 1×10^7 cells/ml (see Note 8).
8. Mix DNA (5 μg DNA per electroporation) and cells (100 μl per electroporation at a density of 1×10^7 cells/ml) in the 1.5 ml microcentrifuge tube.

9. Add 3 ml of Neon Electrolytic buffer into the Neon transfection tubes (MPT100) and place the tube in the Neon Transfection system device.
10. Use Gold Tip to aspirate 100 μl of the DNA-cell mixture and place pipet and Gold Tip to the station.
11. Click start and then the flashing arrow button to porate cells.
12. Immediately transfer electroporated cells to a 60 mm dish with MitCMEF seeded on Geltrex-coated dishes and add 5 ml pre warmed MEF conditioned media (see Note 9).

3.5 Drug Selection and Establishment of Stable Modified ESC

1. Allow the electroporated ESC to recover on MitCMEF in MEF conditioned media.
2. Monitor transfected cells at 24–48 h post transfection to determine the toxicity and efficiency of the transfection by observing the positive and negative control cells.
3. At about 72 h post transfection, when cells appear to form colonies, add drug to the culture media. Summary of the media and culture conditions used during transfection and subsequent selection is tabulated in Table 1 (see Note 10).
4. Use at least three different drug concentration, one at optimal and two lesser and higher ranges during selection (see Note 11).
5. Change media with fresh MEF-CM containing drug everyday (see Note 12).
6. If the transfected vector contains a visible marker, monitor expression of the visible marker (Fig. 2).
7. Scan plates regularly after 1–2 weeks after initiation of drug selection. If the transfected ESC are seeded on Feeder MEF cells that are resistant to the drug used to select, hESC media containing drug can be used instead of MEF-CM.
8. Distinct drug resistance colonies begin to appear by week 2–3 after initiating drug selection. Carefully mark the colonies by circling the colonies from the bottom of the culture dish using an inverted microscope and monitor the colonies regularly (Fig. 2).
9. When the colonies are large enough for passage, carefully dislodge the colonies after a brief (5–10 min) collagenase treatment. Chop the colonies into small pieces using the sharp knife edge of a Pasteur pipette.
10. Transfer the cells to a single well of a 24-well dish with drug resistant MEFs plated on Matrigel-coated dishes and containing conditioned media.

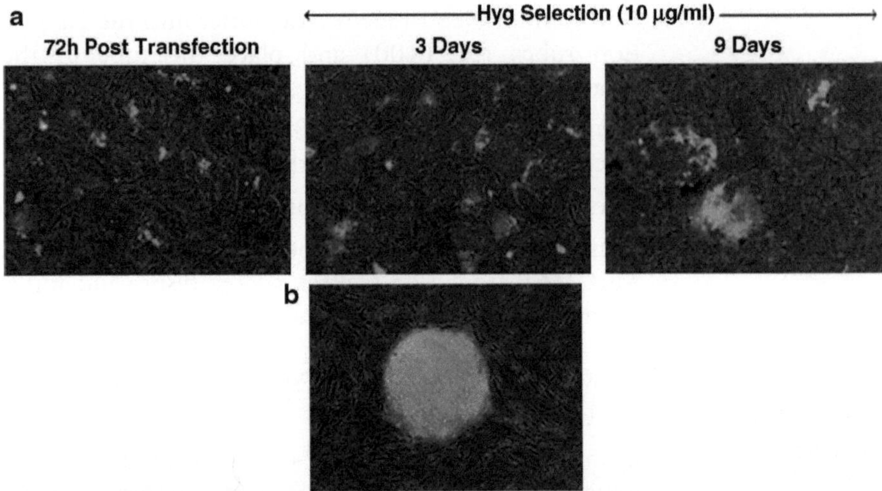

Fig. 2 Monitoring transgene expression in cells post transfection. (**a**) Cells transfected with the episomal vector pEBNA-Ef1a-GFP vector into H9 ESC was monitored daily to observe morphology of cells and emergence of GFP positive cells and their progression to colonies with drug selection. (**b**) Resulting drug resistant clones were pooled and stably expressing GFP shown on iMEF feeder cells established

11. Change the media the following day after the transferred colonies have attached. Begin drug selection with appropriate concentration of the drug in MEF conditioned media.
12. Expand and maintain cells under selection during expansion and create early passage cell banks before proceeding to further analysis and applications.

4 Notes

1. It is important to eliminate feeder fibroblasts since they can compete for the DNA during transfection.
2. Incubation time may vary among different batches of collagenase; therefore, you need to determine the appropriate incubation time by examining the colonies.
3. Avoid making single cell suspensions.
4. Presence of endotoxin in DNA results in poor transfection and survival of transfected cells.
5. It is best to use DNA stocks with concentration of 1 mg/ml.
6. Pipette the cells gently after TrypLE to ensure single cell suspension without damaging the cells.
7. The protocol described is using 100 μl Tip (Cat#MPK10096).
8. On an average most hESC cell line yield $3–5 \times 10^6$ cells per 60 mm dish.

9. It is important to include an "unmanipulated negative control" and a "no DNA electroporation control" to determine any toxicity associated with electroporation. In the absence of a visible marker on the plasmid containing the gene of interest, it is also important to include a positive control where cells are transfected with a visible reporter such as GFP to determine transfection efficiency.
10. 10 μg/ml Hygromycin is optimal for H9 ESC. However, kill curves need to be determined for each ESC cell line to determine the optimal drug concentration to use.
11. Drug resistance may also vary depending on the promoter driving the expression of the resistance gene.
12. If the transfected ESC is seeded on feeder MEF cells that are resistant to the drug used for selection, hESC media containing drug can be used instead of MEF-CM.

Acknowledgment

This work was supported by Life Technologies.

References

1. Pfeifer A, Ikawa M, Dayn Y, Verma IM (2002) Transgenesis by lentiviral vectors: lack of gene silencing in mammalian embryonic stem cells and preimplantation embryos. Proc Natl Acad Sci U S A 99:2140–2145
2. Smith-Arica JR, Thomson AJ, Ansell R, Chiorini J, Davidson B, McWhir J (2003) Infection efficiency of human and mouse embryonic stem cells using adenoviral and adeno-associated viral vectors. Cloning Stem Cells 5:51–62
3. Vetere A, Marsich E, Di Piazza M, Koncan R, Micali F, Paoletti S (2003) Neurogenin3 triggers beta-cell differentiation of retinoic acid-derived endoderm cells. Biochem J 371:831–841
4. Ward CM, Stern PL (2002) The human cytomegalovirus immediate-early promoter is transcriptionally active in undifferentiated mouse embryonic stem cells. Stem Cells 20:472–475
5. Kim JH, Do HJ, Choi SJ, Cho HJ, Park KH, Yang HM, Lee SH, Kim DK, Kwack K, Oh SK, Moon SY, Cha KY, Chung HM (2005) Efficient gene delivery in differentiated human embryonic stem cells. Exp Mol Med 37:36–44
6. Kobayashi N, Rivas-Carrillo JD, Soto-Gutierrez A, Fukazawa T, Chen Y, Navarro-Alvarez N, Tanaka N (2005) Gene delivery to embryonic stem cells. Birth Defects Res C Embryo Today 75:10–18
7. Eiges R, Schuldiner M, Drukker M, Yanuka O, Itskovitz-Eldor J, Benvenisty N (2001) Establishment of human embryonic stem cell-transfected clones carrying a marker for undifferentiated cells. Curr Biol 11:514–518
8. Zwaka TP, Thomson JA (2003) Homologous recombination in human embryonic stem cells. Nat Biotechnol 21:319–321
9. Costa M, Dottori M, Ng E, Hawes SM, Sourris K, Jamshidi P, Pera MF, Elefanty AG, Stanley EG (2005) The hESC line Envy expresses high levels of GFP in all differentiated progeny. Nat Methods 2:259–260
10. Costa M, Dottori M, Sourris K, Jamshidi P, Hatzistavrou T, Davis R, Azzola L, Jackson S, Lim SM, Pera M, Elefanty AG, Stanley EG (2007) A method for genetic modification of human embryonic stem cells using electroporation. Nat Protoc 2:792–796
11. Weaver JC (1995) Electroporation theory. Concepts and mechanisms. Methods Mol Biol 55:3–28
12. Lakshmipathy U, Buckley S, Verfaillie C (2007) Gene transfer via nucleofection into adult and embryonic stem cells. Methods Mol Biol 407:115–126

13. Liu Y, Lakshmipathy U, Ozgenc A, Thyagarajan B, Lieu P, Fontes A, Xue H, Scheyhing K, MacArthur C, Chesnut JD (2010) hESC engineering by integrase-mediated chromosomal targeting. Methods Mol Biol 584:229–268
14. Liu Y, Thyagarajan B, Lakshmipathy U, Xue H, Lieu P, Fontes A, MacArthur CC, Scheyhing K, Rao MS, Chesnut JD (2009) Generation of platform human embryonic stem cell lines that allow efficient targeting at a predetermined genomic location. Stem Cells Dev 18:1459–1472
15. Thyagarajan B, Liu Y, Shin S, Lakshmipathy U, Scheyhing K, Xue H, Ellerstrom C, Strehl R, Hyllner J, Rao MS, Chesnut JD (2008) Creation of engineered human embryonic stem cell lines using phiC31 integrase. Stem Cells 26:119–126
16. Thyagarajan B, Scheyhing K, Xue H, Fontes A, Chesnut J, Rao M, Lakshmipathy U (2009) A single EBV-based vector for stable episomal maintenance and expression of GFP in human embryonic stem cells. Regen Med 4:239–250

Chapter 22

Site-Specific Integration in Human ESC Using Jump-In™ TI™ Technology

Chad C. MacArthur

Abstract

Engineering of human embryonic stem cells (hESC) offers a great potential tool for the study of human gene function. There are many techniques that can be used to engineer human cells, but most are lacking in either specificity or efficiency. Jump-In™ TI™ technology utilizes two bacteriophage recombinases (PhiC31 and R4) to specifically, efficiently, and stably introduce genetic elements into the genome of human ESCs. The techniques described here allow the user to first deliver a targeting site to a defined locus, and second to deliver the genetic elements of interest to that targeting site, allowing for stable, single copy integration into the genome. These integrated elements show high levels of expression in the pluripotent state, as well as in multiple differentiated lineages.

Key words hESC, Engineering, Recombinase, Gateway®, Jump-In™

1 Introduction

Many different methods exist to achieve stable integration of transgenes into a cellular genome. Viral delivery methods offer the advantage of stable and typically robust expression of transgenes, but integration occurs randomly and at multiple sites. Homologous recombination offers the advantage of precise control over the site of integration and copy number, but in human cells this method typically has very low efficiency. Bacteriophage recombinases like PhiC31 and R4 can be utilized to deliver transgenes in a more specific and efficient manner. These recombinases catalyze the integration of genetic elements at specific recognition sites. The recombinase targets a donor plasmid containing an *attB* recognition site into an *attP* recognition site on the host genome. In addition to the high specificity, integration is also unidirectional, which means that loss of the integrated elements from the host genome is less likely.

In the human genome, there are a number of pseudo-*attP* sites recognized by PhiC31 integrase (1). It has been previously shown

Uma Lakshmipathy and Mohan C. Vemuri (eds.), *Pluripotent Stem Cells: Methods and Protocols*, Methods in Molecular Biology, vol. 997, DOI 10.1007/978-1-62703-348-0_22, © Springer Science+Business Media New York 2013

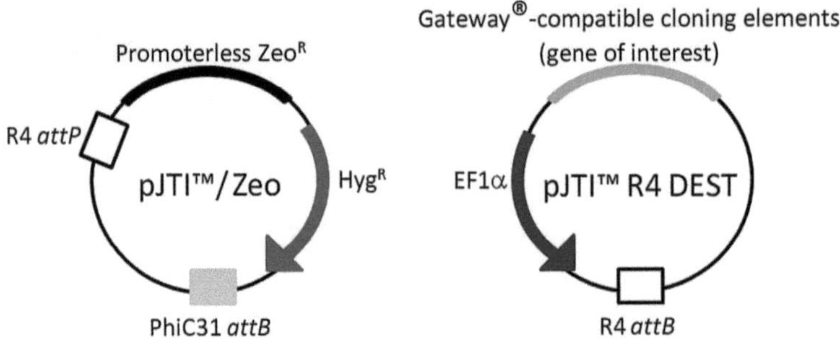

Fig. 1 Jump-In TI™ Vectors. The major components of the pJTI™/Zeo vector are an R4 *attP* site, a Hygromycin B resistance gene for selection of cells, a promoterless Zeocin resistance gene, and a PhiC31 *attB* site. The pJTI™ R4 DEST is equipped with Gateway® compatible cloning elements for simple insertion of your gene(s) of interest, the R4 *attB* site, and a constitutive human EF1α promoter

that many of these pseudo-*attP* sites are located in chromosomal regions that are transcriptionally active and also at intronic regions of genes (2). This means that genetic elements inserted at these sites are likely to be highly expressed and will be less apt to cause deleterious gene disruption. Previous studies have characterized one such PhiC31 pseudo-*attP* site at the chromosomal locus 13q32.3 (3). Genetic elements that were inserted at this locus in hESC were shown to have robust and sustained expression, even after differentiation into multiple lineages (4, 5).

Jump-In TI™ technology utilizes the bacteriophage recombinases PhiC31 and R4 in two subsequent steps to facilitate efficient and specific integration of genetic elements. In the first step, a PhiC31 integrase vector is co-transfected with a pJTI™/Zeo vector (see Fig. 1) that will deliver an R4 *attP* target site. PhiC31 catalyzes integration of the contents of the pJTI™/Zeo vector into the host genome, allowing for targeted delivery of the final genetic target to the R4 *attP* site. In the second step ("retargeting"), a R4 integrase vector is co-transfected with a pJTI™ DEST vector (see Fig. 1) that contains the genetic elements of interest. The R4 integrase catalyzes integration of the pJTI™ DEST vector at a single specific location, targeting the R4 *attP* site that was delivered in the first step. Upon proper integration of the R4 DEST vector, a constitutive EF1α promoter will be placed upstream of the Zeocin resistance gene that was inserted previously as part of the pJTI™/Zeo vector, and Zeocin resistance will only be active in cells where the R4 DEST vector's contents have correctly inserted at the R4 site (Fig. 1).

2 Materials

2.1 hESC Culture Components

DMEM/F-12 (Life Technologies Catalog # 10565).

Knockout Serum Replacement (KSR) (Life Technologies Catalog # 10828).

Table 1
hESC medium composition

Component	Final concentration	For 100 mL
DMEM/F-12	–	79 mL
KSR	20%	20 mL
NEAA (10 mM)	1×(10 μM)	1 mL
BME (55 mM)	1×(55 μM)	100 mL
FGF	4 ng/mL	–

Table 2
MEF medium composition

Component	Final concentration	For 100 mL
DMEM	–	89 mL
ES FBS	10%	10 mL
NEAA (10 mM)	1×(10 μM)	1 mL

2.1.1 hESC Medium (Table 1)

Nonessential amino acids (NEAA) (Life Technologies Catalog # 11140).

2-mercaptoethanol (BME) (Life Technologies Catalog # 21985).

FGF-2 (Life Technologies Catalog # 13256).

2.1.2 Mouse Embryonic Fibroblast Medium (Table 2)

DMEM (Life Technologies Catalog # 10569).

ES-Qualified Fetal Bovine Serum (Life Technologies Catalog # 10439).

NEAA (Life Technologies Catalog # 11140).

2.1.3 Miscellaneous hESC Culture Components

1. Attachment Factor Solution (Life Technologies Catalog # S006100).
2. Collagenase IV (Life Technologies Catalog # 17104019).
3. hESC-qualified Geltrex (Life Technologies Catalog # A1413301).
4. Hygromycin-resistant inactivated MEF (Millipore Catalog# PMEF-H).
5. D-PBS w/o Ca^{2+} Mg^{2+} (Life Technologies Catalog # 14190250).
6. Hygromycin B (Life Technologies Catalog # 10687010).
7. Zeocin™ (Life Technologies Catalog # R25001).

2.2 Transfection Components

1. pJTI™ PhiC31 Integrase Vector (component of the Jump-In™ TI™ Platform Kit, Life Technologies Catalog # A10897).
2. pJTI™/Zeo Vector (component of the Jump-In™ TI™ Platform Kit, Life Technologies Catalog # A10897).
3. pJTI™ R4 Integrase Vector (component of the Jump-In™ TI™ Gateway® Vector Kit, Life Technologies Catalog # A10896).
4. pJTI™ R4 DEST Vector (component of the Jump-In™ TI™ Gateway® Vector Kit, Life Technologies Catalog # A10896).
5. TrypLE™ Express (Life Technologies Catalog # 12604).

2.2.1 Transfecting with the Neon™ Transfection System

1. Neon™ Transfection System.
2. Neon™ Transfection System 100 μL Kit (Life Technologies Catalog # MPK10025).

2.2.2 Transfecting with BTX or Other Electroporator

1. BTX Electroporator.
2. Opti-MEM (Life Technologies Catalog # 12309).
3. Cuvettes (0.4 cm gap).

2.3 Other Reagents

1. CellsDirect Resuspension and Lysis Buffers (Life Technologies Catalog # 11739–010).
2. DNAZol® (Life Technologies Catalog # 10503027).

3 Methods

3.1 Preparing MEF-Conditioned Medium

1. Coat T175 flask(s) with Attachment Factor (0.1% gelatin) for 60 min at 37°C (see Note 1). Warm ~50 mL of MEF medium in a 37°C water bath.
2. Thaw MEF vial(s) by placing in a 37°C water bath for 90–120 s, until there is just a small ice crystal left (see Note 2).
3. Transfer vial(s) to the culture hood (thoroughly spray with ethanol first), and transfer the contents of the vial(s) to a 50 mL conical tube with a P1000. Slowly (drop-wise) add 10 mL of pre-warmed MEF medium to the cells, gently swirling the tube as medium is added.
4. Transfer cells to a 15 mL conical tube, and centrifuge at 200 × *g* for 4 min.
5. Aspirate supernatant, resuspend cell s in an appropriate volume of MEF medium, and count the cells.
6. Plate MEF at a density of 5.4×10^4 cells/cm^2 (approximately nine million cells per T175).
7. The following day, remove the MEF medium and add 90 mL of pre-warmed complete hESC medium to each T175.

Table 3
Amount of DNA to use for transfections

	Creating platform line		Retargeting	
	μg pJTI™/Zeo vector	μg pJTI™ PhiC31 integrase vector	μg R4 DEST vector	μg R4 integrase vector
Neon	5	5	5	5
BTX	15	15	15	15

8. The following day, collect the medium (now MEF-CM) from the flasks and filter-sterilize to remove any MEF cells. Add another 90 mL of pre-warmed hESC to each T175 flask. Collect MEF-CM in this manner for a total of 7 days.
9. MEF-CM can be stored at 4°C for up to 2 weeks, or frozen at −20°C and stored for up to 4 months. FGF-2 should be added to the MEF-CM at a concentration of 4 ng/mL immediately before use.

3.2 Culture of hESC for Transfection

1. Culture and expand hESCs for transfection using standard hESC culture methods (see Note 3).
2. For each transfection reaction, between 1×10^6 cells (for Neon™) and 2×10^6 cells (for BTX or other electroporator) are required (see Note 4).
3. At least 1–2 days prior to transfection, plate the necessary number of 60 mm dishes with Hygromycin-resistant MEF (one 60 mm dish per reaction) (see Note 5).

3.3 Transfection of hESC with the Neon™ Transfection System

1. Turn on the Neon™ instrument; adjust parameters to 850 V, 30 ms, 1 pulse.
2. Prepare DNA in 1.5 mL microcentrifuge tubes, typically 10 μg total DNA for each reaction (see Table 3).
3. For each electroporation (1×10^6 cells), add 5 mL pre-warmed MEF-CM to one 60 mm Hyg-resistant MEF dish and put it back into incubator (see Note 6).
4. Harvest cells to be transfected; rinse cells once with 2 mL of D-PBS, then add 2 mL of TrypLE™ Express per 60 mm dish. After 4 min incubation at room temperature, gently dislodge cells and transfer to a 15 mL conical tube, wash dish 3× with 2 mL of D-PBS and transfer to the same tube. Take a small aliquot of cells to count cell number and viability (see Notes 7 and 8).
5. Centrifuge cells at $200 \times g$ for 4 min. Remove supernatant (see Note 9).

6. Resuspend cells in 10 mL of D-PBS and centrifuge again. Remove supernatant.
7. Resuspend cells using Resuspension buffer R at a density of 1×10^7 cells/mL (see Note 10).
8. Mix DNA (10 μg total) and cells (100 μL, 1×10^6 cells) in the 1.5 mL microcentrifuge tube that contains the DNA (prepared in step 2).
9. Add 3 mL of Electrolytic buffer E2 to the Neon™ Tube and place the tube in the pipette station.
10. Use a 100 μL Neon™ Tip to aspirate 100 μL of the DNA–cell mixture (be careful to avoid introducing any bubbles) and place the Neon™ Pipette (with Neon™ Tip) in the pipette station.
11. Press "Start" on the touch screen to initiate electroporation. After delivering the electric pulse, "Complete" is displayed on the touch screen to indicate that electroporation is complete.
12. Immediately transfer electroporated cells to the MEF dish prepared in step 3, and incubate in the 37°C cell incubator (see Notes 11 and 12).

3.4 Transfection of hESC with a BTX or Other Electroporator

1. Prepare DNA in a 1.5 mL microcentrifuge tube, typically 30 μg total DNA per electroporation (see Table 3).
2. For each electroporation (2×10^6 cells), add 3 mL of pre-warmed MEF-CM to one 60-mm Hyg-resistant MEF dish, and put it back into the incubator.
3. Pre-warm 3 mL MEF-CM for each EP reaction in a 15 mL conical tube in a 37°C water bath.
4. Harvest cells to be electroporated; rinse cells once with 2 mL of D-PBS, then add 2 mL of TrypLE™ Express per 60 mm dish. After 4 min incubation at room temperature, gently dislodge cells and transfer to a 15 mL conical tube, wash dish 3× with 2 mL of D-PBS and transfer to the same tube. Take a small aliquot of cells to count cell number and viability (see Notes 7 and 8).
5. Centrifuge cells at $200 \times g$ for 4 min. Remove supernatant (see Note 9).
6. Resuspend cells in 10 mL D-PBS and centrifuge again. Remove supernatant.
7. Resuspend the cells with pre-warmed Opti-MEM™ to a concentration of 2.5×10^6 cells/mL.
8. Add 800 μL of cells to the 1.5 mL microcentrifuge tube that contains the DNA (prepared in step 1), and transfer the mixture to an electroporation cuvette with a gap of 0.4 cm.
9. Electroporate the cells using the BTX ECM630 electroporator (or Bio-Rad Gene pulser II or Xcell system) with the following

conditions: 500 V, 250 μF. If using the BTX ECM830 electroporator, the conditions are: 200 V, 10 ms, 2 pulses.

10. Immediately transfer the cells into the 15 mL conical tube (prepared in step 3) using a Pasteur pipette or 1 mL serological pipette, and incubate at room temperature for 5 min.
11. Transfer the mixture from step 10 into the 60 mm dish (prepared in step 2) and incubate in the 37°C cell incubator (see Note 11).

3.5 Creating the R4 Platform Line

3.5.1 Culture of hESCs for Transfection

Culture hESCs for transfection as described above (see Notes 3 and 4).

3.5.2 Transfection

Transfect cells as described above, using one of the two methods. For creating the R4 platform line, use pJTI™ PhiC31 integrase and the pJTI™/Zeo platform vectors (see Note 13).

3.5.3 Selection of Transfected Cells

1. Carefully monitor cells every day after transfection. Selection drugs are typically added between 48 and 96 h after transfection (see Note 14).
2. When cells are ready for selection, add Hygromycin B to complete MEF-CM at a concentration of 10 μg/mL.
3. Replace medium every day with complete MEF-CM (supplemented with 10 μg/mL Hygromycin B) and monitor cells for the emergence of hESC colonies (see Note 15).
4. When selection is complete, cells are ready to be passaged onto fresh MEF. 1–2 days prior to passage, plate 12-well plates with MEF (see Note 16).
5. Manually passage each colony onto a separate well of a 12-well MEF plate (see Notes 17 and 18).

3.5.4 Characterization of Platform Line

Determine Integration Site in the Genome

1. Continue to replace medium every day. Passage cells when they reach confluence.
2. At the next passage, collect 1.0–3.0 × 10^4 cells for PCR (see Notes 19 and 20).
3. Run nested PCR to verify the presence of the inserted elements at the Chromosome 13 locus, using the sequences listed in Table 4. Eliminate any clones which do not have integration at the Chromosome 13 locus (see Note 21).

Verify Single Copy Integration of pJTI™/Zeo Vector

1. Continue to expand clones using standard hESC culture methods, passaging onto Hyg-resistant MEF and keeping cells under Hygromycin selection.

Table 4
PCR primer sequences for detection of integration of the pJTI™/Zeo vector at the 13q32.3 locus

Primer name	Sequence	Size (bp)	PCR conditions
CHR13PF (primary)	AGTTAAGCCAGCCCCGACAC	592	58°C, 35 cycles
CHR13PR	TTTTGGCTACCAGTACTAGGCAGG		
CHR13SF (secondary)	CTTGTCTGCTCCCGGCATCC	467	55°C, 30 cycles
CHR13SR	CCCAGCAATGGAGCCTGATT		

2. To verify that only a single copy of the platform vector has integrated into the genome, a Southern Blot should be performed. For the Southern Blot, harvest genomic DNA from one 60 mm dish (at 80–90% confluence) using DNAzol, or preferred method of choice (see Notes 22–24).
3. Perform Southern Blot using standard methods (see Note 25).

3.6 Retargeting the R4 Platform Line

3.6.1 Culture of R4 Platform hESC for Transfection

Culture the R4 platform line for transfection as described above. It is important to keep the R4 platform line under Hygromycin selection during regular culture and expansion (see Notes 26 and 27).

3.6.2 Transfection

Transfect cells as described above using one of the two methods. For retargeting of the R4 platform line, use pJTI™R4 integrase and pJTI™ R4 DEST (containing the user's genetic target).

3.6.3 Selection of Transfected Cells

1. Carefully monitor cells every day after transfection. Selection drugs are typically added between 48 and 96 h after transfection (see Note 14).
2. When cells are ready for selection, add Zeocin™ to complete MEF-CM at a concentration of 1.0 μg/mL (see Note 28).
3. Replace medium every day with complete MEF-CM (supplemented with 1.0 μg/mL Zeo) and monitor cells for the emergence of hESC colonies (see Note 29).
4. When selection is complete, cells are ready to be passaged onto fresh MEF. 1–2 days prior to passage, plate 12-well plates with MEF (see Note 16).
5. Manually passage each colony onto a separate well of a 12-well MEF plate. Ideally, passage as many colonies as possible (see Note 30).

Table 5
PCR primer sequences for detection of proper integration of the insert at the R4 site

Target	Primer name	Sequence	Size (bp)
EF1a-Zeo	EF1AZF EF1AZR	GCCTCAGACAGTGGTTCAAAGTTT TGATGAACAGGGTCACGTCGT	534
Hyg^R gene	HYGF HYGR	ATGAAAAAGCCTGAACTCACC ATTGACCGATTCCTTGCG	430

3.6.4 Characterization of Retargeted Cells

Verify Proper Orientation of Genetic Target at the R4 Site

1. Continue to replace medium every day. Passage cells when they reach confluence.
2. At the next passage, collect $1.0–3.0 \times 10^4$ cells to run PCR (see Note 19).
3. Run PCR to check for proper integration of the insert, using the sequences listed in Table 5 (see Note 31).
4. Eliminate any clones which do not return a positive PCR result.

Verify Single Copy Integration of Genetic Target

1. Continue to expand clones, passaging onto MEF and keeping cells under selection.
2. To verify that only a single copy of the genetic target has integrated into the genome, a Southern Blot should be performed. For the Southern Blot, harvest genomic DNA from one 60 mm dish (at 80–90% confluence) using DNAzol, or preferred method of choice (see Note 22).
3. Perform Southern Blot using standard methods (see Note 32).

4 Notes

1. MEF-CM production can be scaled up or down to suit the user's needs. If using T175 flasks, approximately nine million MEF are required per flask, and each flask will yield 90 mL per day, for a total of 720 mL.
2. It is best to thaw no more than three vials at a time, to ensure optimal health of the cells.
3. hESC can be expanded prior to transfection on MEF, or on feeder-free conditions. For at least one passage immediately prior to transfection, culture hESCs on feeder-free conditions, to avoid MEF carry-over during transfection. It is critical to ensure that cells used for transfection are undifferentiated and healthy. It is best to use cells that are at 80–90% confluence, as over-confluent cells may be less likely to survive transfection.

4. Typical yield for a 60 mm dish at 80–90% confluence is between 3 and 5×10^6 cells.
5. It is strongly recommended to plate transfected cells onto MEF. Transfected cells are less likely to survive if they are plated onto feeder-free conditions.
6. MEF-CM is used here to help survival of hESC after transfection. Continue to culture cells in MEF-CM at least until the first passage after transfection.
7. Make sure to completely dissociate cells by gently pipetting up and down.
8. It is best to work with one to two dishes at a time, at most. This will minimize the time between cell harvest and transfection, and ensure optimal cell health.
9. If, after centrifugation, the pellet is loose and not fully compact, resuspend the pellet by gently pipetting up and down, and then centrifuge again.
10. Resuspension Buffer R should be at room temperature.
11. Be sure to evenly distribute the cells across the plate.
12. For further experiments, if the DNA will be the same, the Neon™ Tip can be used up to two times, and the Neon™ Tube and E2 buffer can be used up to ten times. If the DNA will be different, change the Neon™ Tip and Neon™ Tube (also E2 buffer) for the next experiment(s).
13. Instructions in this chapter are for use of the pJTI™/Zeo platform vector. There are also two other pJTI™ platform vectors available that contain either Bsd (Blasticidin) or Neo (Neomycin), if a selection drug other than Zeocin is required.
14. Timing the beginning of drug selection is critical. If drug is added too early after transfection, the cells will not be fully recovered and may die. If drug is added too late, many non-drug-resistant colonies may have arisen, and it will be difficult to kill them all off with the drug. Selection drug should be added when colonies have just started to form, and are still small.
15. Selection may take up to 14 days to completely eliminate any non-drug-resistant cells. In the first few days of selection, there may be substantial cell death, this is normal. Continue to maintain cells on drug selection, and do not lower the concentration. Continue to feed the cells with MEF-CM, as the MEF layer may deteriorate during the selection period, and MEF-CM will be critical to maintain pluripotency of the hESC.
16. Prior to plating MEF, make a rough estimate of the number of distinct colonies on each plate. Each colony will be transferred

to a single well of the 12-well plates, so make sure to prepare enough plates with MEF.

17. Passage as many colonies as possible to maximize the chances of isolating the desired clone, as only a fraction of those colonies will have integration at the desired locus.
18. At this stage, MEF-CM is no longer necessary, since cells are passaged onto a fresh MEF layer. Complete hESC medium supplemented with 10 μg/mL Hygromycin can be used from this point forward.
19. Preparation of Genomic DNA using CellsDirect Resuspension and Lysis Buffers.
    (a) Pellet a total of 10,000–30,000 cells.
    (b) Wash the cells with 500 μl PBS.
    (c) Centrifuge cells to pellet and remove PBS.
    (d) Resuspend the cell pellet in a mixture of 20 μl of Resuspension Buffer and 2 μl of Lysis Solution.
    (e) Incubate the cell suspension at 75°C for 10 min.
20. The PCR screens will potentially eliminate a significant number of clones that the user has to carry. In the interest of a streamlined workload, it is best to run the PCR screens as soon as possible (e.g., the first passage after selecting clones).
21. PhiC31 can catalyze integration at a number of possible pseudo *attP* sites in the human genome. The sequence given here is used to verify integration of the contents of the pJTI™/Zeo vector at a specific Chromosome 13 locus (13q32.3). This locus has been well characterized to give robust expression of subsequently retargeted genetic elements, in both undifferentiated and differentiated hESCs. To find integration at any other locus, a plasmid rescue assay must be performed.
22. Lysis of Cells using DNAzol.
    (a) Remove medium from cells.
    (b) Wash 2× with 3 mL of D-PBS.
    (c) Add 2 mL of DNAzol (for 60 mm dish, ~20 cm^2).
    (d) Slowly and gently pipette across the surface of the dish to collect the cells and lyse them.
    (e) Transfer DNAzol with cells lysate to 15 mL conical tube, proceed to genomic DNA extraction.
23. An appropriate probe to use for the Southern Blot is a ~1 kb fragment of the Hyg resistance gene, which can be amplified from the pJTI™/Zeo vector using appropriate primers. The probe can be labeled using a standard random priming kit (e.g., Ambion DECAprime™ II Kit).

24. If using a Southern Blot probe targeted to the Hyg resistance gene, the presence of genomic DNA from Hyg resistant MEF can give a false-positive result. To avoid this false-positive, platform hESC must be cultured on either non-Hyg-resistant MEF or feeder-free conditions for at least two passages prior to collection in DNAzol.
25. After the Southern Blot has been performed, there will probably be a small number of clones remaining which meet the desired criteria (integration at the Chromosome 13 locus, single copy integration). At this point, it is best to expand these remaining clones and create frozen stocks as soon as possible using standard hESC culture methods. Once the frozen stocks have been established, move on to retargeting.
26. During regular culture of the platform line, do not add selection drugs during the following: On the day of passage (return cells to selection the day after passage), and for the first 3–5 days after thawing cells.
27. Platform hESCs can be expanded prior to transfection on MEF, or on feeder-free conditions. For at least one passage immediately prior to transfection, it is best to culture hESCs on feeder-free conditions, to avoid MEF carry-over during transfection. Keep cells under Hyg selection during culture and expansion, but stop selection one to two passages immediately before transduction, to ensure that cells are healthy.
28. Depending on how well the cells recover from transfection, it may be necessary to start selection with a higher Zeocin concentration (2.0 μg/mL) for the first 1–3 days of selection.
29. Selection may take up to 14 days to completely eliminate any non-drug-resistant cells. In the first few days of selection, there may be substantial cell death, this is normal. Continue to feed the cells with MEF-CM, as the MEF layer will deteriorate during the selection period.
30. At this stage, MEF-CM is no longer necessary, since cells are passaged onto a fresh MEF layer. Complete hESC medium can be used from this point forward. Cells should be returned to Hygromycin selection at this point. Complete hESC medium should be supplemented with Hygromycin (10 μg/mL) and Zeocin (1 μg/mL).
31. The Hyg resistance gene can be used here as a positive control and the EF1a-Zeo primers will indicate proper integration of the insert.
32. Generate a probe for the Southern Blot by amplifying a fragment of the pJT™ R4 DEST Vector used for retargeting.

References

1. Thyagarajan B, Olivares EC, Hollis RP, Ginsburg DS, Calos MP (2001) Site specific genomic integration in mammalian cells mediated by phage phiC31 integrase. Mol Cell Biol 21:3926–3934
2. Chalberg TW, Portlock JL, Olivares EC, Thyagarajan B, Kirby PJ, Hillman RT et al (2006) Integration specificity of phage phiC31 integrase in the human genome. J Mol Biol 357:8–48
3. Thyagarajan B, Liu Y, Shin S, Lakshmipathy U, Scheyhing K, Xue H, Ellerström C, Strehl R, Hyllner J, Rao MS, Chesnut JD (2008) Creation of engineered human embryonic stem cell lines using phiC31 integrase. Stem Cells 26(1): 119–126
4. Liu Y, Thyagarajan B, Lakshmipathy U, Xue H, Lieu P, Fontes A, MacArthur CC, Scheyhing K, Rao MS, Chesnut JD (2009) Generation of platform human embryonic stem cell lines that allow efficient targeting at a predetermined genomic location. Stem Cells Dev 18(10):1459–1472
5. Macarthur CC, Xue H, Van Hoof D, Lieu PT, Dudas M, Fontes A, Swistowski A, Touboul T, Seerke R, Laurent LC, Loring JF, German MS, Zeng X, Rao MS, Lakshmipathy U, Chesnut JD, Liu Y (2012) Chromatin insulator elements block transgene silencing in engineered human embryonic stem cell lines at a defined chromosome 13 locus. Stem Cells Dev 21(2): 191–205

Index

Uma Lakshmipathy and Mohan C. Vemuri (eds.), *Pluripotent Stem Cells: Methods and Protocols*, Methods in Molecular Biology, vol. 997, DOI 10.1007/978-1-62703-348-0, © Springer Science+Business Media New York 2013

I

J

K

L

MIX
Papier aus verantwortungsvollen Quellen
Paper from responsible sources
FSC® C105338

If you have any concerns about our products, you can contact us on
ProductSafety@springernature.com

In case Publisher is established outside the EU, the EU authorized representative is:
Springer Nature Customer Service Center GmbH
Europaplatz 3, 69115 Heidelberg, Germany

Printed by Libri Plureos GmbH
in Hamburg, Germany